Springer Desktop Editions in Chemistry

L. Brandsma, S.F. Vasilevsky, H.D. Verkruijsse
Application of Transition Metal Catalysts in Organic Synthesis
ISBN 3-540-65550-6

H. Driguez, J. Thiem (Eds.)
Glycoscience, Synthesis of Oligosaccharides and Glycoconjugates
ISBN 3-540-65557-3

H. Driguez, J. Thiem (Eds.)
Glycoscience, Synthesis of Substrate Analogs and Mimetics
ISBN 3-540-65546-8

K. Faber (Ed.)
Biotransformations
ISBN 3-540-66949-3

W.-D. Fessner (Ed.)
Biocatalysis, From Discovery to Application
ISBN 3-540-66970-1

S. Grabley, R. Thiericke (Eds.)
Drug Discovery from Nature
ISBN 3-540-66947-7

H.A.O. Hill, P.J. Sadler, A.J. Thomson (Eds.)
Metal Sites in Proteins and Models, Iron Centres
ISBN 3-540- 65552-2

H.A.O. Hill, P.J. Sadler, A.J. Thomson (Eds.)
Metal Sites in Proteins and Models, Phosphatases, Lewis Acids and Vanadium
ISBN 3-540-65553-0

H.A.O. Hill, P.J. Sadler, A.J. Thomson (Eds.)
Metal Sites in Proteins and Models, Redox Centres
ISBN 3-540-65556-5

F.J. Leeper, J.C. Vederas (Eds.)
Biosynthesis, Polyketides and Vitamins
ISBN 3-540-66969-8

A. Manz, H. Becker (Eds.)
Microsystem Technology in Chemistry and Life Sciences
ISBN 3-540-65555-7

P. Metz (Ed.)
Stereoselective Heterocyclic Synthesis
ISBN 3-540-65554-9

H. Pasch, B. Trathnigg
HPLC of Polymers
ISBN 3-540-65551-4

J. Rohr (Ed.)
Bioorganic Chemistry, Deoxysugars, Polyketides and Related Classes: Synthesis, Biosynthesis, Enzymes
ISBN 3-540-66971-X

T. Scheper (Ed.)
New Enzymes for Organic Synthesis, Screening, Supply and Engineering
ISBN 3-540-65549-2

F.P. Schmidtchen (Ed.)
Bioorganic Chemistry, Models and Applications
ISBN 3-540-66978-7

Springer

Berlin
Heidelberg
New York
Barcelona
Hong Kong
London
Milan
Paris
Singapore
Tokyo

K. Faber (Ed.)

Biotransformations

Professor Dr. Kurt Faber
Institute of Organic Chemistry
University of Graz
Heinrichstr. 28
A-8010 Graz, Austria
E-mail: Kurt.Faber@kfunigraz.ac.at

Description of the Series

The Springer Desktop Editions in Chemistry is a paberback series that offers selected thematic volumes from Springer chemistry series to graduate students and individual scientists in industry and academia at very affordable prices. Each volume presents an area of high current interest to a broad non-specialist audience, starting at the graduate student level.

Formerly published as hardcover edition in the review series
Advances in Biochemical Engineering/Biotechnology (Vol. 63) ISBN 3-540-64496-2

Cataloging-in-Publication Data applied for

ISBN 3-540-66949-3
Springer-Verlag Berlin Heidelberg New York

Die Deutsche Bibliothek - CIP-Einheitsaufnahme
Biotransfomations / K. Faber (ed.).- Berlin; Heidelberg; New York; Barcelona, Hong Kong; London; Milan; Paris; Singapore; Tokyo: Springer 2000
(Springer desktop editions in chemistry)
ISBN 3-540-66949-3

Cover: design & production, Heidelberg
Typesetting: Fotosatz-Service Köhler OHG, Würzburg
Printed on acid-free paper SPIN: 10720505 02/3020 hu - 5 4 3 2 1 0

Preface

The use of enzymes – employed either as isolated enzymes, crude protein extracts or whole cells – for the transformation of non-natural organic compounds is not an invention of the twentieth century: they have been used for more than one hundred years. However, the object of most of the early research was totally different from that of the present day. Whereas the elucidation of biochemical pathways and enzyme mechanisms was the main driving force for the early studies, in contrast it was mainly during the 1980s that the enormous potential of applying natural catalysts to transform non-natural organic compounds was recognized. This trend was particularly well enhanced by the recommendation of the FDA-guidelines (1992) with respect to the use of chiral bioactive agents in enantiopure form.

During the last two decades, it has been shown that the substrate tolerance of numerous biocatalysts is often much wider than previously believed. Of course, there are many enzymes which are very strictly bound to their natural substrate(s). They play an important role in metabolism and they are generally not applicable for biotransformations. On the other hand, an impressive number of biocatalysts have been shown to possess a wide substrate tolerance by keeping their exquisite catalytic properties with respect to chemo-, regio- and, most important, enantio-selectivity. This made them into the key tools for biotransformations. As a result of this extensive research during the last two decades, biocatalysts have captured an important place in contemporary organic synthesis, which is reflected by the fact that ~8% of all papers on synthetic organic chemistry contained elements of biotransformations as early as in 1991 with an ever-increasing proportion. It is now generally accepted, that biochemical methods represent a powerful synthetic tool to complement other methodologies in modern synthetic organic chemistry.

Whereas several areas of biocatalysis – in particular the use of easy-to-use hydrolases, such as proteases, esterases and lipases – are sufficiently well research to be applied in every standard laboratory, other types of enzymes are still waiting to be discovered with respect to their applicability in organic-chemistry transformations on a preparative scale. This latter point is stressed in this volume, which concentrates on the "newcomer-enzymes" which show great synthetic potential.

February 1998

Kurt Faber
Graz, University of Technology

Contents

Biocatalytic Asymmetric Decarboxylation
H. Ohta 1

Biocatalytic Applications of Hydroxynitrile Lyases
D.V. Johnson, H. Griengl 31

Stereoinversions Using Microbial Redox-Reactions
A. J. Carnell 57

Biotransformations with Peroxidases
W. Adam, M. Lazarus, C. R. Saha-Möller, O. Weichold
U. Hoch, D. Häring, P. Schreier 73

Production of Chiral C3- and C4-Units by Microbial Enzymes
S. Shimizu, M. Kataoka 109

Polyamino Acids as Man-Made Catalysts
J. V. Allen, S. M. Roberts, N. M. Williamson 125

Epoxide Hydrolases and Their Synthetic Applications
R. V. A. Orru, A. Archelas, R. Furstoss, K. Faber 145

Microbial Models for Drug Metabolism
R. Azerad 169

Subject Index 219

Biocatalytic Asymmetric Decarboxylation

Hiromichi Ohta

Department of Chemistry, Keio University, 3-14-1 Hiyoshi, Kohoku-ku, Yokohama 223-0061, Japan. *E-mail: hohta@chem.keio.ac.jp*

Biocatalytic decarboxylation is a unique reaction, in the sense that it can be considered to be a protonation reaction to a "carbanion equivalent" intermediate in aqueous medium. Thus, if optically active compounds can be prepared via this type of reaction, it would be a very characteristic biotransformation, as compared to ordinary organic reactions. An enzyme isolated from a specific strain of *Alcaligenes bronchisepticus* catalyzes the asymmetric decarboxylation of α-aryl-α-methylmalonic acid to give optically active α-arylpropionic acids. The effect of additives revealed that this enzyme requires no biotin, no co-enzyme A, and no ATP, as ordinary decarboxylases and transcarboxylases do. Studies on inhibitors of this enzyme and spectroscopic analysis made it clear that the Cys residue plays an essential role in the present reaction. The unique reaction mechanism based on these results and kinetic data in its support are presented.

Keywords: Asymmetric decarboxylation, Enzyme, Reaction mechanism, α-Arylpropionic acid.

1 Introduction 2

2 Screening and Substrate Specificity 3

2.1 Method of Screening 4
2.2 Metabolic Path 4
2.3 Substrate Specificity 6

3 Isolation of the Enzyme and the Gene 7

3.1 Isolation of the Enzyme 8
3.2 Cloning and Heterologous Expression of the Gene 9

4 Effect of Additives on the Enzyme Activity 11

5 Active Site Directed Inhibitor and Point Mutation 12

5.1 Screening of an Active Site-Directed Inhibitor 12
5.2 Titration of SH Residue in the Active Site 14
5.3 Spectroscopic Studies of Enzyme-Inhibitor Complex 15
5.4 Site-Directed Mutagenesis 16

6 Kinetics and Stereochemistry 18

6.1 Effect of Substituents on the Aromatic Ring 18
6.2 Reaction of Chiral α-methyl-α-phenylmalonic Acid 20

Advances in Biochemical Engineering/
Biotechnology, Vol. 63
Managing Editor: Th. Scheper

7 **Effect of Substrate Conformation** 22

7.1 Effect of *o*-Substituents 22
7.2 Theoretical Calculation of the Potential Energy Surface 25
7.3 Indanedicarboxylic Acid 26
7.4 Effect of Temperature on the Rate of Reaction 28

8 **Reaction Mechanism** 29

9 **References** 29

1 Introduction

Biochemical reactions include several types of decarboxylation reactions as shown in Eqs. (1)–(5), because the final product of aerobic metabolism is carbon dioxide. Amino acids result in amines, pyruvic acid and other *α*-keto acids form the corresponding aldehydes and carboxylic acids, depending on the cooperating coenzymes. Malonyl-CoA and its derivatives are decarboxylated to acyl-CoA. *β*-Keto carboxylic acids, and their precursors (for example, the corresponding hydroxy acids) also liberate carbon dioxide under mild reaction conditions.

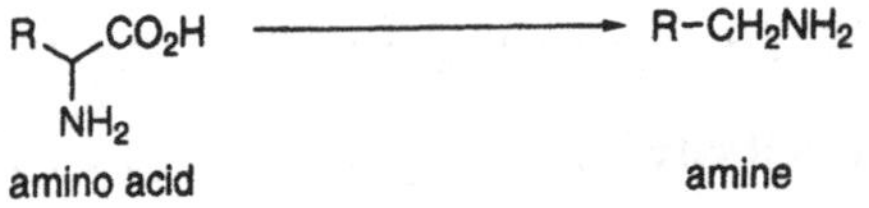

$$R\text{-}CH(NH_2)\text{-}CO_2H \longrightarrow R\text{-}CH_2NH_2 \quad (1)$$

amino acid → amine

$$H_3C\text{-}CO\text{-}CO_2H \longrightarrow H_3C\text{-}CHO \quad (2)$$

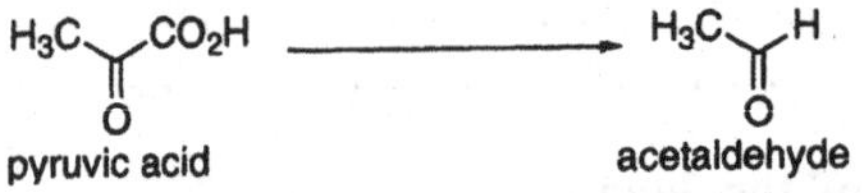

pyruvic acid → acetaldehyde

$$H_3C\text{-}CO\text{-}CH(CO_2H)\text{-}CO\text{-}SCoA \longrightarrow H_3C\text{-}CO\text{-}CH_2\text{-}CO\text{-}SCoA \quad (3)$$

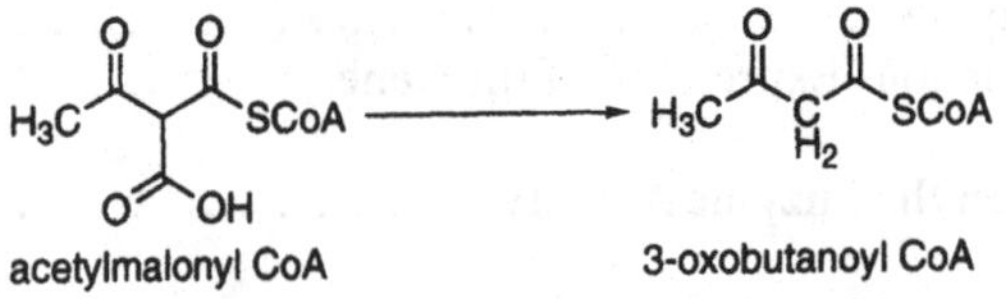

acetylmalonyl CoA → 3-oxobutanoyl CoA

$$HO_2C\text{-}CH_2\text{-}CH(CO_2H)\text{-}CH(OH)\text{-}CO_2H \longrightarrow HO_2C\text{-}CH_2\text{-}CH_2\text{-}CO\text{-}CO_2H \quad (4)$$

iIsocitric acid → α-ketoglutaric acid

$$R\text{-}CO\text{-}OH \rightleftharpoons R\text{-}CO\text{-}O^- \longrightarrow [\,R^-\,] \longrightarrow R\text{-}H \quad (5)$$

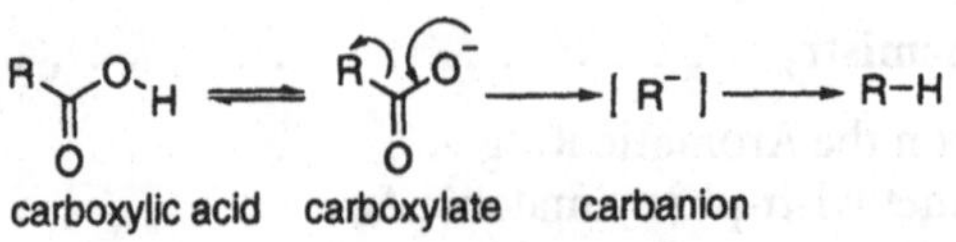

carboxylic acid carboxylate carbanion

The most interesting point from the standpoint of organic chemistry is that the intermediate of the decarboxylation reaction should be a carbanion. The first step is the abstraction of the acidic proton of a carboxyl group by some basic amino-acid residue of the enzyme. C–C bond fission will be promoted by the large potential energy gained by formation of carbon dioxide, provided that the other moiety of the intermediate, i.e. the carbanion, is well stabilized by a neighboring carbonyl group, by the inductive effect of the sulfur atom of coenzyme A, or by some functional group of another other coenzyme. Protonation to the carbanion gives the final product. The characteristic feature of this reaction is the fact that a carbanion is formed in aqueous solution. Nonetheless, at least in some cases, the reactions are enantioselective as illustrated in Eqs. (6) and (7).

$$\underset{\alpha\text{-amino-}\alpha\text{-methylmalonic acid}}{H_3C\text{–}C(NH_2)(CO_2H)CO_2H} \xrightarrow[\text{transferase}]{\text{serine hydroxymethyl}} \underset{\alpha\text{-aminopropionic acid}}{H_3C\text{–}C(NH_2)(H)CO_2H} \quad (6)$$

$$\underset{\alpha\text{-methylmalonyl-CoA}}{H_3C\text{–}C(H)(COSCoA)CO_2H} \xrightarrow[\text{decarboxylase}]{\text{malonyl-CoA}} \underset{(R)\text{-}^3H\text{-propionyl-CoA}}{H_3C\text{–}C(H)(COSCoA)\,^3H} \quad (7)$$

Serine hydroxymethyl transferase catalyzes the decarboxylation reaction of α-amino-α-methylmalonic acid to give (*R*)-α-aminopropionic acid with retention of configuration [1]. The reaction of methylmalonyl-CoA catalyzed by malonyl-coenzyme A decarboxylase also proceeds with perfect retention of configuration, but the notation of the absolute configuration is reversed in accordance with the CIP-priority rule [2]. Of course, water is a good proton source and, if it comes in contact with these reactants, the product of decarboxylation should be a one-to-one mixture of the two enantiomers. Thus, the stereoselectivity of the reaction indicates that the reaction environment is highly hydrophobic, so that no free water molecule attacks the intermediate. Even if some water molecules are present in the active site of the enzyme, they are entirely under the control of the enzyme. If this type of reaction can be realized using synthetic substrates, a new method will be developed for the preparation of optically active carboxylic acids that have a chiral center at the α-position.

2
Screening and Substrate Specificity

At the start of this project, we chose α-arylpropionic acids as the target molecules, because their *S*-isomers are well established anti-inflammatory agents. When one plans to prepare this class of compounds via an asymmetric decarboxylation reaction, taking advantage of the hydrophobic reaction site of an enzyme, the starting material should be a disubstituted malonic acid having an aryl group on its α-position.

2.1
Method of Screening

To screen a microorganism which has an ability to decarboxylate α-aryl-α-methylmalonic acids, a medium was used in which phenylmalonic acid was the sole source of carbon, because we assumed that the first step of the metabolic path would be decarboxylation of the acid to give phenylacetic acid, which would be further metabolized via oxidation at the α-position. Thus, a

$$\underset{\text{phenylmalonic acid}}{\text{Ph-CH}(CO_2H)_2} \xrightarrow{-CO_2} \underset{\text{phenylacetic acid}}{\text{Ph-}CH_2\text{-}CO_2H} \longrightarrow \underset{\text{mandelic acid}}{\text{Ph-CH(OH)-}CO_2H} \longrightarrow \underset{\text{benzoylformic acid}}{\text{Ph-C(=O)-}CO_2H} \longrightarrow \underset{\text{benzoic acid}}{\text{Ph-C(=O)-OH}} \rightleftharpoons \text{further metabolism} \quad (8)$$

microorganism that has an ability to grow on this medium would be expected to decarboxylate α-aryl-α-methylmalonic acids, as the only difference in structure between the two molecules is the presence or absence of a methyl group in the α-position. If the presence of a methyl group inhibits the subsequent oxidation, then the expected monoacid would be obtained. Many soil samples and type cultures were tested and a few strains were found to grow on the medium. We selected a bacterium identified as *Alcaligenes bronchisepticus*, which has the ability to realize the asymmetric decarboxylation of α-methyl-α-phenylmalonic acid [3]. The decarboxylation activity was observed only when the microorganism was grown in the presence of phenylmalonic acid, indicating that the enzyme is an inducible one.

2.2
Metabolic Path of Phenylmalonic Acid

To elucidate the metabolic pathway of phenylmalonic acid, the incubation broth of *A. bronchisepticus* on phenylmalonic acid was examined at the early stage of cultivation. After a one-day incubation period, phenylmalonic acid was recovered in 80% yield. It is worthy of note that the supposed intermediate, mandelic acid, was obtained in 1.4% yield, as shown in Eq. (8). The absolute configuration of this oxidation product was revealed to be *S*. After 2 days, no metabolite was recovered from the broth. It is highly probable that the intermediary mandelic acid is further oxidized via benzoylformic acid. As the isolated mandelic acid is optically active, the enzyme responsible for the oxidation of the acid is assumed to be *S*-specific. If this assumption is correct, the enzyme should leave the intact *R*-enantiomer behind when a racemic mixture of mandelic acid is subjected to the reaction. This expectation was nicely realized by adding the racemate of mandelic acid to a suspension of *A. bronchisepticus* after a 4-day incubation [4].

As shown in Eq. (9), optically pure (*R*)-mandelic acid was obtained in 47% yield, as well as 44% of benzoylformic acid. Benzoic acid was also isolated, although in very low yield, probably as a result of oxidative decarboxylation of benzoylformic acid.

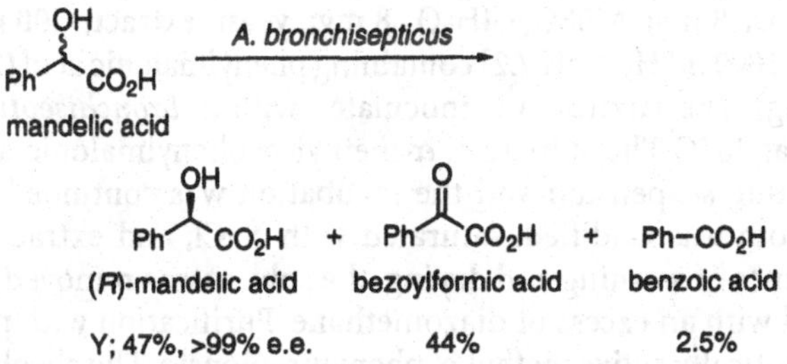

Ph–CH(OH)–CO_2H (mandelic acid) —*A. bronchisepticus*→ (*R*)-mandelic acid + bezoylformic acid + benzoic acid (Ph–CO_2H) (9)

Y; 47%, >99% e.e. 44% 2.5%

The present reaction was proven to occur even when the microorganism had been grown on peptone as the sole carbon source. These results lead to the conclusion that this enzyme system is produced constitutively. In the case of mandelate-pathway enzymes in *Pseudomonas putida*, (*S*)-mandelate dehydrogenase was shown to be produced in the presence of an inducer (mandelic acid or benzoylformic acid) [5]. Thus, the expression of the present oxidizing enzyme of *A. bronchisepticus* is different from that of *P. putida*.

When the resulting mixture of benzoylformic acid and (*R*)-mandelic acid was treated with a cell free extract of *Streptomyces faecalis* IFO 12964 in the presence of NADH, the keto acid can be effectively reduced to (*R*)-mandelic acid (Fig. 1). Fortunately the presence of *A. bronchisepticus* and its metabolite had no influence on the reduction of the keto acid. The regeneration of NADH was nicely achieved by coupling the reaction with reduction by formic acid with the aid of formate dehydrogenase. As a whole, the total conversion of racemic mandelic acid to the *R*-enantiomer proceeded with very high chemical and optical yields. The method is very simple and can be performed in a one-pot procedure [6].

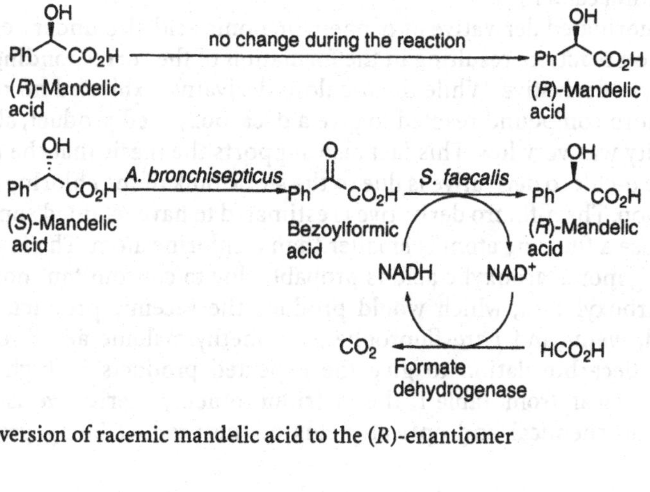

Fig. 1. Conversion of racemic mandelic acid to the (*R*)-enantiomer

2.3 Substrate Specificity

To a 500-ml. Sakaguchi flask were added 50 ml of the sterilized inorganic medium [$(NH_4)_2HPO_4$, 10 g; K_2HPO_4, 2 g; $MgSO_4 \cdot 4H_2O$, 300 mg; $FeSO_4 \cdot 7H_2O$, 10 mg; $ZnSO_4 \cdot 7H_2O$, 8 mg; $MnSO_4 \cdot 4H_2O$, 8 mg; yeast extract, 200 mg and D-biotin, 0.02 mg in 1000 ml H_2O, pH 7.2] containing phenylmalonic acid (250 mg) and peptone (50 mg). The mixture was inoculated with *A. bronchisepticus* and shaken for 4 days at 30 °C The substrate, α-methyl-α-phenylmalonic acid was added to the resulting suspension and the incubation was continued for five more days. The broth was acidified, saturated with NaCl, and extracted with ether. After a sequence of washing and drying, the solvent was removed and the residue was treated with an excess of diazomethane. Purification with preparative TLC afforded optically active methyl α-phenylpropionate. The absolute configuration proved to be *R* by its optical rotation and the enantiomeric excess was determined to be 98% by HPLC, using a column with an optically active immobile phase (Table 1). Since there is no other example of asymmetric decarboxylation, we decided to investigate this new reaction further, although the absolute configuration of the product is opposite to that of anti-inflammatory agents, which is S. First, the alkyl group was changed to ethyl instead of methyl, and it was found that the substrate was not affected at all. This difference in reactivity must come from the difference in steric bulk. Thus it is clear that the pocket of the enzyme for binding the alkyl group is very narrow and has little flexibility. Next, variation of the aromatic part was examined. When the aryl group is 4-chlorophenyl or 6-methoxy-2-naphthyl, the substrate was decarboxylated smoothly to afford the optically active monoacid. Also, α-methyl- α-2-thienylmalonic acid was accepted a good substrate. When the substituent on the phenyl ring was 4-methoxy, the rate of reaction was slower than those of other substrates and the yield was low, although there was no decrease in the enantioselectivity. On the other hand, no decarboxylation was observed when the aryl group was 2-chlorophenyl, 1-naphthyl or benzyl. It is estimated that steric hindrance around the prochiral center is extremely severe and the aryl group must be directly attached to the prochiral center [7].

The α-fluorinated derivative of α-phenylmalonic acid also underwent a decarboxylation reaction, resulting in the formation of the corresponding α-fluoroacetic acid derivative. While the *o*-chloro derivative exhibited no reactivity, the *o*-fluoro compound reacted to give a decarboxylated product, although the reactivity was very low. This fact also supports the thesis that the unreactivity of the *o*-chloro derivative is due to the steric bulk of the chlorine atom at the *o*-position. The *o*-fluoro derivative is estimated to have retained some reactivity because a fluorine atom is smaller than a chlorine atom. The low e.e. of the resulting monocarboxylic acid is probably due to concomitant non-enzymatic decarboxylation, which would produce the racemic product. On the other hand, *meta*- and *para*-fluorophenyl-α-methylmalonic acids smoothly underwent decarboxylation to give the expected products in high optical purity. As is clear from Table 1, the *m*-trifluoromethyl derivative is a good substrate, the chemical and optical yields of the product being practically

Table 1. Asymmetric decarboxylation of α-Alkyl-α-arylmalonic

R, Ar, CO_2H, CO_2H — *Alcaligenes bronchisepticus* → R, Ar, H, CO_2H

α-alkyll-α-aryllmalonic acid → α-ary-lα-alkylacetic acid

Ar	R	[Sub] (%)	Yield (%)	e.e. (%)
	CH_3	0.5	80	98
	C_2H_5	0.2	0	–
Cl–	CH_3	0.5	95	98
CH_3O	CH_3	0.5	96	>95
CH_3O–	CH_3	0.1	48	99
S	CH_3	0.3	98	95
	CH_3	0.2	0	–
Cl	CH_3	0.2	0	–
–CH_2–	CH_3	0.2	0	–
	F	0.1	64	95
F	CH_3	0.3	12	54
F	CH_3	0.5	75	97
F–	CH_3	0.5	54	98
F_3C	CH_3	0.5	54	98

quantitative. This can be attributed to the strong electron-withdrawing effect of this substituent.

3 Isolation of the Enzyme and the Gene

The bacterium isolated from a soil sample was shown to catalyze the decarboxylation of disubstituted malonic acid as described above. Although the configuration of the product was opposite to that of physiologically active anti-inflammatory agents, the present reaction is the first example of an asymmetric synthesis via decarboxylation of a synthetic substrate. The reaction was further investigated from the standpoint of chemical and biochemical conversion. As the first step, the enzyme was isolated from the original bacterium, *Alcaligenes bronchisepticus*, and purified. The gene coding the enzyme was also isolated and overexpressed, using a mutant of *E. coli.*

3.1 Isolation of the Enzyme

A. bronchisepticus was cultivated aerobically at 30 °C for 72 h in an inorganic medium (vide supra) in 1 liter of water (pH 7.2) containing 1% of polypeptone and 0.5% of phenylmalonic acid. The enzyme was formed intracellularly and induced only in the presence of phenylmalonic acid. All the procedures for the purification of the enzyme were performed below 5 °C. Potassium phosphate buffer of pH 7.0 with 0.1 mM EDTA and 5 mM of 2-mercaptoethanol was used thoughout the experiments. The enzyme activity was assayed by formation of pheylacetic acid from phenylmalonic acid. The summary of the purification procedure is shown in Table 2. The specific activity of the enzyme increased by 300-fold to 377 U/mg protein with a 15% yield from cell-free extract [9]. One unit was defined as the amount of enzyme which catalyzes the formation of 1 mmol of phenylacetic acid from phenylmalonic acid per min.

The enzyme was judged to be homogeneous to the criteria of native and SDS-PAGE, HPLC with a TSK gel G-3000 SWXL gel-filtration column, and isoelectric focusing, all of the methods giving a single band or a single peak. The enzyme was tentatively named arylmalonate decarboxylase, AMDase in short.

The molecular mass of the native AMDase was estimated to be about 22 kDa by gel filtration on HPLC. Determination of the molecular mass of denatured protein by SDS-PAGE gave a value of 24 kDa. These results indicate that the purified enzyme is a monomeric protein. The enzyme had an isoelectric point of 4.7. The amino acid sequence of the NH_2-terminus of the enzyme was determined to be Met1-Gln-Gln-Ala-Ser5-Thr-Pro-Thr-Ile-Gly10-Met-Ile-Val-Pro-Pro15-Ala-Ala-Gly-Leu-Val20-Pro-Ala-Asp-Gly-Ala25.

A few examples of decarboxylation reaction using isolated enzyme are shown in Table 3. The most important point is that the reaction proceeded smoothly

Table 2. Purification Table of AMDase from *A. bronchisepticus*

Purification step	Total protein (mg)	Total activity (unit)	Specific activity (unit/mg protein)	Yield (%)
Cell-free extract	8630	10950	1.26	100
Heat fractionation	4280	9540	2.22	87
Ammonium sulfate	2840	10350	3.64	95
DEAE-Toyopearl	244	5868	24.1	54
Butyl-Toyopearl	14.7	3391	231	31
QAE-Toyopearl	4.32	1627	377	15

Table 3. Synthesis of optically active α-arylpropionates using AMDase[a]

$$\mathrm{Ar{-}C(CH_3)(CO_2H)_2} \xrightarrow{\text{AMDase}} \mathrm{Ar{-}CH(CH_3)CO_2H}$$

Ar	Time (h)	Yield (%)	*e.e.* (%)	config.
phenyl	20	100	>99	R
p-methoxyphenyl	20	99	>99	R
2-thienyl	1	97	>99	S

[a] The reaction was carried out in 0.5 M phosphate buffer (pH 6.5) containing 5 mM 2-mercaptoethanol, 22.6 units/ml AMDase (45.2 units/ml for run 2) and 0.1 M substrate (semi-Na salt) 30 °C.

without any aid of the cofactors which are usually required by other decarboxylases and transcarboxylases.

3.2 Cloning and Heterologous Expression of the Gene

For the further investigation of this novel asymmetric decarboxylation, the DNA sequence of the gene should be clarified and cloned in a plasmid for gene engineering. The genomic DNA of *A. bronchisepticus* was digested by *Pst*I, and the fragments were cloned in the *Pst*I site of a plasmid, pUC 19. The plasmids were transformed in an *E. coli* mutant, DH5α-MCR and the transformants expressing AMDase activity were screened on PM plates by the development, of

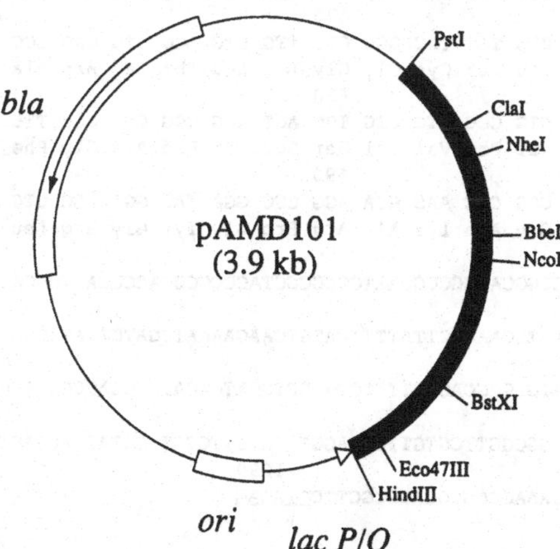

Fig. 2. Partial restriction enzyme map of plasmid pAMD 101. The blackened segment shows *A. bronchisepticus* DNA of 1.2 kb

```
                                                                                -481
                        TCGCCACCTCTGACTTGACCGTCGATTTTTGTGATGCTCGTCAAGGGGGCGGAGCCTATGGAAA
                                                                                -401
AAACGCCAGCAACGCGGCCTTTTTACGGTTCCTGGCCTTTTGCTGGCCTTTTGCTCACATGTTCTTTCCTGCGTTATCCC
                                                                                -321
CTGATTCTGTGGATAACCGTATTACCGCCTTTGAGTGAGCTGATACCGCTCGCCGCAGCCGAACGACCGAGCGCAGCGAG
                                                                                -241
TCAGTGAGCGAGGAAGCGGAAGAGCGCCCAATACGCAAACCGCCTCTCCCCGCGCGTTGGCCGATTCATTAATGCAGCTG
                                                                                -161
GCACGACAGGTTTCCCGACTGGAAAGCGGGCAGTGAGCGCAACGCAATTAATGTGAGTTAGCTCACTCATTAGGCACCCC
                                                                                 -81
AAGCTTTCATCGCGCAGGGTTCGCTGGTGCGCTACCAGCCCGCCGACGCGCTCGCCGCGTGGCTGCCCGCGCAGGAGCAG
HindIII                                                                           -1
CGCTGGGTCGACCTGATCTCGCGCGCCAAGCTGACCTTCGCCCCTTGAGATTTTCGGTATCCACAGTAGGAGAACTTTTC
                                            30                                  60
ATG CAG CAA GCA AGC ACT CCC ACC ATC GGC ATG ATC GTG CCG CCC GCC GCG GGT CTG GTG
Met Gln Gln Ala Ser Thr Pro Thr Ile Gly Met Ile Val Pro Pro Ala Ala Gly Leu Val
                                            90                                 120
CCG GCG GAT GGG GCG CGG CTC TAT CCC GAT CTG CCC TTC ATT GCC AGC GGG CTG GGG CTG
Pro Ala Asp Gly Ala Arg Leu Tyr Pro Asp Leu Pro Phe Ile Ala Ser Gly Leu Gly Leu
                                           150                                 180
GGC TCC GTC ACG CCG GAA GGC TAT GAC GCC GTG ATC GAA TCG GTG GTG GAC CAT GCG CGC
Gly Ser Val Thr Pro Glu Gly Tyr Asp Ala Val Ile Glu Ser Val Val Asp His Ala Arg
                                           210                                 240
CGC CTG CAA AAG CAG GGC GCG GCG GTG GTT TCG CTG ATG GGC ACC TCG CTC AGC TTC TAC
Arg Leu Gln Lys Gln Gly Ala Ala Val Val Ser Leu Met Gly Thr Ser Leu Ser Phe Tyr
                                           270                                 300
CGG GGC GCG GCC TTC AAT GCC GCG TTG ACC GTA GCG ATG CGC GAA GCC ACG GGA CTG CCA
Arg Gly Ala Ala Phe Asn Ala Ala Leu Thr Val Ala Met Arg Glu Ala Thr Gly Leu Pro
                                           330                                 360
TGC ACG ACC ATG AGC ACG GCG GTC CTG AAC GGA TTG CGC GCC CTG GGC GTG CGC CGC GTC
Cys Thr Thr Met Ser Thr Ala Val Leu Asn Gly Leu Arg Ala Leu Gly Val Arg Arg Val
                                           390                                 420
GCG TTG GCG ACG GCC TAT ATC GAC GAT GTG AAC GAG CGC CTG GCG GCA TTC CTG GCC GAA
Ala Leu Ala Thr Ala Tyr Ile Asp Asp Val Asn Glu Arg Leu Ala Ala Phe Leu Ala Glu
                                           450                                 480
GAG AGC CTG GTT CCC ACC GGC TGC CGC AGC CTT GGC ATC ACG GGC GTG GAG GCC ATG GCG
Glu Ser Leu Val Pro Thr Gly Cys Arg Ser Leu Gly Ile Thr Gly Val Glu Ala Met Ala
                                           510                                 540
CGC GTG GAT ACG GCC ACG CTG GTC GAC CTG TGC GTG CGT GCC TTC GAA GCG GCG CCC GAT
Arg Val Asp Thr Ala Thr Leu Val Asp Leu Cys Val Arg Ala Phe Glu Ala Ala Pro Asp
                                           570                                 600
AGC GAC GGC ATC CTG CTG TCT TGC GGC GGC TTG CTG ACG CTG GAC GCC ATA CCC GAA GTC
Ser Asp Gly Ile Leu Leu Ser Cys Gly Gly Leu Leu Thr Leu Asp Ala Ile Pro Glu Val
                                           630                                 660
GAG CGC CGC CTG GGC GTG CCG GTG GTG TCG AGT TCG CCG GCG GGG TTC TGG GAC GCC GTG
Glu Arg Arg Leu Gly Val Pro Val Val Ser Ser Ser Pro Ala Gly Phe Trp Asp Ala Val
                                           690                                 720
CGG CTT GCG GGG GGA GGG GCC AAG GCA AGG CCC GGA TAC GGC CGG CTG TTC GAC GAG TCC
Arg Leu Ala Gly Gly Gly Ala Lys Ala Arg Pro Gly Tyr Gly Arg Leu Phe Asp Glu Ser
                                                                               800
TGACGATAGCGGCTCGCCGGCGGCAAGGCCGCAATCGCGCGCTAGCCGCGCACCGCAGTGACATGAAACCCGCGCAAGCG
***                                                                            880
GGTTTCTTACTTTGTCCTTAATGCAAAACTTATTTTCATGTCAAGAAAATCGATCTGATATGGTCTGTTGAGCGTATCAG
                                                                               960
ATGGCTTAAAAAGCGCATAATGTGCGCACCCTTTCCCAAGCTGCATGTCATGGGGATCGGGTCGATTTTCTGTAAGCGAA
                                                                              1040
GAGCGTGTTGTCACTAGCTCGGCGGTTCGTGTTCCCAGGTTCATGTTCGGTAGCTAGGACAACTTGGAGTATTCACGGAG
                                            1090
ATGTCTATGACAGAAGACGCAAAGCCGCGCAGGCGGTTCCTGCAG
                                         PstI
```

Fig. 3. Nucleoside sequence of the DNA fragment containing the AMDase gene

the blue color of bromothymol blue, due to the pH change. The transition interval of bromothymol blue is pH 6.0 (yellow) to 7.6 (blue). The expected decarboxylation reaction of the dibasic acid, to form a monobasic carboxylic acid, results in an increase in the pH of the medium. In this way, the decarboxylase activity was easily detected as a blue halo around the colony. One of approximately 700 transformants was found to have AMDase activity. The plasmid (pAMD 100) contained an insert of about 2.8 kb. This insert DNA was further digested with *Pst*I and *Hind*III. The *Pst*I-*Hind*III fragment was subcloned in pUC 19 to generate a new plasmid, pAMD 101(Fig. 2).

The *Pst*I fragment from pAMD 100 was subcloned into the *Pst*I site of pUC 119 in both orientations. Various deletion mutants with the 2.8 kb insert were prepared and sequenced. The DNA fragment and the deduced amino acid sequence are shown in Fig. 3. An open reading frame encoding 240 amino acids showed the same NH_2-terminal amino acid sequence as that of AMDase obtained from *A. bronchisepticus*. Based on the sequence, the mass number of the encoded protein was calculated to be 24734. This is in good agreement with that of the enzyme purified from A. bronchisepticus determined by SDS polyacrylamide gel electrophoresis [10]. *E. coli* DH5a-MCR was cultured in TB medium supplemented with 0.01% thiamine. The amount of the enzyme in the cell-free extract was elevated to 1800 units/l culture broth. We conclude that the enzyme is soluble and maintains its active form in the cells, because whole cells of the *E. coli* transformant showed the same activity as cell-fee extracts. The sequence of ten amino acids at the *N*-terminal of the enzyme isolated from *E. coli* was revealed to be completely consistent with that of the enzyme isolated from *A. bronchisepticus*.

The DNA sequence of the encoding AMDase and the amino acid sequence deduced from it was compared with the data base using DNASIS (Hitachi). No significant homologies were observed with any of the sequences searched.

4
Effect of Additives on Enzyme Activity

To investigate the cofactor requirement and the characteristics of the enzyme, the effects of additives were examined using phenylmalonic acid as the representative substrate. The addition of ATP or ADP to the enzyme reaction mixtures, with or without coenzyme A, did not enhance the rate of reaction. From these results, it is concluded that these co-factors are not necessary for this decarboxylase. It is well established that avidin is a potent inhibitor of the biotin-enzyme complex [11–14]. In the present case, addition of avidin has no influence on the decarboxylase activity, indicating that the AMDase is not a biotin enzyme. Thus, the co-factor requirements of AMDase are entirely different from those of known analogous enzymes, such as acyl-CoA carboxylases [15], methylmalonyl-CoA decarboxylases [11] and transcarboxylases [15, 16].

The effects of various compounds and inhibitors on the enzyme activity were tested. The activity was measured in the presence of 1–10 mM of various metal ions and compounds. The enzyme activity was inhibited by sulfhydryl reagents (concentration 1 mM) such as $PbCl_2$ (relative activity, 18%), $SnCl_2$ (17%), $HgCl_2$

(0%), HgCl (8%), $AgNO_3$ (3%), 5,5′-dithiobis(2-nitrobenzoate) (2%), iodoacetate (3%) and *p*-chloromercuribenzoate (PCMB) (0%). *N*-Ethyl maleimide (at 10 mM) causes 72% inhibition of the decarboxylase activity. Thus the AMDase appears to be a thiol decarboxylase. As is clear from the DNA sequence described above that this enzyme contains four Cys residues. At least, some of them exhibit a free SH group which plays an essential role in the active site of the enzyme. The activity of the enzyme was not lost upon incubation with the following agents: several divalent metal cations, such as $NiCl_2$, $CoCl_2$, $BaCl_2$, $MgCl_2$ and $CaCl_2$, carbonyl reagents, such as NaN_3, NH_2OH, KCN, metal chelating agents such as EDTA, 8-quinolinol, bipyridyl, 1,10-phenanthroline, serine inhibitors such as phenylmethanesulfonyl fluoride (at 10 mM). It is estimated that AMDase does not contain metal ions.

5 Active Site-Directed Inhibition and Point Mutation

In the previous studies using inhibitors and additives, it became clear that AMDase requires no cofactors, such as biotin, coenzyme A and ATP. It is also suggested that at least one of four cysteine residues plays an essential role in asymmetric decarboxylation. One possibility is that the free SH group of a cysteine residue activates the substrate in place of coenzyme A. Aiming at an approach to the mechanism of the new reaction, an active site-directed inhibitor was screened and its mode of interaction was studied. Also, site-directed mutagenesis of the gene coding the enzyme was performed in order to determine which Cys is located in the active site.

5.1 Screening of an Active Site-Directed Inhibitor

We screened for a potent inhibitor against the AMDase-catalyzed decarboxylation of α-methyl-α-phenylmalonic acid to give α-phenylpropionic acid. Among the compounds shown in Fig. 4 which have structures similar to the substrate,

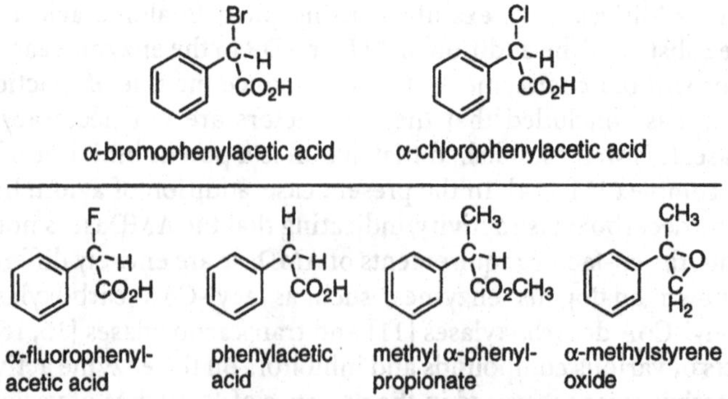

Fig. 4. Compounds tested as inhibitor: the top two have an inhibitory effect while the bottom four do not

(±)-α-halophenylacetic acids remarkably inhibited the reaction. Especially, α-bromophenylacetic acid showed a striking inhibitory effect on the AMDase-catalyzed decarboxylation reaction [17]. On the contrary, other substrate analogues showed no inhibitory effects. A Lineweaver-Burk plot (Fig. 5) for α-bromophenylacetic acid indicated that this compound was a competitive inhibitor [18], with a K_i value of 3.6 mM at 24 °C.

Since the mode of inhibition is competitive and the K_i value is extraordinarily small compared to the K_m value of the substrate (25 mM), it is strongly suggested that this inhibitor blocks the active site and prevents approach of the substrate to the catalytic site of the enzyme. It is assumed that α-bromophenylacetic acid interacts with a cysteine residue at the active site in some way. The high electron-withdrawing effect of the bromine atom would have an important role in the inhibition mechanism. Thus, the mode of binding of the inhibitor to the active site of the enzyme is presumed to resemble that of the substrate closely. Accordingly, disclosure of the way the inhibitor interacts with the enzyme would provide important information on how the enzyme activates the substrate.

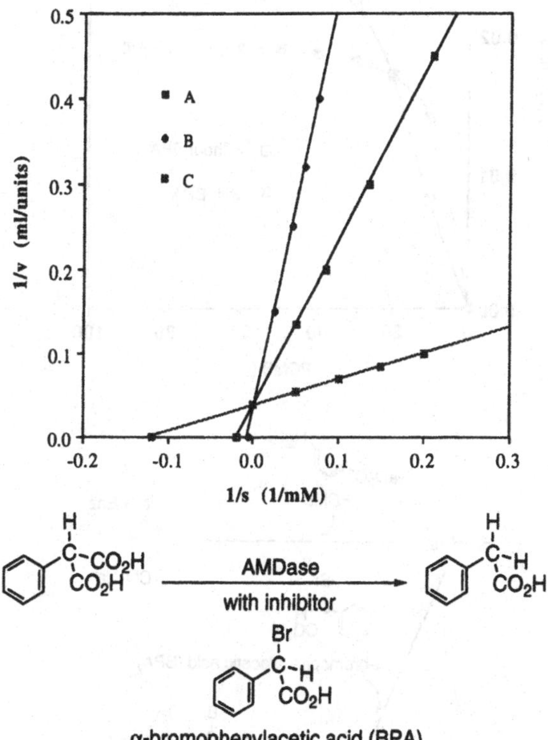

Fig. 5. Inhibition mode of α-bromophenylacetic acid against AMDase-catalyzed decarboxylation. Lineweaver-Burk plot in the presence of the acid; A, 100 μM; B, 20 μM; C, 0 μM

5.2 Titration of the Active Site SH Residue

As deduced from the DNA sequence of the gene, AMDase contains four cysteine residues. Since α-halocarboxylic acids are generally active alkylating agents there is a possibility that α-bromophenylacetic acid reacts with several cysteine residues of the enzyme. Therefore, we tried to clarify how many cysteine residues react with this inhibitor. It is well established that when *p*-chloromercuribenzoate (PCMB) binds to a cysteine residue, the absorbance at 255 nm increases due to the formation of an aryl-Hg–S bond. Thus it is possible to estimate the number of free S-H residues of the enzyme by titration with PCMB solution (Fig. 6). When the native enzyme had reacted with PCMB, the absorbance at 255 nm increased by 0.025. On the other hand, when PCMB solution was added to the enzyme solution after the enzyme was incubated with α-bromophenyl-

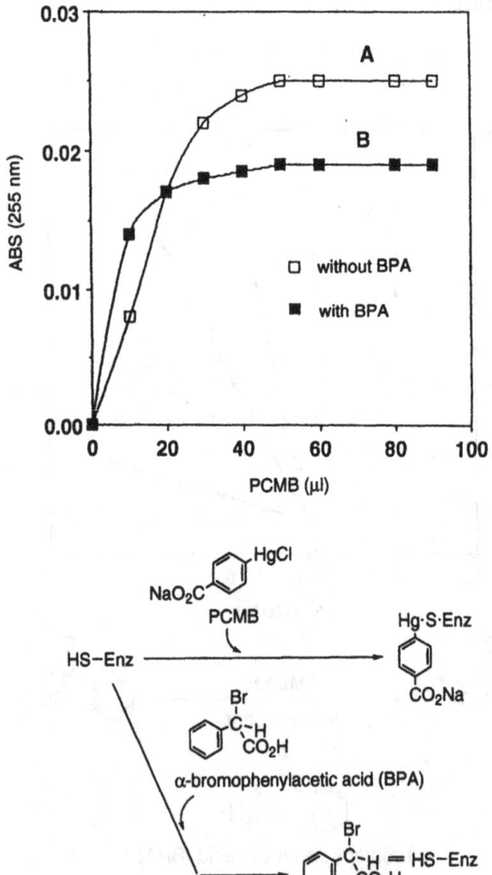

Fig. 6. Titration of cysteine residues with PCMB and BPA

acetic acid, the increase of absorbance was 0.018, just three fourths of the value for the free enzyme. These results clearly show that one fourth of cysteine residues were blocked by α-bromophenylacetic acid and could not react with PCMB. As described above, this enzyme has four cysteine residues. Thus, it is concluded from this titration measurement that all four cysteines are in the free SH form and that one of them, that which reacted with α-bromophenylacetic acid, should be located at the catalytic site. How, then, does the bromo acid block the cysteine residue?

5.3
Spectroscopic Studies of Enzyme-Inhibitor Complex

There are at least three possibile ways in which the inhibitor can bind to the active site: (1) formation of a sulfide bond to a cysteine residue, with elimination of hydrogen bromide [Eq. (10)], (2) formation of a thiol ester bond with a cysteine residue at the active site [Eq. (11)], and (3) formation of a salt between the carboxyl group of the inhibitor and some basic side chain of the enzyme [Eq. (12)]. To distinguish between these three possibilities, the mass numbers of the enzyme and enzyme-inhibitor complex were measured with matrix-assisted laser desorption ionization time-of-flight mass spectrometry (MALDI). The mass number of the native AMDase was observed as 24766, which is in good agreement with the calculated value, 24734. An aqueous solution of α-bromophenylacetic acid was added to the enzyme solution, and the mass spectrum of the complex was measured after 10 minutes. The peak is observed at mass number 24967. If the inhibitor and the enzyme bind to form a sulfide with elimination of HBr, the mass number should be 24868, which is smaller by about one

$$\underset{\text{MW = 24734}}{\text{Enzyme(NH}_2\text{)–SH}} + \underset{\text{MW = 215}}{\text{PhCHBrCO}_2\text{H}} \xrightarrow{-\text{HBr}} \underset{\text{MW = 24868}}{\text{Enzyme(NH}_2\text{)–S–CHPh(CO}_2\text{H)}} \qquad (10)$$

$$\xrightarrow{-\text{H}_2\text{O}} \underset{\text{MW = 24931}}{\text{Enzyme(NH}_2\text{)–S–C(=O)–CHBr–Ph}} \qquad (11)$$

$$\xrightarrow{\text{salt formation}} \underset{\text{MW = 24949}}{\text{Enzyme(SH)–NH}_3^+\ {}^-\text{OOC–CHBr–Ph}} \qquad (12)$$

hundred units than the observed value. On the other hand, if the binding mode is formation of a thiol ester or a salt, the mass number is expected to be 24931 and 24949, respectively. Accordingly, sulfide formation is very unlikely; this conclusion is consistent with the fact that the mode of inhibition is reversible. However, it is difficult to distinguish between the formation of a thiol ester and that of a salt by mass spectroscopy alone [17].
From the kinetics and mass measurement results, α-bromophenylacetic acid was found to bind with AMDase in a reversible manner, and the resulting bond was estimated to be a thiol ester or a simple salt. In the former case, some thiol compounds would be expected to attack the carbonyl group and liberate the free enzyme, resulting in the recovery of AMDase activity, whereas thiol compounds should have no effect on the dissociation of the carboxylate-enzyme complex. In this way, the two possibilities could be distinguished from one another. This prediction turned out to be true. When a large excess of 2-mercaptoethanol was added to the enzyme-inhibitor complex, the activity of the enzyme gradually increased until 100% of its activity was finally restored. Even when the thiol was added 24 hours after the inhibition experiment, recovery of the activity of AMDase was observed. This result clearly shows that α-bromophenylacetic acid was released from the active site of AMDase when 2-mercaptoethanol was added to the AMDase-inhibitor complex. Thus, it is highly probable that the potent inhibitory effect of α-bromophenylacetic acid is due to competitive formation of a thiol ester with a cysteine residue that is present at the active site of the enzyme. These results suggest that the first event which occurs between the substrates and the enzyme is also a formation of a thiol ester. If both the carboxyl group of the substrate and the thiol group of the enzyme are activated by the presence of other amino acids at the active site, formation of a thiol ester without the aid of ATP can become possible. In case of α-bromophenylacetic acid, the strong electron-withdrawing character of the bromine atom at the α-position is considered to activate the carboxyl group for a nucleophilic attack of a thiol group of the enzyme.

In addition to mass spectroscopic studies, we have been able to observe an absorbance which can be assigned to the deformation vibration of a C–S bond ($1103\ cm^{-1}$) by FT-IR spectroscopy of the complex [19].

5.4
Site-Directed Mutagenesis

Site-directed mutagenesis is one of the most powerful methods of studying mechanisms of enzyme-catalyzed reactions. Since this technique makes it possible to replace a specific amino acid residue of an enzyme by an arbitrary one, it is particularly useful to specify the amino acid residue(s) which is responsible for the activity [20–22]. In the case of AMDase, one of four cysteine residues was presumed to be involved in the catalytic site by the titration experiments. To determine which Cys is located at the active site, preparation of four mutant enzymes, in each of which one of the cysteines is replaced another amino acid, and kinetic studies on them, are expected to be most informative. Which amino acid should be introduced in place of cysteine? To decide on the best candidate, the mechanism of the reation should be considered. In the present decarboxylation reaction, the enzyme contains a cysteine residue at the catalytic site, and requires no coenzyme A and ATP. One possible explanation for these unique characteristics is that the cysteine residue of the enzyme itself plays the role of coenzyme A. This assumption leads to a conclusion that the key to the activation of the substrate by Cys will be its nucleophilicity and anion-stabilizing effect. In fact, kinetic studies of this decarboxylation reaction, described in the next chapter, revealed that the transition state has a negative charge.

Thus, it is estimated that a substitution of cysteine in the active site for serine would greatly decrease the rate of reaction, because of the relatively weak nucleophilicity and anion-stabilizing effect of a hydroxy group compared to a thiolate functionality. In this way, if the mutant enzyme partially retains its catalytic activity, even when the essential cysteine in the active site is replaced by serine, the k_{cat} value would greatly decrease whereas the K_m value would not be seriously affected. On the other hand, if a cysteine residue other than the catalytic one is replaced by serine, the effect on reactivity will be moderate: because the steric bulk of serine resembles that of cysteine, it will more or less retain the hydrogen bonding pattern of the wild type enzyme. We therefore prepared four mutant genes, in which one of four codons corresponding to cysteine was replaced by that of serine via site-directed mutagenesis. Four AMDase mutants, expressed in a mutant of *E. coli*, were purified to homogeneity by SDS-PAGE electrophoresis and used in kinetic studies.

First, the activity of the enzyme was measured and kinetic parameters were determined by Lineweaver-Burk plots, using phenylmalonic acid as the substrate. The results are summarized in Table 4. Among four mutants, C188S showed a drastic decrease in the activity (k_{cat}/K_m). The low activity was due to a decrease in the catalytic turnover number (k_{cat}) rather than in affinity for the substrate (K_m).

The CD spectrum of the C188S mutant is essentially the same as that of the wild type enzyme, which reflects the fact that the tertiary structure of this mutant changed little compared to that of the wild-type enzyme. Calculated values of the secondary structure content of the mutant enzymes, based on the J-600S Secondary Structure Estimation system (JASCO), are shown in Table 5. These data also show that there is no significant change in the tertiary structure of the C188S mutant. The fact that the k_{cat} value of this mutant is extremely small, despite little change in conformation, clearly indicates that Cys^{188} is located at the active site. Another mutant which showed only a small change in confor-

Table 4. Relative activities and Kinetic parameters of the wild type and four mutant enzymes

	K_m (mM)	k_{cat} (s^{-1})	k_{cat}/K_m
Wild type	13.3	366	27.5
C101S	4.3	248	57.6
C148S	11.5	100	8.7
C171S	9.1	62.3	6.8
C188S	4.9	0.62	0.13

Table 5. Secondary-structure content of the wild type and four mutant enzymes

	α-Helix (%)	β-Sheet (%)	Turn (%)
Wild type	39.6	4.8	11.1
C101S	38.0	3.2	5.0
C148S	29.6	5.7	5.6
C171S	21.5	4.2	5.0
C188S	32.9	4.1	9.1

mation was C101S, which exhibited higher activity than the wild enzyme. The higher activity is attributed to a smaller K_m value. The specific reason for higher activity is not clear at present, but it is assumed either that Cys^{101} is located near the catalytic site, or that mutation brought about some increase in flexibility, which made the induced a tighter fit of the enzyme to the substrate.

On the other hand, the catalytic activity of mutants C148S and C171S decreased in spite of the smaller K_m values than that of the wild type enzyme. It can be assumed that the decrease in α-helix structure caused a decrease in k_{cat} value. The distance between the catalytic amino acid and the binding substrate would become longer because of the change in conformation. It was thus concluded that $Cysteine^{188}$ is located in the catalytic site of the enzyme [23].

6 Kinetics and Stereochemistry

So far, it has become clear that Cys^{188} plays an essential role in the asymmetric decarboxylation of disubstituted malonic acids. It follows that studies of reaction kinetics and stereochemistry will serve to disclose the role of the specific cysteine residue and the reaction intermediate.

6.1 Effect of Substituents on the Aromatic Ring

Because this reaction proceeds only when the aromatic ring is directly attached to the α-carbon atom, the electronic effect of the substituents on the phenyl ring will be significant controlling factors, and quantitative studies on the rate of reaction will provide important information concerning the mechanism of the reaction. Fortunately, since AMDase accepted a wide range of *meta*- and *para*-substituted phenylmalonic acids, it is possible to evaluate the electronic effects on the rate of reaction quantitatively. The kinetic experiments were performed using 15 mM of monosubstituted phenylmalonic acids in 50 mM Tris-HCl buffer at 30 °C [9]. The results are summarized in Table 6. It is clear that the reactivity of the compounds tested is more strongly dependent on k_{cat} than on K_m.

If the values of the fluorine-substituted substrate are excluded, the Michaelis constants of various arylmalonic acids can be regarded as nearly the same. Then it can be concluded that the difference in k_{cat} values is mainly due to the dif-

Table 6. Kinetic constants of AMDase for substituted phenylmalonic acis

X-C_6H_4-$CH(CO_2H)_2$ —AMDase→ X-C_6H_4-CH_2CO_2H

X	K_m (mM)	V_{max} (U/mg)	k_{cat} (s^{-1})	Relative activity
H	13.9	882.7	353	100
p-CH_3	8.89	311.9	125	55
p-CH_3O	9.36	229.9	92.0	39
p-F	13.5	488.1	195	57
m-F	35.7	5451	2180	241
o-F	27.5	1666	666	95
p-Cl	8.32	1954	782	370
m-Cl	9.84	3554	1422	569
o-Cl	12.6	2413	1085	338

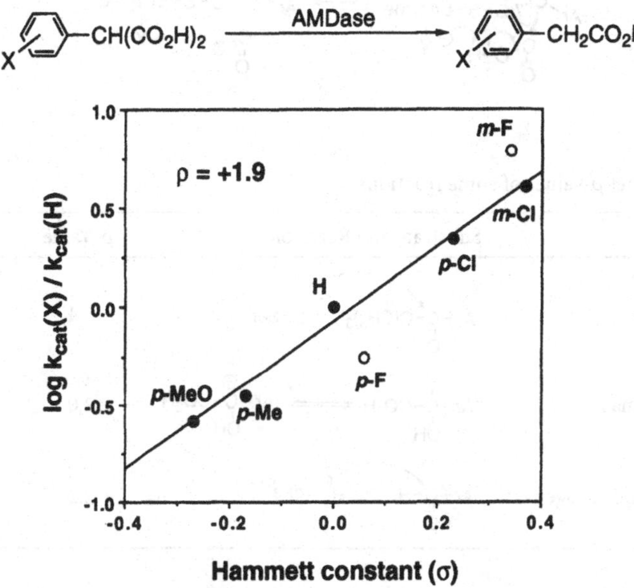

Fig. 7. Hammett plot of k_{cat} for the AMDase catalyzed decarboxylation of a series of X-phenylmalonic acids

ference in the electronic effect of the substituents. Accordingly, it will be instructive to examine the relationship between k_{cat} and Hammett's ρ-values for the substituents. As shown Fig. 7, the logarithm of k_{cat} for the compounds listed in Table 6 over that for the nonsubstituted substrate clearly shows a linear correlation, the ρ-value being + 1.9. The fact that the ρ-value is positive means that the transition state has a negative charge. Some similar values that have been obtained in other reactions are listed in Table 7. The rate-determining step of the chymotrypsin catalyzed hydrolysis of benzoic acid ester is nucleophilic attack of the activated serine residue at the carbonyl group [24]. The key step in racemization of mandelic acid is assumed to be abstraction of the α-proton to form an enolate [25, 26]. In the case of reduction of substituted benzaldehyde, the transition state is produced by the attack of hydride (or an electron) at the carbonyl carbon atom [27]. Consideration of these reaction mechanisms and their ρ-values leads to the conclusion that the key step in the present decarboxylation reaction is probably either nucleophilic attack by the enzyme (possibly Cys[188]) at one of the carboxyl groups or formation of an enolate, as illustrated in Eq. 13.

$$Ar\text{-}C(CH_3)(CO_2H)CO_2H \;\; HS\text{-}Enzyme \;(\ominus X) \longrightarrow Ar\text{-}C(CH_3)(COOH)\text{-}C(O^{\ominus})(X)\text{-}S\text{-}Enzyme \longrightarrow Ar\text{-}C(CH_3){=}C(O^{\ominus})\text{-}S\text{-}Enzyme \;\; (C{=}O{=}O) \;\; HX \qquad (13)$$

Table 7. Hammet ρ-values of some reactions

Enzyme	Substrate and Reaction	ρ-Value	Ref.
Chymotrypsin	$Ar\text{-}C(=O)\text{-}C(CH_3)_3 \;\; {}^{\ominus}O\text{-}Ser$	+ 1.4	[24]
Madelate racemase	$Ar\text{-}CH(OH)\text{-}CO_2H \rightleftharpoons Ar\text{-}C^{\ominus}(OH)\text{-}CO_2H$	+ 2.0	[25, 26]
Yeast alcohol dehydrogenase	$Ar\text{-}C(=O)\text{-}H \;\; H^{\ominus}$ (or $e^{\ominus}$)	+ 2.2	[27]

6.2
Reaction of Chiral α-Methyl-α-phenylmalonic Acid

In order to obtain more information on the reaction intermediate, the stereochemical course of the reaction was investigated. The absolute configuration of the product from α-methyl-α-phenylmalonic acid was unambiguously determined to be *R*, based on the sign of specific rotation. Then, which carboxyl group remains in the propionic acid and which is released as carbon dioxide? To solve this problem we have to distinguish between two prochiral carboxyl groups. The most effective way to do this would be to prepare both enantiomers of chiral α-methyl-α-phenylmalonic acid, each of known configuration, containing ^{13}C on either of the two carboxyl groups. Preparation of both enantiomers of chiral α-methyl-α-phenylmalonic acid was carried out starting from ^{13}C containing phenylacetic acid via resolution of racemic α-methyl-α-hydroxymethylphenylacetic acid, as illustrated in Fig. 8. Since the optical rotations of both enantiomers of this acid are known, the absolute configurations of (+)- and (–)-enantiomers were unambiguously determined. Jones' oxidation gave the desired chiral ^{13}C-labeled malonic acid, although they could not be distinguished by optical rotation.

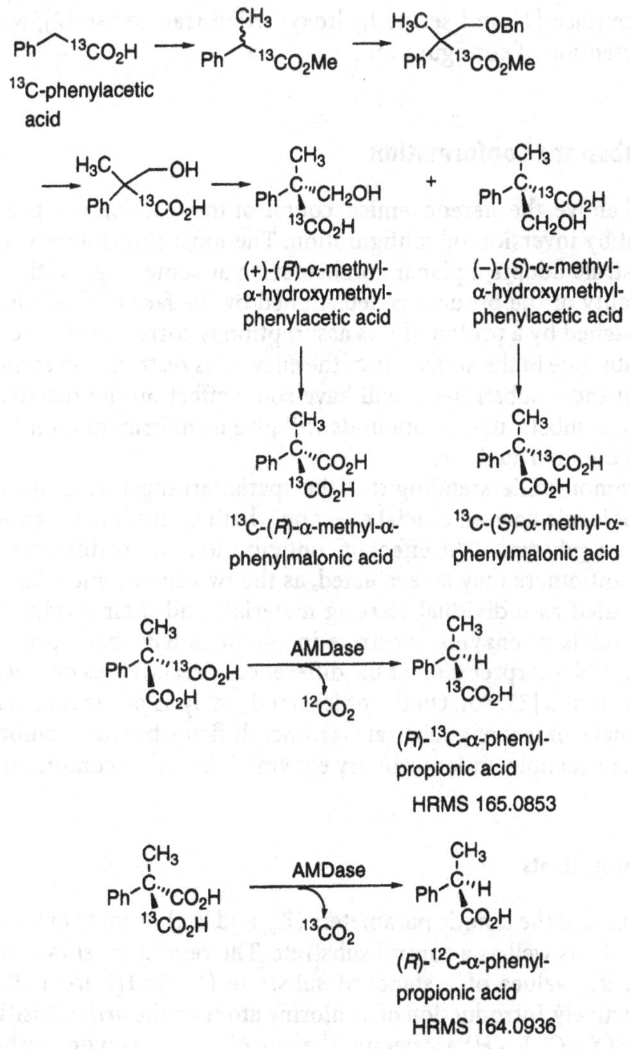

Fig. 8. Preparation and reaction of ^{13}C-containing chiral α-methyl-α-phenylmalonic acid

The enzymatic reaction was performed at 30 °C for 2 hours in a volume of 1 ml of 250 mM phosphate buffer (pH 6.5) containing 50 mM of KOH, 32 U/ml of the enzyme, and [1-^{13}C]-substrate. The product was isolated as the methyl ester. When the (*S*)-enantiomer was employed as the substrate, ^{13}C remained completely in the product, as confirmed by ^{13}C NMR and HRMS. In addition, spin-spin coupling between ^{1}H and ^{13}C was observed in the product, and the frequency of the C–O bond-stretching vibration was down-shifted to 1690 cm^{-1} (cf. 1740 cm^{-1} for ^{12}C–O). On the contrary, reaction of the (*R*)-enantiomer resulted in the formation of (*R*)-monoacid containing ^{13}C only within natural abundance. These results clearly indicate that the *pro-R* carboxyl group of malonic acid is eliminated to form (*R*)-phenylpropionate with inversion of configuration [28]. This is in sharp contrast to the known decarboxylation reaction by malonyl CoA decarboxylase [1] and serine hydroxymethyl transferase [2], which proceeds with retention of configuration.

7
Effect of Substrate Conformation

As described above, the stereochemical course of the reaction was proven to be accompanied by inversion of configuration. The most probable explanation is that the substrate adopts a planar conformation at some stage of the reaction, and the chirality of the product is determined by the face of this intermediate that is approached by a proton. If this assumption is correct and the conformation of the substrate in the active site of the enzyme is restricted in some way, the steric bulk of the *o*-substituents will have some effect on the reactivity. Thus, studies of the *o*-substituted compounds will give us information on the stereochemistry of the intermediates.

It is a common understanding that the spatial arrangements of the substituents of a molecule have an crucial effect on whether an enzyme can accept the compound as a substrate. The effect of configuration on the difference of reactivities of enantiomers may be evaluated, as the two enantiomers can be separated and treated as individual starting materials and their products. In fact, promising models of enzyme-substrate interactions have been proposed that permit successful interpretation of the difference of reactivities between a given pair of enantiomers [29, 30]. On the other hand, analysis of the reactivity of the conformational isomers of a substrate is rather difficult, because conformers are readily interconvertible under ordinary enzymatic reaction conditions.

7.1
Effect of *o*-Substituents

First, we examined the kinetic parameters (K_m and k_{cat}) of some *ortho*-substituted compounds, as well as a control substrate. The results are shown in Table 8. The K_m and k_{cat} values of a standard substrate (X, R=H) are 13.9 mM and 353 s^{-1}, respectively. Introduction of a chlorine atom on the *ortho*-position of the benzene ring (X=Cl, R=H) accelerates the rate of reaction obviously because of its electron-withdrawing property. The steric effect of this substituent is con-

Table 8. Effect of *o*-substituents

X	R	k_m (mM)	k_{cat} (sec^{-1})
H	H	13.9	353
Cl	H	12.6	1085
H	CH_3	25.5	30
Cl	CH_3	No Reaction	
CH_3	CH_3	No Reaction	

sidered to be small, as the K_m value is nearly the same as that of unsubstituted compound (X, R=H). On the other hand, substitution of the α-hydrogen with a methyl group (X=H, R=CH_3) decreases the k_{cat} value less than one tenth (30 s^{-1}). This can be accounted for by the direct binding of an electron-donating group near the reaction site. The steric effect of a methyl group is considered to be small, judging from the K_m values of this compound and phenylmalonic acid. Taking into consideration all of these results together, substitution of a chlorine atom at the *ortho* position of an α-methyl compound is reasonably expected to bring about some rate enhancement, due to its electron-withdrawing effect. However, what happened was entirely different: the α-(*o*-chlorophenyl)-α-methyl derivative (X=Cl, R=CH_3) does not undergo decarboxylation at all by incubation with AMDase. The starting material was recovered intact. It is noteworthy that the corresponding *p*-chlorophenylmalonic acid was smoothly decarboxylated to give the expected monocarboxylic acid. As is clear from the case of *o*-chlorophenylmalonic acid (X=Cl, R=H), a chlorine atom alone is not bulky enough to inhibit the reaction, so it is concluded that the inactivity of the disubstituted compound (X=Cl, R=CH_3) is caused by the presence of two substituents at the *ortho* and α-positions. Further evidence that the steric repulsion between the *ortho*- and α-substituents is the crucial factor inhibiting the enzymatic reaction is also demonstrated by the *o*-methyl derivative (X, R=CH_3), which is not affected by the enzyme.

The most probable interpretation of the above results is that the conformation disfavored by steric repulsion between the *ortho*- and α-substituents is the same conformation that is required for the substrate to be bound in the active site of the enzyme. Undoubtedly it is conformation A (*syn*-periplanarwith respect to the *ortho*- and α-substituents) illustrated in Fig. 9. If the substrate could undergo the reaction via the other planar conformation (B), the expected product would have been obtained, because conformation (B) is free from steric repulsion between the two substituents, and the substrate would have had no difficulty to take up this conformation. The actual inactivity of the two compounds (X, R=Cl, CH_3 and CH_3, CH_3) suggests that, for some reason, conformation (B) is disfavored in the pocket of the enzyme. Then, how much is the energy

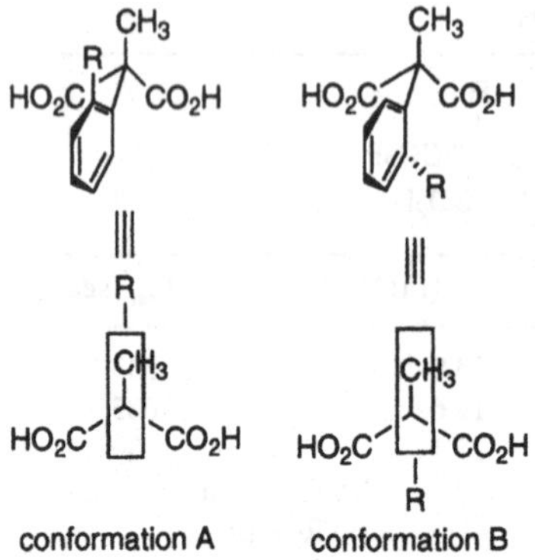

conformation A conformation B

Fig. 9. Possible planar conformations: *A* *syn*-periplanar; *B* *anti*-periplanar

difference between the two conformers A and B? Apparently the *syn*-periplanar conformer A is less favored than B. However, if the binding energy with the enzyme overcomes the difference in potential energy between free A and B, and the enzyme forces the substrate to adopt conformation A, decarboxylation of the substrate will be able to proceed. Thus, whether a compound can react smoothly or not will depend on a balance between the enzyme-substrate binding energy and the potential energy of the "reactive conformer". The free energy of formation of the enzyme-substrate complex is easily calculated, based on K_m values, according to Eqs. 14–17.

$$E + S \underset{}{\overset{K^{\ddagger}_{ES}}{\rightleftharpoons}} [ES] \xrightarrow{k_{cat}} E + P \tag{14}$$

$$\Delta G^{\ddagger}_{ES} = \Delta H^{\ddagger}_{ES} - T\Delta S^{\ddagger}_{ES} \tag{15}$$

$$\ln K^{\ddagger}_{ES} = -\Delta G^{\ddagger}_{ES} / RT \quad \text{(where } R = 1.986 \text{ cal / °K·mol, } T = 298 \text{ °K)} \tag{16}$$

$$K^{\ddagger}_{ES} = K_m^{-1} \tag{17}$$

Of course, the K_m value of α-(*o*-chlorophenyl)-α-methylmalonic acid is not available because of its inactivity. If we suppose that its "imaginary K_m value" is not very different from that of *o*-chlorophenylmalonic acid (X=Cl, R=H, 12.6 mM), then the free energy (ΔG) of formation of enzyme-substrate complex can be calculated as −2.6 kcal/mol at 25 °C. In addition, supposing that the ΔS values for formation of the enzyme-substrate complex are not very different between these two compounds because the ligands around the prochiral centers are similar, then the difference in ΔH between the favored and disfavored conformation will be the key to the interpretation of the reactivity difference between the compounds with and without a methyl group at the α-position.

7.2
Theoretical Calculation of the Potential Energy Surface

To explore the potential energy surfaces for two types of arylmalonic acid, we employed the ab initio molecular orbital method on the internal rotation of the benzene ring [31]. The theoretical calculations were carried out with the Gaussian 2 program [32]. The molecular structures for various rotational angles were optimized by using the 3–21G* basis set with the Hartree-Fock method [33]. The results are shown in Figs. 10 and 11. The potential energy diagram for *o*-chlorophenylmalonic acid is shown in Fig. 10. We obtained two stable structures, that correspond to the *syn*- and anti-periplanar conformers A and B in Fig. 9, respectively. The energy difference between these two conformers is calculated to be 0.8 kcal/mol. This energy difference is far smaller than the binding energy of this substrate to the enzyme, and indicates that steric repulsion between the α-methyl hydrogen atoms and the chlorine atom is not so significant. Thus, it is possible for *o*-chlorophenylmalonic acid to bind the enzyme by taking up the *syn*-periplanar conformation A.

The rotational energy diagram for α-(*o*-chlorophenyl)-α-methylmalonic acid is shown in Fig. 11. In contrast to a simple potential curve for the non-methylated compound, we obtained four energy minima for the forms at the dihedral angles 24.6°, 73.0°, 178.2°, and 278.8°. The most stable conformer adopts a dihedral angle of 178°, which corresponds to the *anti*-periplanar conformer B, whereas the potential energy of the *syn*-periplanar conformation is about 5.5 kcal/mol higher. This potential-surface curve clearly indicates that the structure corresponding to the *syn*-periplanar conformation is unstable due to the

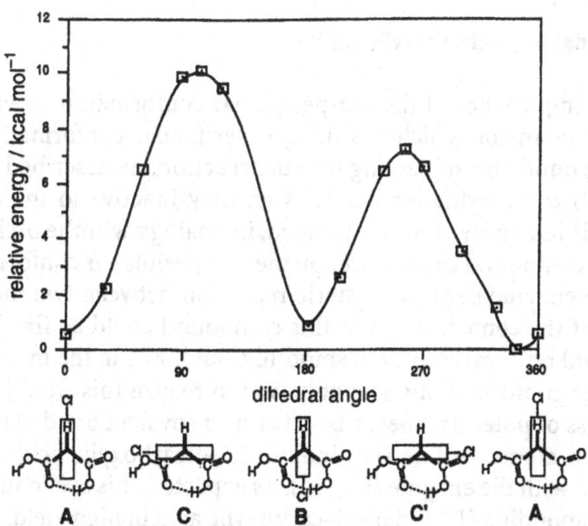

Fig. 10. The potential energy diagram for C–C bond rotation in *o*-chlorophenylmalonic acid calculated with the HF/3–21G* method

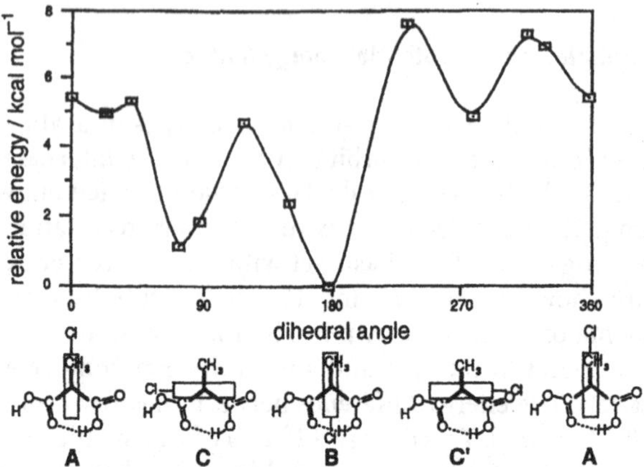

Fig. 11. The potential energy diagram for C–C bond rotation in α-(*o*-chlorophenyl)-α-methylmalonic acid calculated with the HF/3–21G* method

steric repulsion between the chlorine atom and the α-methyl group. Accordingly, the chlorophenyl derivative can be incorporated in the active site of the enzyme in *syn*-periplanar form, whereas the α-(*o*-chlorophenyl)-α-methylmalonic acid is unable to compensate for the energy loss suffered by adopting the *syn*-conformation during binding with the enzyme. This result is consistent with the actual reactivities of the two *o*-chlorophenyl derivatives. The most probable explanation of the results as a whole would be that the decarboxylation reaction will proceed only when the substrates can take up the *syn*-periplanar conformation in the active site of the enzyme.

7.3. Reaction of Indane-1,1-dicarboxylic Acid

The essential importance of the *syn*-periplanar conformation A was confirmed by designing an analog which has the *syn*-periplanar conformation of an unreactive compound and subjecting it to the reaction. As described earlier, α-(*o*-methylphenyl)-α-methylmalonic acid is entirely inactive to the enzyme. The reason for this inactivity is now estimated, in analogy withthe *o*-chloro-derivative, that this compound cannot occupy the *syn*-periplanar conformation in the pocket of the enzyme because of steric repulsion between two methyl groups. Accordingly, if the conformation of this compound could be fixed to *syn*-periplanar, it would be decarboxylated smoothly. But how can the molecule be fixed in an unstable conformation? The only way to realize this would be compensatingfor a loss of potential energy by creating a covalent bond between the two methyl carbon atoms. In this way, indane-1,1-dicarboxylic acid was prepared and incubated with the enzyme (Fig. 12). As expected, this cyclic substrate afforded the corresponding (*R*)-indane-1-carboxylic acid in high yield. The k_{cat} value (1.56 s^{-1}) is smaller than that of phenylmalonic acid (353 s^{-1}) and α-methyl-α-phenylmalonic acid (92 s^{-1}) because of the electron-donating property of two

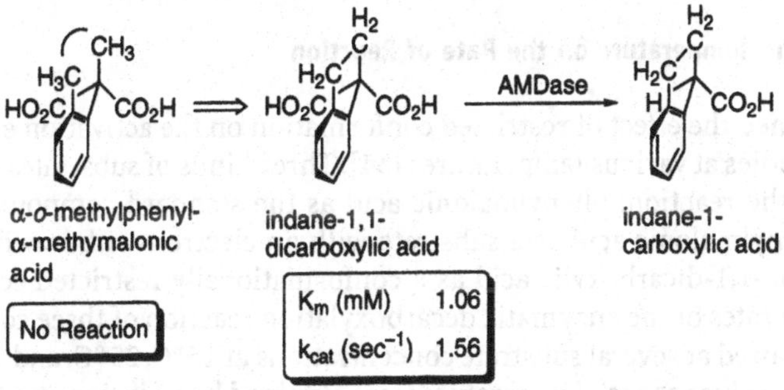

Fig. 12. Reaction of indane-1,1-dicarboxylic acid, the mimic of non-reactive dimethyl derivative

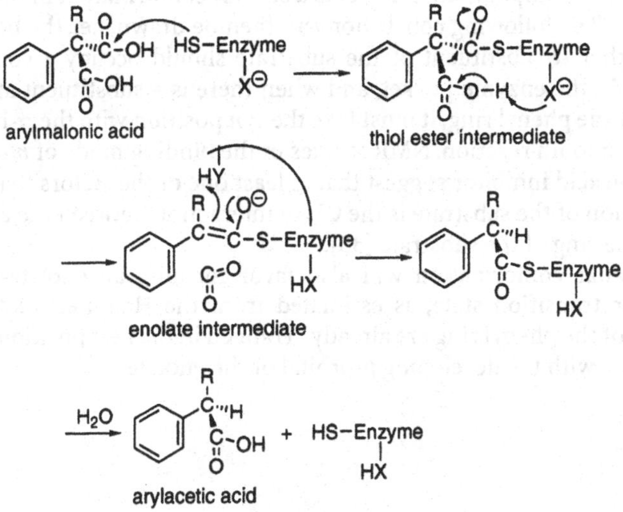

Fig. 13. Proposed reaction mechanism of AMDase-catalyzed decarboxylation

methylene groups on the *ortho*- and *α*-positions. The marked contrast between the reactivities of the dimethyl derivative and indane-1,1-dicarboxylic acid is clearly due to the difference in their freedom of conformation. It is noteworthy that the K_m value of indane-1,1-dicarboxylic acid is smaller by one order of magnitude than those of acyclic compounds. Evidently, it is due to the fact that its conformation is already arranged in the form that fits the binding site of the enzyme; in other words, probably because of the decrease of activation entropy. If the $\Delta H^{\ddagger}$ values for *α*-methyl-*α*-phenylmalonic acid and indane-1,1-dicarboxylic acid are assumed to be the same, the difference in $\Delta S^{\ddagger}$ between two compounds is calculated to be 6.3 cal/K · mol, based on the difference in K_m values.

7.4 Effect of the Temperature on the Rate of Reaction

We examined the effect of restricted conformation on the activation entropy by kinetic studies at various temperatures [34]. Three kinds of substrates were subjected to the reaction: phenylmalonic acid as the standard compound, *ortho*-chlorophenylmalonic acid as a substrate with an electron-withdrawing group, and indane-1,1-dicarboxylic acid as a conformationally restricted compound. The initial rates of the enzymatic decarboxylation reaction of three compounds were measured at several substrate concentrations at 15 °C, 25 °C, and 35 °C. The k_{cat} and K_m values at each temperature were obtained by a Lineweaver-Burk plot, and an Arrhenius plot was made based on these data. The values of the activation enthalpy and entropy are summarized in Table 9. The activation entropy of indane-1,1-dicarboxylic acid is smaller than the others by 9 to 11 cal degree^{-1}mol^{-1}. It can be deduced that the rotation of the benzene ring of this compound is fixed at the most favorable angle for the substrate to bind to the enzyme. Other compounds should take a similar conformation in the active site of AMDase. The following conclusion can then be drawn, i.e., the benzene ring and the other α-substituent of the substrate should occupy a coplanar conformation in the enzyme pocket, and when there is a substituent at the *ortho*-position of the phenyl ring, it must take the *syn* position with the α-hydrogen to undergo a smooth reaction. NMR studies of the binding mode of *m*-fluorinated phosphonic acid inhibitor suggest that at least one of the factors that freeze the conformation of the substrate is the CH-π interaction between the enzyme and the benzene ring of the substrate [35].

This planar conformation will also favor forming an enolate-type intermediate or transition state, as estimated from the Hammett plot, since the *p*-orbitals of the phenyl ring are already arranged in the best positions to be able to conjugate with the developing *p*-orbital of the enolate.

Table 9. Activation parameters for AMDase-catalyzed decarboxylation

Substrate	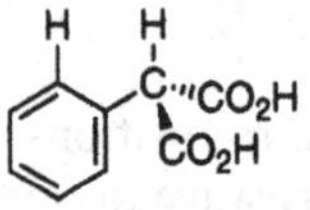 phenylmalonic acid	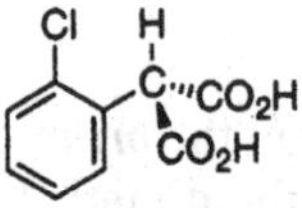 *o*-chlorophenyl-malonic acid	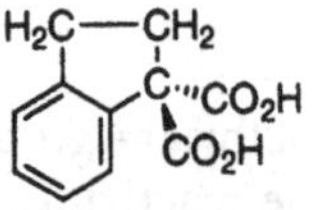 indane-1-1-dicarboxylic acid
k_{cat} (s^{-1})	250	934	1.9
K_m (mM)	11.9	12.9	0.92
$\Delta S^{\neq}$ (cal mol^{-1} K^{-1}	−38.5	−36.8	−27.6
$\Delta S^{\neq}$ (kcal mol^{-1} K^{-1}	2.7	2.4	8.9
$\Delta H^{\neq}$ (kcal mol^{-1} K^{-1}	14.1	13.3	17.1

8 Reaction Mechanism

Our present conclusions as to the mechanism of the novel enzymatic decarboxylation described above are as follows.

The characteristic points that have to be taken into consideration are:

(i) AMDase requires no coenzymes, such as biotin, coenzyme A, and ATP
(ii) this reaction is inhibited by thiol specific reagents, such as PCMB and iodoacetate
(iii) replacement of Cys^{188} with Ser greatly decreases the value of k_{cat}, whereas K_m is not affected
(iv) the transition state of this reaction has a negative charge; (v) the prochirality of the substrate dicarboxylic acid is strictly differentiated and the reaction proceeds with inversion of configuration
(vi) the α-substituent and the *o*-substituent on the benzene ring are likely to occupy a coplanar conformation in the enzyme pocket
(vii) spectroscopic studies on an enzyme-inhibitor complex suggest that the substrate binds to the enzyme via a thiol ester bond.

The Cysteine residue at 188 from the *N*-terminal will be activated by some basic amino acid residue, and attack the *pro*-S carboxyl group to form a thiol ester intermediate. The carboxyl group will probably be protonated and the reactivity to a nucleophile will be higher than usual. Then, a basic amino acid will abstract the proton from another free carboxyl group, and an enolate and carbon dioxide will be generated by C–C bond fission.The high electronegativity of a sulfur atom bound as a thiol ester will facilitate the formation of enolate anion. In this way, Cys^{188} plays an essential role in this reaction in place of coenzyme A. Enantioface-differentiating protonation from an acidic part of the enzyme will give an optically active monocarboxylic acid-enzyme thiol ester, which – in turn – will be hydrolyzed to give the final product and liberate the free enzyme. An X-ray crystallographic analysis of the tertiary structure of AMDase, is now in progress, with the object of elucidating the mechanism more precisely.

9 References

1. Kim YS, Kolattukudy PE (1980) J Biol Chem 255:686
2. Thomas NR, Rose JE, Gani D (1991) J Chem Soc Chem Commun:908
3. Miyamoto K, Ohta H (1990) J Am Chem Soc 112:4077
4. Miyamoto K, Ohta H (1992) Biotech Lett 14:363
5. Hegeman GD (1966) J Bacteriol 91:1140, 1155, 1161
6. Tsuchiya S, Miyamoto K, Ohta H (1992) Biotech Lett 14:1137
7. Miyamoto K, Ohta H (1991) Biocatalysis 5:49
8. Miyamoto K, Tsuchiya S, Ohta H (1992) J Fluorine Chem 59:225
9. Miyamoto K, Ohta H (1992) Eur J Biochem 210:475
10. Miyamoto K, Ohta H (1992) Appl Microbiol Biotech 38:234
11. Galvian J H, Allen S H G (1968) Arch Biochem Biophys 126:838

12. Hoffmann A, Hilpert W, Dimroth P (1989) Eur J Biochem 94:23c
13. Green NM (1965) Biochem Z 118:67
14. Green NM, Tomes E J (1970) Biochem J 118:67
15. Boyer PD (1972) The enzyme, vol 6. Academic Press, New York, pp 37–115
16. Wood HG, Lochmuller H, Riepertinger C, Lynen F (1963) Biochem Z 337:247
17. Kawasaki T, Watanabe M, Ohta H (1995) Bull Chem Soc Jpn 68:2017
18. Segel IH (1975) Biochemical calculations, John Wiley & Sons, New York, p 235
19. Kawasaki T, Fujioka Y, Saito K, Ohta H (1996) Chem Lett:195
20. Bahattacharyya K, Leomte M, Rieke CJ, Garavito RM, Smith W (1996) J Biol Chem 271:2179
21. Mohamedali K, Kurz LC, Rudolph FB (1996) Biochem 35:1672
22. Hashimoto Y, Yamada K, Motoshima H, Omura T, Yamada H, Yasuochi T, Miki T, Ueda T, Imoto T (1996) J Biochem 119:145
23. Miyazaki M, Kakidani H, Hanzawa S, Ohta H (1997) Bull Chem Soc Jpn 70: in press
24. Bender ML, Nakamura K (1962) J Am Chem Soc 84:2577
25. Hegeman GD, Rosenberg EY, Kenyon GL (1970) Biochem 9:4029
26. Kenyon GL, Hegeman GD (1970) Biochem 9:4036
27. Klinman JP (1972) J Biol Chem 247:7977
28. Miyamoto K, Tsuchiya S, Ohta H (1992) J Am Chem Soc 114:6256
29. Jones JB, Jakovac IJ (1982) J Can Chem 60:19
30. Toone EJ, Werth MJ, Jones JB (1990) J Am Chem Soc 112:4946
31. Miyamoto K, Ohta H, Osamura (1994) Bioorg Med Chem 2:469
32. Frisch MJ, Trucks GM, Hed-Gordon M, Gill PMW, Wong MW, Foresman JES, Gomperts R, Andres JL, Raghavachri K, Binkley JS, Baker JJ, Stewart P, Pople JA (1992) Gaussian 92, Revision C. Gaussian, Inc., Pittsburgh PA
33. Hehre WJ, Radom L, Schleyer PvR, Pople JA Ed (1986) Ab initio molecular orbital theory, John Wiley, New York
34. Kawasaki T, Horimai E, Ohta H (1996) Bull Chem Soc Jpn 69:3591
35. Kawasaki T, Saito K, Ohta H (1997) Chem Lett:351

Received February 1998

Biocatalytic Applications of Hydroxynitrile Lyases

Dean V. Johnson · Herfried Griengl

Institut für Organische Chemie der Technischen Universität Graz, Stremayrgasse 16, A-8010 Graz, Austria. *E-mail: Dvj@orgc.tu-graz.ac.at*

Hydroxynitrile lyases (Hnls) are enzymes that catalyse the stereoselective addition of hydrocyanic acid to aldehydes and ketones and the reverse reaction, the decomposition of cyanohydrins. This biotransformation (in the synthesis direction) can generate a product with a new chiral centre which possesses geminal difunctionality, i.e. a hydroxyl and nitrile moiety at a single carbon atom, and which also represents a versatile synthetic intermediate in organic chemistry. Currently, this area of research is sufficiently well established that enzymes for the synthesis of either (*R*)- or (*S*)-cyanohydrins are available. The Hnl from almonds, *Prunus amygdalus* (PaHnl), provides an easy access to (*R*)-cyanohydrins. Further, recent advances in cloning and overexpression techniques have provided two of these (*S*)-Hnl enzymes, those from *Hevea brasiliensis* (HbHnl) and *Manihot esculenta* (MeHnl), in sufficient quantities for potential application to industrial syntheses. Further, the crystallisation of the Hnl from HbHnl has revealed new information about their 3D structure and a tentative reaction mechanism for cyanohydrin cleavage has been postulated.

This review surveys the practical sources of Hnls and the development of their use in aqueous, organic and biphasic systems to yield enantiomerically enriched cyanohydrins.

The potential of cyanohydrins as synthetic intermediates in organic chemistry will also be presented.

Keywords: Molecular cloning, Overexpression, Reaction mechanism, Transhydrocyanation, Bioactive products.

1 **Introduction** 32

2 **Hydroxynitrile Lyases** 33

2.1 Their Distribution in Nature, Role and Biochemical Characterisation 33

2.2 Availability of the Enzymes, Molecular Cloning and Overexpression 36

2.3 Three Dimensional (3D) Structure 37

3 **Enzyme Catalysed Cyanohydrins Reactions** 39

3.1 Reaction Mechanism 39

3.2 Biocatalytic Transformations, Scope and Limitations 40

3.2.1 (*R*)-Cyanohydrins 40

3.2.2 (*S*)-Cyanohydrins 41

3.3 Procedures 44

3.3.1 Synthesis in Aqueous, Organic and Biphasic Mixtures 44

3.3.2 Transhydrocyanation 45

Advances in Biochemical Engineering/
Biotechnology, Vol. 63
Managing Editor: Th. Scheper

3.3.3 Stereoselective Decomposition of Racemic Cyanohydrins (Dehydrocyanation) 47
3.3.4 Immobilisation Techniques and On-Line Production of Cyanohydrins 48

4 Enantiopure Cyanohydrins as Intermediates for the Synthesis of Bioactive Compounds 49

4.1 Chemical Transformations 49
4.2 Unsaturated Cyanohydrins 51

5 Conclusions and Outlook 52

6 References 53

1 Introduction

As the pharmaceutical industry searches for alternatives to resolution techniques for the production of drugs in an enantiomerically pure form [1a–c], superior methods that can deliver an chiral intermediate as a single isomer, e.g. biotransformations [1a–c, 2], will be very advantageous. In this respect, the hydroxynitrile lyases (Hnl) and their catalytic production of chiral cyanohydrins may have a significant role to play.

The Hnls catalyse the asymmetric addition of hydrogen cyanide (HCN) to the carbonyl moiety of an aldehyde or ketone (Scheme 1) to yield a chiral cyanohydrin (1) (where R^1 and R^2 R^1 = Alkyl or Aryl and R^2 = H or Alkyl). Reflecting their role in nature, the cyanohydrin is also cleaved by the HNL to yield HCN and the parent carbonyl compound. From nature complementary (*R*)- and (*S*)-Hnls are available which show distinct differences from each other, particularly in terms of substrate specificity and product selectivity. The synthetic reaction was first applied by Rosenthaler in 1908 [3] using a (*R*)-hydroxynitrile lyase (EC 4.1.2.10) from almonds with benzaldehyde to yield a chiral cyanohydrin, mandelonitrile (2). Over a half a century later (in 1966) Becker and Pfeil [4] employed the same enzyme (though in an immobilized form) to yield optically pure madelonitrile on a multigram scale using a continuous process [5]. This early pioneering work laid the foundation for some of the research that was to follow over the next 30 years. During this period, the discovery of enzymes with Hnl activity was ongoing, although it is only in the last decade that most advances in

$$R^1C(=O)R^2 \underset{HCN}{\overset{Hnl}{\rightleftharpoons}} R^1C(OH)(CN)R^2 \quad \mathbf{1}$$

Scheme 1. Enzymatic cyanohydrin formation

their synthetic application and the understanding of their selectivity have been realised. The object of this review is to outline the progress that has been made in this area of biocatalysis and, wherever possible, give an indication of where its future may lie.

2
Hydroxynitrile Lyases

2.1
Their Distribution in Nature, Role and Biochemical Characterisation

Some 3000 plant species exhibit the ability to release HCN from their tissues, a process which is known as cyanogenesis. This common phenomenon is well recognised in mainly plant sources, as well as a few non-plant sources [6].

In a healthy state the HCN is compartmentalised (at the tissue level) as cyanogenic glycosides or cyanolipids as a means to prevent premature cyanogenesis, although upon plant tissue damage the release of HCN will occur. The catabolism of these glycosides is initiated by one (or more) glycosidases which cleave the *O*-glycosidic bond of the cyanogenic glycoside to yield the cyanohydrin, which subsequently decomposes to yield the appropriate carbonyl compound and HCN (for plant defence). This decomposition occurs spontaneously (base catalysed), but is considerably accelerated by the action of a hydroxynitrile lyase [6–9] (Scheme 2). Alternatively the cyanogenic glycosides can be utilised by *β*-cyanoalanine synthetase, a process in which they are refixed by reaction with L-cysteine to form *β*-cyanoalanine. This is then hydrolysed by *β*-cyanoalanine hydrolase to L-asparagine and hence in this sequence the HCN can be considered as a nitrogen source for amino acid synthesis [10, 11].

Presently, the Hnls from eleven cyanogenic plants (from six plant families) have been purified and characterised and the properties of a selection of them are outlined in Table 1. These may be separated into those with (flavoproteins)

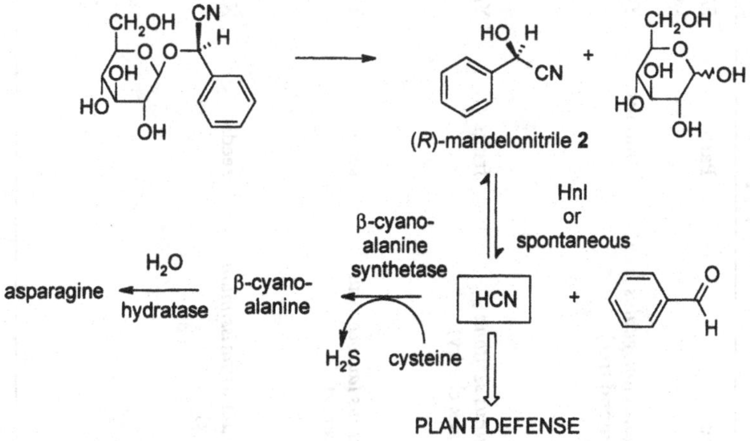

Scheme 2. Catabolism of cyanogenic glycosides

Table 1. The biochemical properties of seven Hnls

Plant	Part	FAD Co-enzyme	Molecular Native	Weight (kDa) SDS-page	Natural substrate (as glycoside)	Ref.
Prunus amygdalus (almond tree)	almonds	YES	72	72	(*R*)-mandelonitrile 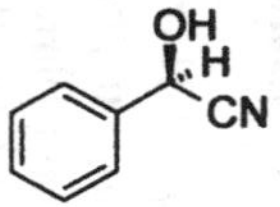	13, 14, 15
Prunus seritona (black cherry)	fruits	YES	50	59	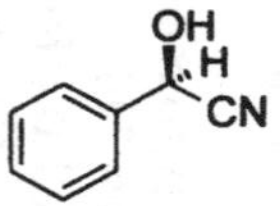	17
Prunus laurocerasus (cherry)	seeds	YES	60	Not given	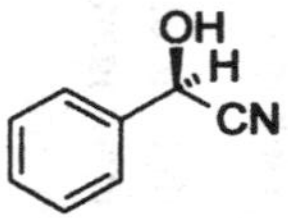	15
Linum usitatissimum (flax)	seeds	NO	82–87	42–43	acetone cyanohydrin  (*R*)-2-butanone cyanohydrin 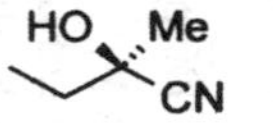	27, 28, 29

Table 1 (continued)

Plant	Part	FAD Co-enzyme	Molecular Native	Weight (kDa) SDS-page	Natural substrate (as glycoside)	Ref.
Manihot esculenta (manioc)	leaves	NO	92–124	28.5–30	acetone cyanohydrin  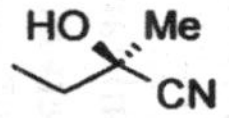	33, 34
Sorghum bicolor (millet)	seedlings	NO	108	55	(*S*)-4-hydroxy-mandelonitrile 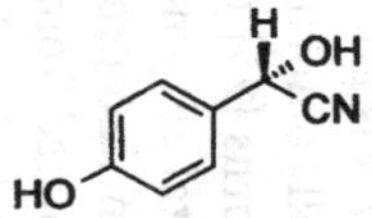	13, 14, 29–32
Hevea brasiliensis (rubber tree)	leaves	NO	60	30		38–40

and without (non-flavoproteins) the flavin adenine dinucleotide (FAD) coenzyme. To date the flavoprotein Hnls have been isolated from one plant family, the *Rosaceae* [12–17], and are highly glycosylated enzymes which show a sequence homology to FAD-dependent oxidoreductases [16]. The monomeric units have molecular weights ranging from 20–42 kDa, occur as homo or heterooligomers and, in general, a 5- to 50-fold purification yields a pure protein. Furthermore, their natural substrate is (*R*)-mandelonitrile. The FAD coenzyme is an important structural feature [18–23] which for Hnl activity is required in its oxidised state [24, 25]; however it is not involved in a redox reaction and may in fact be required for the overall conformational structure of the active enzyme [24–26]. It was suggested by Jorns [25] that the FAD-containing Hnls may have evolved from a single ancestoral enzyme.

In contrast, the non-flavoprotein Hnls are less uniform in their biochemical properties and are isolated from a variety of plant families, such as the seedlings of *Linum usitatissimum* (flax) [27–29] and *Sorghum bicolor* (millet) [13, 14, 29–32] and the leaves of *Manihot esculenta* [33, 34] (manioc), *Phlebodium aureum* (fern) [35], *Ximenia americana* (sandalweed) [36, 37] and *Hevea brasiliensis* (rubber tree) [38–40]. Furthermore, it is only after a higher degree of purification (100- to 150-fold with respect to specific activity) that a homogeneous protein is obtained. The diversity of the biochemical properties within this group of Hnls is demonstrated in the variation of their selectivity and substrate acceptance, e.g. the Hnls from *Linium usitatissimum* (LuHnl) cleave (*R*)-cyanohydrins as their substrate [27] whilst those from *Hevea brasiliensis* (HbHnl) operate only on (*S*)-cyanohydrins [39,40]. This trend in variation is also reflected in the wide ranging molecular weight found within this group (Table 1).

2.2
Availability of the Enzymes, Molecular Cloning and Overexpression

Though the Hnls from 11 cyanogenic plants have been studied, only 4 have found useful applications in synthesis, those from *Sorghum bicolor* (SbHnl), *Manihot esculenta* (MeHnl), *Prunus amygdalus* (almond tree, (PaHnl)) and *Hevea brasiliensis* (HbHnl). The latter three of these Hnls are available in preparative quantities, the PaHnl (a (*R*)-Hnl) is commercially available from almonds, whilst more recently the genes for the Me- and Hb-(*S*)-Hnls have been cloned and overexpressed [39, 41]. This represents a significant advantage for these Hnls as large quantities of enzymes are now readily available for synthetic applications.

The almond meal PaHnl can be obtained from commercial sources or alternatively can be prepared [42, 43] by grinding almonds and defatting the powder three times with ethyl acetate. These procedures makes this Hnl an attractive enzyme to use on a multigram scale and hence it has been widely applied in organic synthesis (as outlined in Sect. 3.3.1).

The HbHnl is obtained from the leaves of the rubber tree plant and a crude extract is easily prepared by homogenisation of the frozen leaves, followed by centrifugation [38–40]. A 5-step purification procedure of this crude extract (with over a 100-fold purification factor) to yield a homogenous HbHnl has been recently reported [38]. The purified enzyme was also cloned and expressed into *Escherichia coli* and *Saccharomyces cerevisiae* organisms and the protein sequence of the cloned HbHnl determined [39] from which the key amino acid residues of the active site were identified. An amino acid replacement study (i.e. cysteine 81 by serine) yielded a mutant with significantly reduced activity and it was suggested that this amino acid (cysteine 81), amongst others, has an important role in the catalytic action of the HbHnl. Detailed sequence homology studies revealed that the HbHnl shows no significant homology to the SbHnl (*S*-Hnl) or to the (*R*)-Hnl from *Pseudomonas serotina* [16], although it is highly homologous to the MeHnl [33], which makes the Hb-Hnl and MeHnl highly competitive sources of enzyme for (*S*)-cyanohydrin production. Recently a more successful and efficient expression system for HbHnl has been developed using methanol-inducible *Pichia pastoris* [44]. The intracellular Hnl protein is produced in high levels (approximately 60% of the total cellular protein) and exhibits a high specific activity (40 U/mg). High-cell-density cultivation yields more than 20 g of pure Hnl protein per litre of culture volume. This expression system is sufficiently well developed to provide simple access to the quantities of enzyme that are required for industrial applications.

In contrast, the (*S*)-Hnl from S. *bicolor* enzyme has been purified [13, 31] and cloned [32] but for functional expression it requires complex posttranslational processing [41] and is still not easily available in sufficient quantities for application to technical processes.

More successfully, the (*S*)-Hnl from *Manihot esculenta* has also been overexpressed in *E. coli* [41] and the lysate of the transformed cells showed an enzyme activity of 0.5 units per ml of the culture. A culture of 80 l volume of the recombinant MeHnl followed by a short purification procedure [41] yielded 40,000 U. To obtain the equivalent amount of enzyme from the parent plant material would require the processing of 100–200 kg of dried cassava leaves and thus this recombinant method for the production of MeHnl is a significant practical development. Hence, this recombinant MeHnl has allowed a study of (*S*)-cyanohydrin production to be performed [41].

2.3 Three Dimensional (3D) Structure

To understand fully the mechanism of operation for an enzyme system the elucidation of the 3D structure is invaluable. Initial crystallisation experiments [45] were followed by a successful structural analysis of the HbHnl [46] which yielded the first 3D structure of a hydroxynitrile lyase (Fig. 1). The structure contains a large β-sheet which is surrounded on both sides by α-helices and a variable "cap" region. This structural arrangement fits the HbHnl into the a larger class of enzymes known as the α,β-hydrolase fold family [47, 48] of which other enzymes within this family which possess carboxypeptidase, lipase, thioesterase and oxidoreductase activity also show similar structural features. Further recent homology and structural investigations using structural prediction algorithms and sequential alignment programs [49] are in accordance with the suggestion that the Hb-Hnl belongs to this hydrolase family. These similarities assist in the

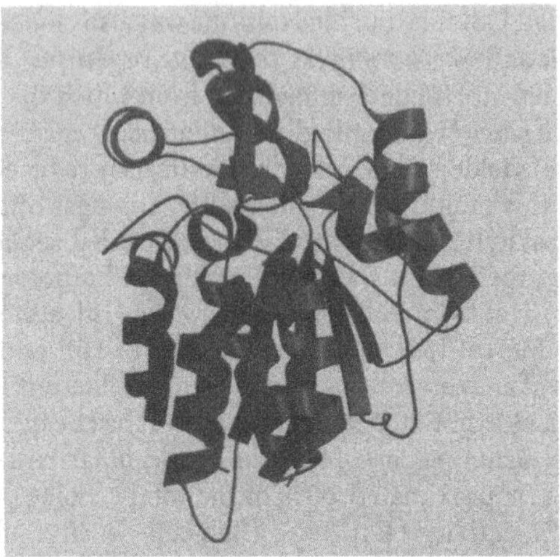

Fig. 1. A ribbon stereorepresentation of the Hnl from *Hevea brasiliensis*

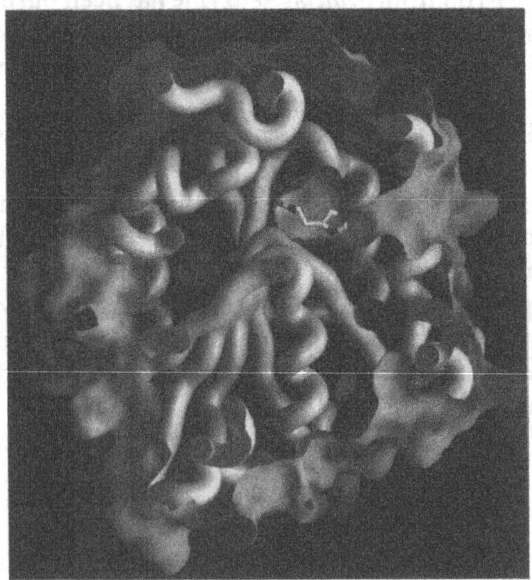

Fig. 2. A section through the surface representation of the Hnl crystal structure. A histidine molecule is shown in the active site

identification of the active site and the determination of some of the residues participating in the enzyme catalysed formation of cyanohydrins by HbHnl. Consequent to this observed similarity, a sharp turn ("nucleophilic elbow") between a β-sheet and an α-helix which contains serine$_{80}$ as the central residue was also identified [46].

Furthermore, it was also established that a narrow channel connects the active site, which is buried deep inside the protein, to the surface [46] (Fig. 2).

This structural feature is in agreement with the proposed Uni Bi mechanism for hydroxynitrile lyases [25, 40], which suggests that reagents enter the active site in a sequential fashion. The residues involved in the catalytic action of the active site and their positive assignment are discussed in Sect. 3.1.

Crystallisation studies using the PaHnl have also been reported [50]. Four isomeric forms of this Hnl were isolated from the aforementioned defatted almond meal and subsequently separated. Three of these isoenzymes were successfully crystallised and shown to belong to the monoclinic space group. Though by this means additional data were included, the clearest picture for the 3D structure of a hydroxynitrile lyase still remains with the HbHnl.

3 Enzyme Catalysed Cyanohydrin Reactions

3.1 Reaction Mechanism

Using site-directed mutagenesis, Wajant and Pfizenmaier [51] have recently identified the amino acids involved in a putative catalytic triad for the MeHnl. The order of these residues are identical to those found in enzymes belonging to the α,β-hydrolase fold family, suggesting that the MeHnl also belongs to this family. From these key residues a mechanism for the cleavage of cyanohydrins by the MeHnl has been postulated [51] (Scheme 3). In the presence of a cyanohydrin molecule it was suggested that the proton of Ser_{80} is transferred to the

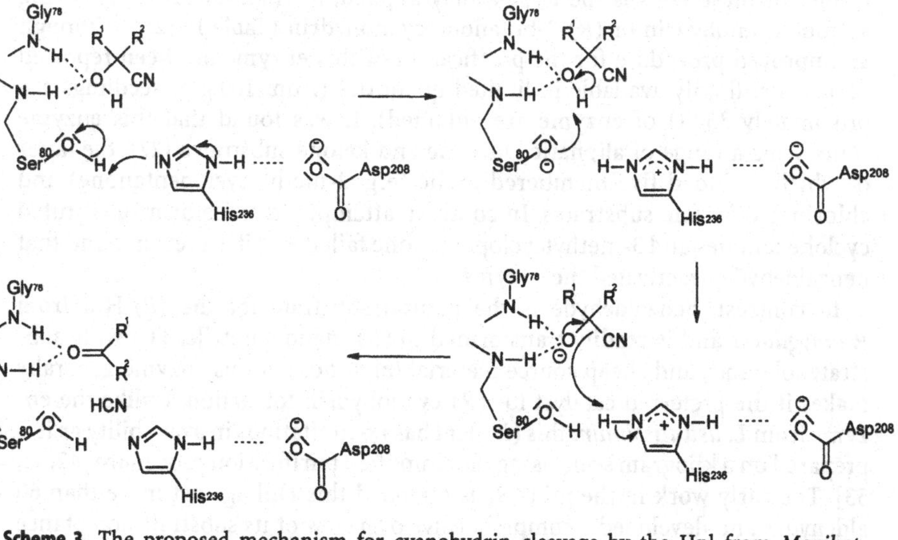

Scheme 3. The proposed mechanism for cyanohydrin cleavage by the Hnl from *Manihot esculenta*

nitrogen of His_{236}. The resultant oxyanion of Ser_{80} acts as a base in abstracting the proton of the hydroxyl group of the cyanohydrin, yielding a negatively charged oxygen on the substrate, which is stablized by the amide backbone of the Ser_{80} and Gly_{78} residues. This charge is the driving force for breaking the C-CN bond to yield the carbonyl compound and HCN.

From X-ray studies (outlined in Sect. 2.3) of the HbHnl and by analogy to the observed catalytic mechanism of structurally related proteins (also of the α,β-hydrolase fold family) another reaction sequence was tentatively postulated [46, 49]. For cyanohydrin formation it is suggested that a (presumably) catalytic triad containing the residues Ser_{80}, His_{235} and Asp_{207} is directing the course of the reaction. The significance of these amino acids within the catalytic triad was supported by their replacement in site directed mutagenesis experiments [46, 49], such that mutants in which the Ser_{80} or His_{235} were replaced by alanine or serine showed no enzyme activity. Also replacing the residues adjacent to the active site (i.e. Glu_{70} to Ala and Cys_{81} to Ser) resulted in significantly reduced levels of measured activity for these mutants. This suggests that they may also have a significant role in the catalytic mechanism. This mechanism is currently an issue of debate and more recently a modified mechanism has been discussed [52].

3.2 Biocatalytic Transformations, Scope and Limitations

3.2.1 *(R)-Cyanohydrins*

The two established Hnls, those from *L. usitatissimum* and *P. amygdalus*, have found biocatalytic applications for the production of (*R*)-cyanohydrins. The former of these Hnls is the least widely applied, the natural substrates being acetone cyanohydrin or (*R*)-2-butanone cyanohydrin (Table 1) [28]. Although an improved procedure for the purification of this enzyme has been reported [27] it is still only available in limited quantities (from 100 g of seedlings approximately 350 U of enzyme are obtained). It was found that this enzyme transforms a range of aliphatic aldehyde and ketone substrates [27], the latter of which included five-membered cyclic (e.g. 2-methylcyclopentanone) and chlorinated ketone substrates. In contrast, attempts to transform substituted cyclohexanones and 3-methylcyclopentanone failed and it was even found that benzaldehyde deactivated the enzyme.

In contrast, benzaldehyde is the natural substrate for the (*R*)-Hnl from *P. amygdalus* and is readily transformed to (*R*)-mandelonitrile. The wide substrate tolerance and cheap source material (almonds) of this enzyme generally makes it the preferred catalyst for (*R*)-cyanohydrin formation. Unlike the enzyme from *L. usitatissimum* this (*R*)-Hnl has no limitations in availability as it is prepared on a kilogram scale using the simplified purification procedure [42, 43, 53]. The early work in the mid-1960s screened this Hnl against more than 60 aldehydes and developed a comprehensive overview of its substrate acceptance [54–56]. Latterly, numerous groups, including those of van der Gen, Effenberger and Kanerva have applied this enzyme in the synthesis of chiral cyanohydrins, although for their extensive applications of the PaHnl the group of van der Gen is the most cited. They have applied benzaldehyde (**3**) to this enzyme and obtained good conversion (95% in 40 min at 0 °C), and the corresponding cyanohydrin was formed with excellent selectivity [57] (Table 2). In comparison to benzaldehyde the reaction of 4-methoxybenzaldehyde (**4**), was found to be slower (Table 2). Subsequently, these initial results were rapidly followed by the development of a broader spectrum of substrates [58] in which a variety of substrates, in an aqueous ethanol medium, including the cyclic aldehyde **5**, saturated and unsaturated aldehydes (**6** and **7**) and heterocyclic substrates (such as **8**) were tested. The range of aldehyde substrates has been further extended by Effenberger to include 3-phenoxy benzaldehyde (**9**) and 3-methylthio propionaldehyde (**10**) [59, 60], but significantly they showed that the more sterically hindered ketone substrates such as **11** and **12** can be transformed with good enantiomeric excess and yield [61, 62]. Although some bulkier substituted aryl aldehydes have been reported to be a hindrance to the enzyme catalysed reaction [63], in general, the (*R*)-oxynitrilase from *P. amygadalus* combines low substrate specificity with high enantioselectivity and provides the ideal system for (*R*)-cyanohydrin production.

Recent work [64] by Kiljunen and Kanerva has been directed towards the search for novel sources of (*R*)-oxynitrilases which may transform bulky aryl aldehydes. For this purpose whole cell preparations (called meal) from apple seeds and cherry, apricot and plum pips were tested for their (*R*)-cyanohydrin activity. In this study a comparison of almond and apple meal showed that they possess similar properties for the formation of the (*R*)-stereogenic centre. However, in certain cases higher enantioselectivity was observed using the apple meal preparation. Additionally, apple meal (*R*)-Hnl has also been applied to transform ketones into their corresponding cyanohydrins [65] thus creating a wider repertoire of substrates for this latest of (*R*)-Hnls. Thus it has only recently been shown that apple meal (*R*)-oxynitrilase is now an additional member of the (*R*)-Hnl family.

Also in the (*R*)-class of Hnls, although of more limited application, is the (*R*)-mandelonitrile lyase from *Phlebodium aureum* [35] which catalyses the addition to some aromatic and heterocyclic carbonyls but only poorly to aliphatic carbonyls.

3.2.2
(S)-Cyanohydrins

For the selective generation of (*S*)-cyanohydrins the enzymes from *S. bicolor*, *M. esculenta* and *H. brasiliensis* are available for use. However the enzyme from *S. bicolor* (from the seedlings), which was first isolated by Bové and Conn in 1960 [66], has two restrictions. First, after a six-step purification from 10 kg of *S. bicolor* plants only a limited quantity (160 mg, with a specific activity of approximately 260 U/mg) of pure enzyme in three isomeric forms is obtained [13, 14, 30, 31, 67] which represents a time-consuming procedure. Second, the substrate specificity of this enzyme is high, as it accepts only aromatic aldehydes as sub-

Table 2. Selected aldehydes transformed by PaHnl

Substrate	Product	e.e. of Cyanohydrin (%)	Conversion	Conditions	Ref.
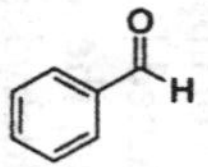 3	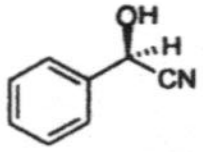 OH, H, CN	99	95	Ethanol/Acetate buffer (pH 5.4), 0 °C, KCN/HOAc	57
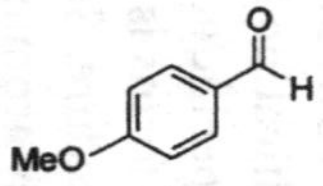 MeO 4	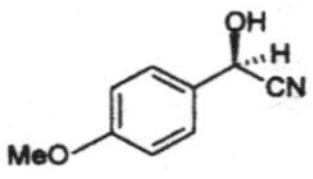 MeO, OH, H, CN	78	78	Ethanol/Acetate buffer (pH 5.4), 0 °C, KCN/HOAc	57
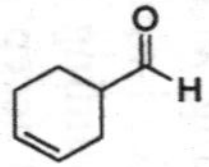 5	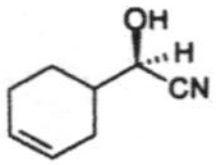 OH, H, CN	55	86	Ethanol/Acetate buffer (pH 5.4), 0 °C, KCN/HOAc	58
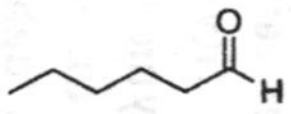 6	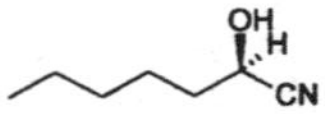 OH, H, CN	67	95	Ethanol/Acetate buffer (pH 5.4), 0 °C, KCN/HOAc	58
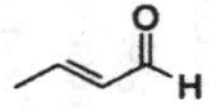 7	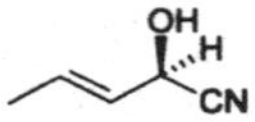 OH, H, CN	69	96	Ethanol/Acetate buffer (pH 5.4), 0 °C, KCN/HOAc	58
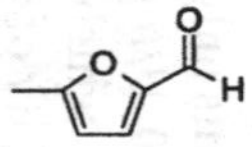 8	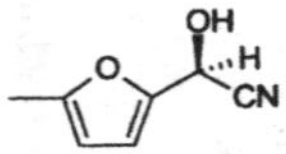 OH, H, CN	78	25	Ethanol/Acetate buffer (pH 5.4), 0 °C, KCN/HOAc	58

Table 2. (continued)

Substrate	Product	e.e. of Cyanohydrin (%)	Conversion	Conditions	Ref.
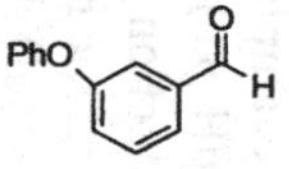 **9**	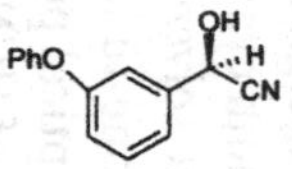	98	99	Ethanol/Acetate buffer (pH 5.4), 0 °C,HCN	59
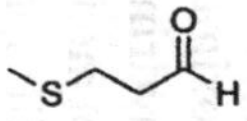 **10**	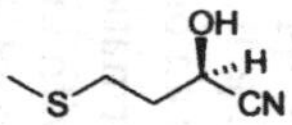	96	75	Ethanol/Acetate buffer (pH 5.4), 0 °C,HCN	59
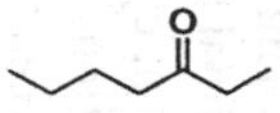 **11**	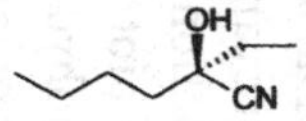	90	21	Diisopropyl ether/ Acetate buffer (pH 4.5), 0°C,HCN	62
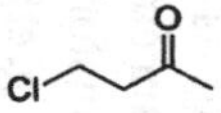 **12**	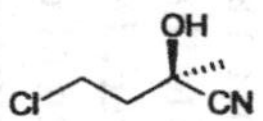	84	87	Diisopropyl ether/ Acetate buffer (pH 4.5), 0 °C,HCN	61
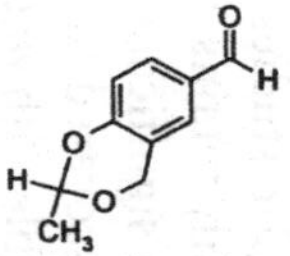 **13**	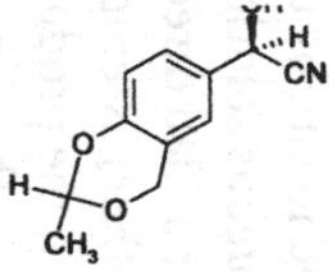	96	60	Diisopropyl ether/ Acetate buffer (pH 4.5), 0 °C,HCN	122

strates. However a wide range of 3- and 4-substituted aromatic aldehydes are converted and the products are often obtained with excellent selectivity [63, 68, 69].

Thus the recent developments made for the (*S*)-Hnls for *M. esculenta* and *H. brasiliensis* are of significant value and have widened the scope of applications in the area of (*S*)-cyanohydrin production.

Following the initial isolation of the Hnl from *M. esculenta* [33] in which the peptide sequence was established, an overexpressed version of this enzyme (in *E. coli*) was prepared [41]. This system is not limited for enzyme quantity (as outlined in Sect. 2.3), and can accept a wide range of aromatic, heterocyclic and aliphatic aldehydes, as well as ketones, as substrates. In practical terms, a measure of the degree of enzyme inhibition by substrates is of significant value and for this system this has been quantified for a range of aldehydes, ketones and alcohols [70]. It was deduced that ketones and alcohols are competitive inhibitors, whilst aldehydes are noncompetitive inhibitors.

Similar progress has also been made for the (*S*)-Hnl from *H. brasiliensis*. The purified protein has been overexpressed in a number of systems [39, 44] (see Sect. 2.3), which adequately fulfills the practical demands required for transformations using this system.

The substrate range of this Hnl is also broad and to date a wide range of aldehydes of varying structural nature [71] and chain length [72, 73] can be easily transformed.

At present, these latter two Hnls represent the most superior catalysts for the production of (*S*)-cyanohydrins and it is expected that their development will rapidly continue to yield processes of industrial and technical relevance.

3.3 Procedures

3.3.1 *Synthesis in Aqueous, Organic and Biphasic Mixtures*

The HCN, an indispensable prerequisite for cyanohydrin formation, may be generated in situ by the addition of an acidic component (e.g. citric acid buffer) to an aqueous solution of an alkali cyanide (e.g. potassium cyanide) [42, 53, 58, 71, 72] or added directly to the reaction when an organic solvent (such as diisopropyl ether) is employed. In all biotransformations one fundamental factor governing the production of cyanohydrins with high optical purity is the suppression of the non-enzymatic (spontaneous) addition of HCN to the carbonyl substrate. In an aqueous system this suppression can be achieved by lowering the pH (to 3.5–4.0) of the reaction medium [68, 71, 74], although working at such low pH values has the disadvantage of significantly reducing the enzyme activity (for studies regarding HbHnl see [75]). For example, at pH 4.0 the HbHnl only retains 20% of its optimal (approximately pH 6.0) catalytic activity [76]. A significant advancement in cyanohydrin production was made when it was discovered that transformations performed in organic media which are not miscible with water (e.g. ethyl acetate or diisopropyl ether) show virtually no spontaneous addition of HCN to the carbonyl moiety [42, 53, 59, 60]. It has been suggested that the use of a water-immiscible organic solvent suppresses the non-enzymatic addition to a greater degree than the enzyme-catalysed addition [59]. In addition, optical purities of the products were significantly enhanced for reactions with predominantly organic reaction media (e.g. ethyl acetate) compared to organic/aqueous mixtures (e.g. ethanol/water) [59]. Furthermore, certain solvents showed reduced levels of inactivation towards the Hnl, and thus for the PaHnl it was shown that in diisopropyl ether enzyme activity is retained for several weeks, which is in contrast to ethyl acetate where it is reduced by greater than 50% in less than 6 h [77]. Undoubtedly, the disadvantage of working in such an organic system is that the HCN cannot be generated in situ and instead requires the direct handling of this highly toxic reagent. It must first be generated in a neat form (for example see [78]) and added either undiluted or diluted with the organic solvent to the reaction mixture. However, the benefits of improved conversion and optical purity outweigh this disadvantage and make this a preferred method for cyanohydrin production.

In a biphasic system a number of reaction variables (e.g. temperature, pH, solvent and solvent/buffer ratio) are involved in a complex interplay to yield the enantiomeric excess and conversion for the reaction. In a recent elegant study directed at the development of an industrial process for the production of cyanohydrins with the PaHnl [79], van der Gen et al. have examined this interplay using an optimal design for the variation of the reaction parameters. A number of crucial observations were made: i) methyl *tert*-butyl ether (MTBE) was the preferred choice of solvent over ethyl acetate as conversions in this solvent were greater; ii) the temperature and pH of the system have a significant effect on the enantiomeric excess of the product, but not on the conversion (this is as a consequence of enhanced non-enzymatic addition); iii) a change in the ratio of organic solvent/buffer towards the buffer increased the conversion but decreased the enantiomeric excess of the product. The implication of this is that the non-enzymatic addition is more prevalent when a large volume of buffer is employed in the reaction system. The optimisation of these reaction parameters has yielded a system in which 9.5 kg of (*R*)-mandelonitrile could be produced in 30 min using 15 g of PaHnl (in 100 l volume) [79]. This methodical approach to Hnl-catalysed reactions has clearly produced an industrially attractive process.

3.3.2
Transhydrocyanation

A method of employing an organic solvent as a reaction medium that incorporates the safe transfer of HCN would be of great advantage to cyanohydrin production. Thus the recent development of enzyme catalysed transhydrocyanation, which employs acetone cyanohydrin [80–84] as a HCN carrier, is of significant value. This method makes the handling process of HCN both easier and less dangerous.

With the aim of developing a biocatalytic method which might reduce the tendency for non-enzymic addition of HCN and be broadly applicable, the group of Kyler et al. first introduced transhydrocyanation [85] using the PaHnl.

Their goal was to introduce an alternative cyanide carrier (donor) which would generate a cyanide-bound form of the PaHnl and/or produce sufficiently low levels of HCN that the reaction would be directed away from non-enzymatic addition and towards the enzymatic pathway. Acetone cyanohydrin (Scheme 4) was selected for this purpose as it had previously been employed in the trans-hydrocyanation of carbonyls [86–88]. This choice of donor proved to be advantageous (compared to e.g. mandelonitrile) for a number or reasons; it is miscible with the buffer used in the system, commercially available and its by-product (acetone) is volatile and has a favourable equilibrium constant. Using this method, more than ten aldehydes were applied and the cyanohydrin products were generally obtained in high optical purity (always greater than 88% e.e.). For this system, the optical purity of the product could be increased by decreasing the pH of the reaction buffer, but below pH 5.0 the enzymatic activity (of the PaHnl) was significantly reduced and hence substrate conversions were low. Most importantly the use of a biphasic system was crucial in obtaining products with optimum enantiomeric purity. It was noted that the partitioning of the substrates between the organic and aqueous phases has a pronounced influence on the enantiomeric purity of the product, i.e. substrates that partition predominantly towards the aqueous phase yield products with significantly decreased optical purity. Conversely, substrates with low water solubility yield products that are almost enantiomerically pure. This trend was also reflected in the values for the logarithms of the partition coefficients (log P) which were calculated for the substrates. For those with positive log Ps high enantiomeric excesses (88–99%) for the products were recorded, whilst negative log Ps yielded compounds with low enantiomeric excess (2–30%) [85]. This finding is crucial to the successful production of cyanohydrins in this biphasic system.

Subsequently, this procedure was applied to the SbHnl [63, 89] and HbHnl [73] with a variety of aldehydes, and also with PaHnl to study the diastereogenic addition of HCN to chiral aryl substituted aldehydes [90].

A systematic study of transhydrocyanation using a series of aliphatic aldehydes (with the PaHnl) with increasing chain length [91] showed that a change towards a greater chain length yielded a significant decrease in the rate of formation and the enantiomeric excess of the cyanohydrin product. It was proposed that the retarded enzymatic reactions (longer than 24 h) observed for the longer chain substrates (i.e. heptyl and nonyl aldehydes) allowed the corresponding non-selective chemical reaction to become more significant over the extended reaction periods.

In a recent development, the kinetics of the transhydrocyanation process between benzaldehyde and acetone cyanohydrin were studied by ^{1}H NMR methods

HO CN
O
R1 R2
HO CN
R1 R2
+
O
Hnl

Scheme 4. The principle of transhydrocyanation

[92] to further elucidate the mechanism of the HCN exchange process. Thus during a time course study the quantities of each component of the transhydrocyanation reaction (i.e. mandelonitrile, benzaldehyde, acetone cyanohydrin and acetone) were quantified. It has been postulated [93] that the transhydrocyanation consists of two separate steps, (i) cleavage of acetone cyanohydrin followed by (ii) synthesis of the new cyanohydrin, both processes being enzyme mediated. Further studies should allow the formulation of a complete kinetic model for this process and the optimisation of all the reactions within this complex system.

3.3.3
Stereoselective Decomposition of Racemic Cyanohydrins (Dehydrocyanation)

The natural reaction of the Hnls is the cleavage of a cyanohydrin to yield HCN and the corresponding carbonyl compound. Thus utilising this catalytic behaviour in organic synthesis, it would be possible to treat a racemic cyanohydrin with a (*R*)- or (*S*)-Hnl in an effort to decompose selectively one enantiomer of this mixture (exemplified in Scheme 5). This would leave one enantiomer unreacted which would thus become enriched in this mixture. This principle (dehydrocyanation) has been applied to the preparation of optically pure cyanohydrins from racemic mixtures. This dehydrocyanation process can be regarded as a kinetic resolution, but critical to the success of this method are the removal of the unwanted carbonyl product (from the decomposed cyanohydrin) and the capture of the HCN liberated during cyanohydrin cleavage, both of which favourably shift the equilibrium of this reaction.

One of the earliest examples in this field [94] was reported by Mao and Anderson who showed that the SbHnl was specific for the dehydrocyanation of *p*-hydroxybenzaldehyde cyanohydrin. In a recent patent [95] the process of dehydrocyanation was successfully demonstrated using the PaHnl as the catalyst and a gas-membrane extraction method to remove the undesired aldehydes and HCN, which yielded a wide range of (*S*)-cyanohydrins, often with excellent enantiomeric excess. In further patent literature, Niedermeyer [96] also generated (*S*)-cyanohydrins via decomposition of a racemic mixture, but the problems of workup resulted in decreased yields.

The preparation of (*S*)-ketone cyanohydrins has also been achieved in a one-pot procedure, using the PaHnl, by decomposition of the corresponding racemic cyanohydrins followed by stereoselective addition of the HCN that is liberated to ω-bromoaldehydes [97]. This tandem reaction process yielded both (*S*)-ketone cyanohydrins (from the decomposition of the racemic cyanohydrin) and (*R*)-ω-bromocyanohydrins (the HCN addition product), the latter being con-

HO H / R CN —(*R*)-Hnl⇌ HO H / R CN + RCHO + HCN

Scheme 5. Selective (*S*)-cyanohydrin formation by enantioselective decomposition of a racemic mixture

verted to 2-cyanotetrahydrofurans, which are of synthetic interest as they are common structural components of terpenoids, pheromones, antibiotics and *C*-glycosides.

Most recently, Effenberger and Schwämmle [98] used sodium hydrogen sulfite in the enzymatic decomposition of racemic aromatic and heterocyclic aldehyde cyanohydrins. This reagent captured the liberated aldehydes as bisulfite adducts, a reaction which is favoured due to their small dissociation constants compared to the corresponding cyanohydrins [99, 100]. This method employed the (*R*)-selective PaHnl for decomposition which generally yielded aromatic or heterocyclic (*S*)-cyanohydrins in moderate e.e. and yield. However, a disadvantage to this system was the discovery that racemic aliphatic cyanohydrins could not be selectively decomposed to yield aliphatic (*S*)-cyanohydrins. Also in this study [98], aldehydes were removed with hydroxylamine (to form oximes) and semicarbazide (to form semicarbazones), the latter of which generally yielded the most superior cleavage results due to the advantageous precipitation of the semicarbazone from the reaction solution. However, these reagents have a very deleterious effect on prolonged enzyme reactions, with a loss of over 70% of the enzyme activity being recorded.

Undoubtedly, the cyanohydrin formation reactions catalysed by the Me- and HbHnls remain the preferred method for (*S*)-cyanohydrin formation.

3.3.4
Immobilisation Techniques and On-Line Production of Cyanohydrins

As early as 1966, Becker and Pfeil applied the methodology of immobilised Hnls, by combining a cellulose-based ion-exchanger (ECTEOLA cellulose) with the PaHnl to form a column containing immobilised enzyme for on-line cyanohydrin production. This example became the successful precedent for the methods outlined in this section. To date, the simplest form of immobilisation for a Hnl is that utilised by Zandbergen et al. [42] in which almond meal (prepared from grinding almonds) provides a support for the PaHnl. This method is economical, cheap and widely applied as the source material (almonds) can be purchased in conventional stores. A number of alternative support materials for this Hnl were examined [59] (including DEAE-cellulose and glass beads) of which cellulose proved the most suitable.

The characteristics of a support material are of great importance to the measured enzyme activity [79, 101]. Hydrophobic carriers have a low ability to attract water, thus leaving more available for the enzyme, hence Wehtje et al. [102, 103] have shown that celite is a suitable carrier for the PaHnl to yield an immobilized form of the enzyme. In contrast, controlled pore glass (CPG) and Sephadex G25 were found to be less well suited to enzyme support as, using these systems, cyanohydrin synthesis was significantly reduced (over 30%). Sephadex also promoted the spontaneous addition of HCN to benzaldehyde [102]. A series of batch experiments showed that if the solvent (diisopropyl ether) surrounding the immobilised PaHnl contained insufficient water (i.e. less than 2%), it would be extracted from the enzyme preparation and consequently enzyme activity was lost [102]. These results were the basis for the production of stainless steel reactors containing celite immobilised PaHnl, through which a water saturated substrate solution was passed, and in 50 h of operation only a minor decrease (approximately 3%) in enzyme activity was observed. At high concentrations of benzaldehyde (1 mol/l) and HCN (2 mol/l) good cyanohydrin productivity (2.20 g/h) was observed.

The PaHnl has also been covalently immobilized to porous silica using glutaraldehyde as the spacer unit [104], which was found to be a superior binding agent. Reactor design is a significant factor in the on-line production of optically pure cyanohydrins and the reactants (aldehyde and HCN) should not be allowed to mix upstream from the enzyme reactor, else problems with spontaneous addition may occur. The reactors of this design were run for almost 200 h and combined good product ((*R*)-mandelonitrile) yield (94%) with good optical purity (greater than 94%). During this time some activity loss (40%) was observed and according to a protein assay this was not attributable to leakage of the enzyme from the system.

An excellent production figure for (*R*)-mandelonitrile (2400 g/l per day) was achieved by Kragl et al. [105] using a continuously stirred tank reactor in which an ultrafiltration membrane enables continuous homogenous catalysis to occur from an enzyme (PaHnl) which is retained within the reaction vessel. In order to quench the reaction the outlet of this vessel was fed into a vessel containing a mixture of chloroform and hydrochloric acid, which allowed for accurate product analysis.

Of alternative immobilisation design is the formation of lyotropic liquid crystals (LC) which provides an immobilisation matrix for the enzyme which is suitable for use in an organic solvent. The catalyst is entrapped in the LC phase and the substrates dissolved in the surrounding organic phase, from which the subsequent products may be isolated [106]. Such a catalyst (LC-PaHnl) was prepared by Miethe et al. [107] using the PaHnl as the protein. In stability tests the LC enzyme gradually lost activity only in the first four days (of a 20 day test period), after which the enzyme activity remained constant for the subsequent period of time. It was suggested that the high viscosity of the LC, which may limit unfavourable conformational changes in the protein, may account for the eventual high stability observed in this system. The LC crystal showed a broad operational range for pH (3.0–7.0) for (*R*)-mandelonitrile production. These properties lend the LC-PaHnl to be a suitable candidate for the continuous process of cyanohydrin production and indeed a maximal figure of 2650 g/l per day for (*R*)-mandelonitrile production was recorded using this catalyst.

4 Enantiopure Cyanohydrins as Intermediates for the Synthesis of Bioactive Compounds

4.1 Chemical Transformations

To the synthetic chemist the chiral cyanohydrins are extremely useful building blocks as they possess a high degree of functionality and are chiral compounds

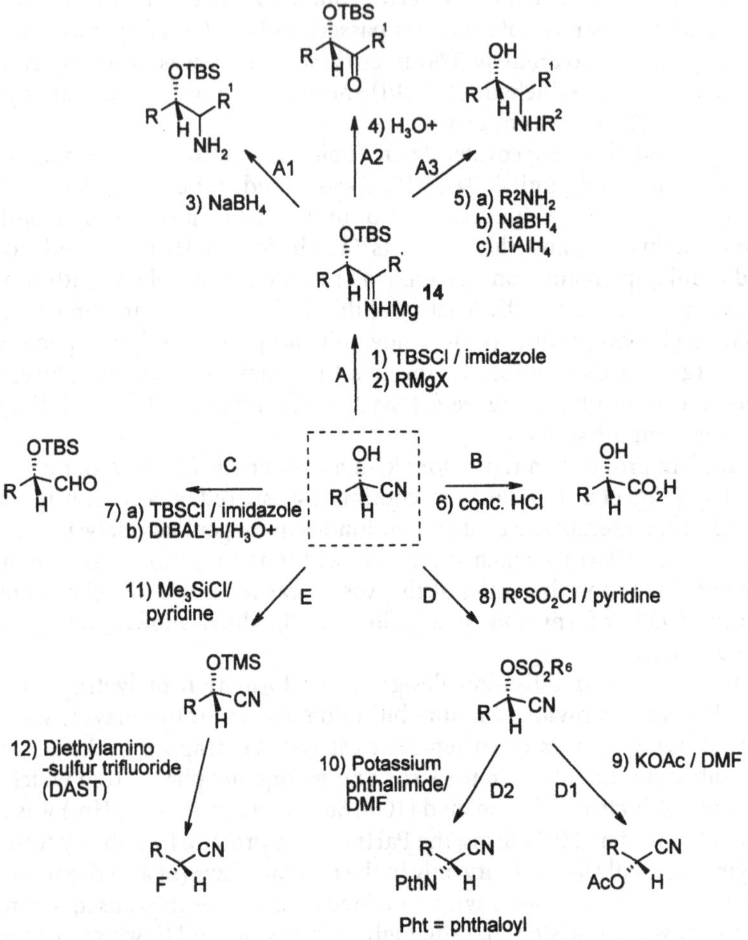

Scheme 6. Further chemical transformations of chiral cyanohydrins

which are (often) enantiomerically pure. As a consequence a number of groups have exploited their synthetic utility for the preparation of useful synthetic intermediates and compounds of biological interest. A number of transformations may be performed directly on the chiral cyanohydrin, the main prerequisite being appropriate protection of the hydroxyl moiety. Suitable protecting groups that have been incorporated in synthesis are ethers [108–110] and silyloxy ethers [57, 111]. Only weak bases (such as pyridine and imidazole) are employed for proton abstraction of the hydroxyl group as the use of stronger bases results in racemisation and decomposition of the cyanohydrin. In the case of THP protection, toluene sulphonic acid is used in a catalytic quantity. Scheme 6 summarises some of the recent synthetic interconversions performed on protected chiral cyanohydrins. These include protection of the hydroxyl group (sequence A) followed by nucleophilic addition to the nitrile moiety of a nucleophilic metallo compound (such as a Grignard reagent). The intermediate imine **14** may undergo further reactions, such as reduction (A1) [112], hydrolysis (A2) [57] or transimination to form a secondary imine followed by reduction and deprotection (A3) [113] to yield a range of products.

The chiral cyanohydrins also lead directly to *α*-hydroxy acids by hydrolysis (sequence B) [69] and to protected *α*-hydroxy aldehydes by first hydroxyl group protection, followed by reduction of the nitrile and hydrolysis of the intermediate imine (not shown) (sequence C) [114]

By selection of an appropriate derivative the hydroxyl group can become activated towards displacement (via an S_N2 exchange process) with a suitably chosen nucleophile. Conversion to the sulfonate ester (SO_2R) [115] promotes displacement by either an acetate (^-OAc) (sequence D1) or phthalimide nucleophile (Sequence D2), whilst the trimethylsilyl derivative facilitates introduction of fluorine (Sequence E) [116].

4.2 Unsaturated Cyanohydrins

The formation of unsaturated cyanohydrins (from *α*,*β*-unsaturated aldehydes) is of further advantage as these products possess an additional synthetic potential. As in the saturated cyanohydrins (above in Scheme 6) they possess the same opportunities for elaboration of the hydroxyl or nitrile moiety, although the presence of the carbon-carbon double bond offers the possibility for additional transformations to be performed such as additions [108], oxidative cleavage [117, 118] and epoxidation [119] (Scheme 7). Thus, these highly functionalised chiral units can be of greater importance to an organic chemist.

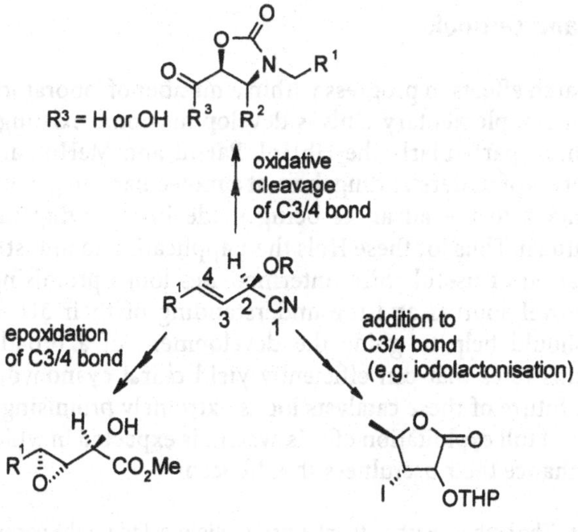

Scheme 7. Some opportunities for transformations of unsaturated cyanohydrins

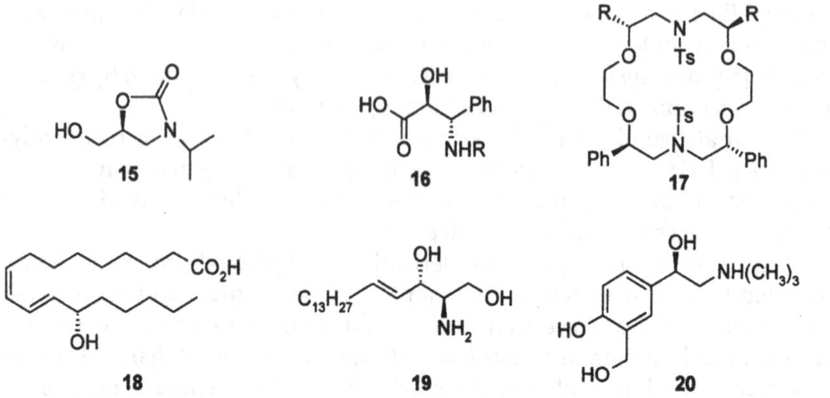

Fig. 3. Examples of bioactive compounds synthesised from cyanohydrins

Therefore, the chiral cyanohydrins are valuable and versatile synthons as their single hydroxyl asymmetric centre is accompanied by at least one other chemical functionality. Thus with careful functional group protection, differential and selective chemical transformations can be performed. Such synthetic techniques lead to production of interesting bioactive compounds and natural products. These products include intermediates of β-blockers **15** [117], β-hydroxy-α-amino acids **16** [118], chiral crown ethers **17** [111], coriolic acid **18** [120], sphingosines **19** [121], and bronchodilators such as salbutamol **20** [122] (Fig. 3).

5 Conclusions and Outlook

From the research efforts in progress within a number of laboratories an expanding library of complementary Hnls is developing. Some existing members in this collection, in particularly the HbHnl, PaHnl and MeHnl, are being fully exploited in terms of understanding them at a molecular and genetic level which is complementary to the advances being made in clarifying their substrate acceptance pattern. Thus for these Hnls their application to industrial syntheses for the generation of useful chiral intermediates looks promising. The determination of novel sources and the understanding of their 3D structure and mechanism should help to guide the development of a broad overlapping network of substrates that can efficiently yield chiral cyanohydrins. Thus, in summary, the future of these catalysts looks extremely promising and over the next time period full exploitation of this system is expected in which new developments to enhance their usefulness should occur.

Acknowledgement. The authors wish to thank Current Biology Ltd for the permission to reproduce of Figs. 1 and 2.

6 References

1. (a) Stinson SC (1995) Chem Eng News 73:44; (b) Collins AN, Sheldrake GN, Crosby J (1997) Chirality in Industry II. Wiley, New York; (c) Sheldon RA (1993) Chirotechnology. Marcel Decker, New York
2. Tanaka A, Tetsuya T, Kobayashi T (1993) Industrial application of immobilized biocatalysts. Marcel Dekker, New York
3. Rosenthaler L (1908) Biochem Z 14:238
4. Becker W, Pfeil E (1966) J Am Chem Soc 88:4299
5. Becker W, Pfeil E (1966) Ger Offen 1,593,260, 25 September 1969 Chem Abstr 1969. 72:P3061t
6. Conn EE (1981) Cyanogenic glycosides. In: Stumpf PK, Conn EE (eds) The biochemistry of plants: a comprehensive treatise. Academic Press, New York, 7:479–500
7. Poulton JE (1990) Plant Physiol 94:401
8. Selmar D, Lieberei R, Biehl B, Conn EE (1989) Physiol Plant 75:97
9. Lieberei R, Selmar D, Biehl B (1985) Pl Syst Evol 150:49
10. Rosenthal GA, Bell EA (1979) Naturally occurring toxic nonprotein amino acids. In: Rosenthal GA, Janzen DH (eds) Herbivores, their interaction with secondary plant metabolites. Academic Press, New York
11. Witthohn K, Nauman CM (1984) Z Naturforsch 39:837
12. Aschoff H-J, Pfeil E (1970) Hoppe-Seyler's Z Physiol Chem 351:818
13. Jansen I, Woker R, Kula M-R (1992) Biotechnol Appl Biochem 15:90
14. Smitskamp-Wilms E, Brussee J, van der Gen A, van Scharrenburg GJM, Sloothaak JB (1991) Recl Trav Chim Pays-Bas 110:209
15. Gerstner E, Kiel U (1975) Hoppe-Seyler's Z Physiol Chem 356:1853
16. Cheng I-P, Poulton JE (1993) Plant Cell Physiol 34:1139
17. Wu H-C, Poulton JE (1991) Plant Physiol 96:1329
18. Jaenicke L, Preun J (1984) Eur J Biochem 138:319
19. Jorns MS (1985) Eur J Biochem 146:481
20. Vargo D, Jorns MS (1979) J Am Chem Soc 101:7625
21. Vargo D, Pokora A, Wang S-W (1981) J Biol Chem 256:6027
22. Pokora A, Jorns MS, Vargo D (1982) J Am Chem Soc 104:5466
23. Jorns MS, Ballenger C, Kinney G, Pokora A, Vargo D (1983) J Biol Chem 258:8561
24. Becker W, Benthin U, Eschenhof E, Pfeil E (1963) Biochem Z 337:156
25. Jorns MS (1979) J Biol Chem 254:12,145
26. Bärwald K-R, Jaenicke L (1978) FEBS Lett 90:255
27. Albrecht J, Jansen I, Kula M-R (1993) Biotechnol Appl Biochem 17:191
28. Xu L-L, Singh BK, Conn EE (1988) Arch Biochem Biophys 263:256
29. Wajant H, Riedel D, Benz S, Mundry K-W (1994) Plant Sci 103:145
30. Wajant H, Mundry K-W (1993) Plant Sci 89:127
31. Wajant H, Böttinger H, Mundry K-W (1993) Biotechnol Appl Biochem 18:75
32. Wajant H, Mundry K-W, Pfizenmaier K (1994) Plant Mol Biol 26:735
33. Hughes J, Carvalho FPDeC, Hughes MA (1994) Arch Biochem Biophys 311:496
34. Wajant H, Förster S, Böttinger H, Effenberger F, Pfizenmaier K (1995) Plant Sci 108:1
35. Wajant H, Förster S, Selmar D, Effenberger F, Pfizenmaier K (1995) Plant Physiol 109:1231
36. Kuroki GW, Conn EE (1989) Proc Natl Acad Sci USA 86:6978
37. van Scharrenberg GJM, Sloothaak JB, Kruse CG, Smitskamp-Wilms E, Brussee J (1993) Ind J Chem 32b:16
38. Wajant H, Förster S (1996) Plant Sci 115:25
39. Hasslacher M, Schall M, Hayn M, Griengl H, Kohlwein SD, Schwab H (1996) J Biol Chem 271:5884
40. Schall M (1996) PhD Thesis, University of Graz
41. Förster S, Roos J, Effenberger F, Wajant H, Sprauer A (1996) Angew Chem Int Ed Engl 35:437, Angew Chem 108:493

42. (a) Zandbergen P, van der Linden J, Brussee J, van der Gen A (1991) Synth Commun 21:1387; (b) Zandbergen P, van der Linden J, Brussee J, van der Gen A (1995) Cyanohydrin formation. In: Roberts SM, Wiggins K, Casy G (eds) Preparative biotransformations: whole cell and isolated enzymes in organic synthesis. Wiley, New York, chap 4:51
43. Vernau J, Kula M-R (1990) Biotechnol Appl Biochem 12:397
44. Hasslacher M, Schall M, Hayn M, Bona R, Rumbold K, Lückl J, Griengl H, Kohlwein SD, Schwab H (1997) Protein Expression and Purif 11:61
45. Wagner UG, Schall M, Hasslacher M, Hayn M, Griengl H, Schwab H, Kratky C (1996) Acta Cryst D52:591
46. Wagner UG, Hasslacher M, Griengl H, Schwab H, Kratky C (1996) Structure 4:811
47. Ollis DL, Goldman A (1992) Protein Eng 5:197
48. Cygler M, Doctor BP (1993) Protein Sci 2:366
49. Hasslacher M, Kratky C, Griengl H, Schwab H, Kohlwein SD (1997) Proteins Struct Funct Genet 27:438
50. Lauble H, Müller K, Schindelin H, Förster S, Effenberger F (1994) Proteins Struct Funct Genet 19:343
51. Wajant H, Pfizenmaier K (1996) J Biol Chem 271:25,830
52. Zuegg J, Gugganig M, Wagner UG, Kratky C (1997) Abstract No C-24, 3rd International Symposium on Biocatalysis and Biotransformations, 22–26 September Le Grand Motte, France Club Bioconversions en Synthese Organique, CNRS Marseille
53. van den Nieuwendijk AMCH, Warmerdam EGJC, Brussee J, van der Gen A (1995) Tetrahedron: Asymmetry 6:801
54. Becker W, Freund H, Pfeil E (1965) Angew Chem 77:1139
55. Becker W, Pfeil E (1966) Biochem Z 346:301
56. Klabunowski JI (1963) Asymmetrische Synthese, Veb Deutscher Verlag der Wissenschaften 2:84
57. Brussee J, Roos EC, van der Gen A (1988) Tetrahedron Lett 29:4485
58. Brussee J, Loos WT, Kruse CG, van der Gen A (1990) Tetrahedron 46:979
59. Effenberger F, Ziegler T, Förster S (1987) Angew Chem 99:491
60. Ziegler T, Hörsch B, Effenberger (1990) Synthesis 575
61. Effenberger F, Hörsch B, Weingart F, Ziegler T, Kühner S (1991) Tetrahedron Lett 32:2605
62. Effenberger F, Heid S (1995) Tetrahedron Asymmetry 6:2945
63. Kiljunen E, Kanerva LT (1996) Tetrahedron Asymmetry 7:1105
64. Kiljunen E, Kanerva LT (1997) Tetrahedron Asymmetry 8:1225
65. Kiljunen E, Kanerva LT (1997) Tetrahedron Asymmetry 8:1551
66. Bové C, Conn EE (1961) J Biol Chem 236:207
67. Woker R, Champluvier B, Kula M-R (1992) J Chromatogr 584:85
68. Niedermeyer U, Kula M-R (1990) Angew Chem Int Ed Engl 29:386, Angew Chem 102:423
69. Effenberger F, Hörsch B, Förster S, Ziegler T (1990) Tetrahedron Lett 31:1249
70. Chueshul S, Chulavatnatol M (1996) Arch Biochem Biophys 334:401
71. Schmidt M, Hervé S, Klempier N, Griengl H, (1996) Tetrahedron 52:7833
72. Klempier N, Pichler U, Griengl H (1995) Tetrahedron: Asymmetry 6:845
73. Klempier N, Griengl H, Hayn M (1993) Tetrahedron Lett 34:4769
74. Kragl U, Niedermeyer U, Kula M-R, Wandrey C (1990) Ann NY Acad Sci 613:167
75. Hickel A, Graupner M, Lehner D, Hermetter A, Glatter O, Griengl H (1997) Enz Microb Technol 21:361
76. Selmar D, Carvalho FJP, Conn EE (1987) Anal Biochem 166:208
77. Bauer B, Strathmann H, Effenberger F (1991) Patent No. 4041 896 Chem Abstr 1970, 73, P3061t
78. Slotta KH (1934) Chem Ber 67:1028
79. Loos WT, Geluk HW, Ruijken MMA, Kruse CG, Brussee J, van der Gen (1995) Biocatalysis and Biotransformations 12:255
80. Kobayashi Y, Hayashi H, Miyaji K, Inoue S (1986) Chem Lett 931
81. Nowitzki O, Münnich I, Stucke H, Hoffmann HMR (1996) Tetrahedron 52:11,799
82. Castejón P, Moyano A, Pericàs MA, Riera A (1996) Tetrahedron 52:7063

83. Cocker W, Grayson DH, Shannon PVR (1995) J Chem Soc Perkin Trans I 1153
84. Burton TS, Caton MPL, Coffee ECJ, Parker T, Stuttle KAJ, Watkins GL (1976) J Chem Soc Perkin Trans I 2550
85. Ognyanov VI, Datcheva VK, Kyler KS (1991) J Am Chem Soc 133:6992
86. Finiels A, Geneste P (1979) J Org Chem 44:1577
87. Okano V, doAmaral L, Cordes EH (1976) J Am Chem Soc 98:4201
88. Young PR, McMahon PE (1979) J Am Chem Soc 101:4678
89. Kiljunen E, Kanerva LT (1994) Tetrahedron: Asymmetry 5:311
90. Danieli B, Barra C, Carrea G, Riva S (1996) Tetrahedron: Asymmetry 7:1675
91. Huuhtanen TT, Kanerva LT (1992) Tetrahedron: Asymmetry 3:1223
92. Hickel A, Gradnig G, Griengl H, Schall M, Sterk H (1996) Spectrochim Acta Part A 52:93
93. Hickel A (1996) PhD Thesis, Technical University Graz
94. Mao C-H, Anderson L (1967) Phytochemistry 6:473
95. van Eikeren P (1993) US Pat 5,241,087, 31st August 1993 Chem Abstr 1994, 120, P8336f
96. Niedermeyer U (1993) Ger Patent DE 41 39 987, 4th December 1991, Chem Abstr 1993, 119, P70564m
97. Menéndez E, Brieva, Rebolledo F, Gotor Y (1995) J Chem Soc Chem Commun 989
98. Effenberger F, Schwämmle A (1997) Biocatalysis and Biotransformations 14:167
99. Svirbely EJ, Roth JF (1953) J Am Chem Soc 75:3106
100. Ching W-M, Kallen RG (1978) J Am Chem Soc 100:6119
101. Reslow M, Adlercreutz P, Mattiasson B (1988) Eur J Biochem 172:573
102. Wehtje E, Adlercreutz P, Mattiasson B (1993) Biotechnol Bioeng 41:171
103. Wehtje E, Adlercreutz P, Mattiasson B (1990) Biotechnol Bioeng 36:39
104. Wehtje E, Adlercreutz P, Mattiasson B (1988) Appl Microbiol Biotechnol 29:419
105. Kragl U, Niedermeyer U, Kula M-R, Wandrey (1990) Ann NY Acad Sci 613:167
106. Voß H, Miethe P (1992) Enzymes entrapped in liquid crystals: a novel approach for biocatalysis in nonaqueous media. In: Tramper J, Vermae MH, Beet HH, Stockar UV (eds) Biocatalysis in non-conventional media, progress in biotechnology. Elsevier, London 8:739
107. Miethe P, Jansen I, Niedermeyer, Kragl U, Haftendorn R, Kula M-R, Mohr K-H, Meyer H-W (1992) Biocatalysis 7:61
108. Schmitz J (1992) PhD Thesis, University of Cologne
109. Zandbergen P, van den Nieuwendijk AMCH, Brussee J, van der Gen A, Kruse CG (1992) Tetrahedron 48:3977
110. Zandbergen P, Willems HMG, van der Marcel GA, Brussee J, van der Gen A (1992) Synth Commun 22:2781
111. de Vries EFJ, Steenwinkel P, Brussee J, Kruse CG, van der Gen A (1993) J Org Chem 58:4315
112. Brussee J, Dofferhof F, Kruse CG, van der Gen A (1990) Tetrahedron 46:1653
113. Brussee J, van der Gen A (1991) Recl Trav Chim Pays Bas 110:25
114. Brussee J (1992) PhD Thesis, University of Leiden
115. Effenberger F, Stelzer U (1991) Angew Chem 103:866
116. Stelzer U, Effenberger F (1993) Tetrahedron: Asymmetry 4:161
117. Wamerdam EGJC, Brussee J, van der Gen A, Kruse CG (1994) Helv Chim Acta 77:252
118. Wamerdam EGJC, van Rijn RD, Brussee J, Kruse CG, van der Gen A (1996) Tetrahedron: Asymmetry 7:1723
119. Wamerdam EGJC, van den Nieuwendijk AMCH, Brussee J, Kruse CG, van der Gen A (1996) Tetrahedron: Asymmetry 7:2539
120. Johnson DV, Griengl H (1997) Tetrahedron 53:617
121. Johnson DV, Griengl H (1997) Abstract No P I-52, 3rd International Symposium on Biocatalysis and Biotransformations, 22–26 September, Le Grand Motte, France Club. Bioconversions en Synthese Organique, CNRS Marseille
122. Effenberger F, Jäger J (1997) J Org Chem 62:3867

Received February 1998

Stereoinversions Using Microbial Redox-Reactions

Andrew J. Carnell

Department of Chemistry, Robert Robinson Laboratories, University of Liverpool, Liverpool L69 7ZD. *E-mail: acarnell@liverpool.ac.uk*

This paper aims to provide a summary of the recent literature on the use of redox enzymes to carry out stereoinversion reactions on chiral secondary alcohols. Emphasis has been placed on biotransformations which result in the deracemization of a racemic substrate to give high value synthetic intermediates in a theoretical 100% yield. Most of the biocatalysts which are competent to carry out such transformations are whole cell systems, which contain the necessary cofactor recycling machinery to facilitate this otherwise entropically disfavoured process. The first section deals with deracemization of compounds such as mandelic acid and pantoyl lactone using two microorganisms which display enantiocomplementary stereospecificity. The deracemization of chiral alcohols such as β-hydroxyesters, aryl ethanols and terminal 1,2-diols with single microorganisms will then be discussed and the influence of growth and reaction conditions on the selectivity observed will be emphasised. Then the ability of several microorganisms to deracemize by double stereoinversion substrates with two stereocentres such as cyclohexan-1,2-diol, *cis* and *trans* indan-1,2-diol and pentan-2,4-diol will be presented and some mechanistic rationale proposed. Lastly enzymes known as epimerases which are important in sugar and deoxysugar biosynthesis will be discussed with reference to some recent work on the mechanism of UDP-glucose epimerase.

Keywords: Microbial, Stereoinversion, Deracemization, Redox, Epimerase.

1 **Introduction** 58

2 **Deracemization Through Stereoinversion Using Two Microorganisms** 59

3 **Deracemization Through Stereoinversion Using a Single Microorganism** 63

3.1 Deracemization of Compounds with One Stereocentre 63

3.2 Deracemization of Compounds with Two Stereocentres 66

4 **Regiospecific Stereoinversions in Sugar Biosynthesis** 69

5 **Conclusions** 71

6 **References** 71

Advances in Biochemical Engineering/
Biotechnology, Vol. 63
Managing Editor: Th. Scheper

1 Introduction

Current regulations governing the introduction of chiral molecules as pharmaceuticals or agrochemicals require that each enantiomer should undergo testing for biological activity prior to development and marketing [1]. In response to this requirement organic chemists have focused a great deal of effort in developing new methods for accessing enantiomerically pure compounds. Two fundamental approaches towards the preparation of homochiral materials are resolution of racemic mixtures and asymmetric synthesis. Catalytic methods are the most effective in terms of economics and environmental impact. Resolution processes involve separations usually based on a difference in physical properties of diastereomeric derivatives, such as fractional crystallization, or a difference in energy between two diastereomeric transition states in a kinetic resolution. There are several potential disadvantages, particularly from an industrial perspective, in using a resolution process:

1) the maximum theoretical yield is 50% and perfect resolutions are rare;
2) separation of the product formed from the remaining substrate in a kinetic resolution may be problematic;
3) only one enantiomer may be required in which case the other may need to be discarded or an alternative use be found; occasionally an enantioconvergent route to the target can be found using the "unwanted enantiomer."

Solutions to overcoming these problems have been developing in recent years and one general approach which has been successfully applied on a large scale is the *dynamic resolution*. This involves rapid in situ racemization of the substrate in a kinetic resolution process in order to present the chiral selector such as an asymmetric chemical reagent or an enzyme [2] with a racemic substrate throughout the course of the reaction, thus optimising the enantiomeric purity of the product. This approach is of course limited to substrates which are capable of being racemized under the reaction conditions. Selective crystallisation of an equilibrating mixture of enantiomers/diasteriomers is also an attractive approach to industry but may require extensive development for any particular example.

Methods in asymmetric synthesis, particularly those which employ transition metal catalysis, have and continue to contribute enormously to the synthetic chemists armoury and, providing both enantiomers of the catalyst ligand are available, one can synthesise either enantiomer of a chiral intermediate at will.

Relatively little attention has been paid to the conversion of racemic compounds into their enantiomerically pure versions in a single process, in other words a *deracemization*. For certain classes of chiral compounds such as secondary alcohols, this approach should provide many benefits, particularly to the pharmaceutical industry. Existing routes to high value intermediates in their racemic form may be modified to provide the equivalent homochiral product, thus reducing the extent of development chemistry required. In addition, the unwanted enantiomer (or enriched mixture) from a resolution process could be converted to the desired antipode. Obviously a deracemization is energetically "uphill" since one is moving from an entropically favoured racemate to a more ordered single enantiomer. Therefore some energy input will be required and enzyme systems are uniquely set up to provide such a driving force through coupled in vivo redox cofactor transformations.

This article will report the use of enzymes contained in whole-cell systems (living or resting cells) for the deracemization of chiral secondary alcohols through irreversible stereoinversion. The discussion will be presented in three sections dealing with (i) sequential oxidation-reduction with two microorganisms leading to deracemization, (ii) single microorganism redox stereoinversions and (iii) regiospecific stereoinversions catalysed by epimerases in sugar biosynthesis.

2
Deracemization Through Stereoinversion Using Two Microorganisms

Enantiomerically pure secondary alcohols are important intermediates for the synthesis of pharmaceuticals, agrochemicals, pheromones, flavours and fragrances, liquid crystals and as chiral auxiliaries in asymmetric synthesis [3, 4]. Methods for their preparation include asymmetric hydrogenation or reduction using chirally modified metal hydrides and enzymatic or microbial reduction of the corresponding ketones. They can also be obtained by enzymatic kinetic resolution of their corresponding carboxylic esters using hydrolytic enzymes such as lipases and esterases. In most cases the stereochemical course of an enzymatic reduction with a dehydrogenase enzyme may be predicted from a simple model generally referred to as Prelog's rule [5]. Baker's yeast alcohol dehydrogenase shows Prelog specificity giving almost exclusively *S*-configured alcohols whereas the dehydrogenase from *Lactobacillus kefir* affords predominantly *R*-alcohols, the anti-Prelog product [6]. In many situations, where a particular enantiomer of a chiral alcohol is required, the corresponding racemate or opposite enantiomer may be more readily available than the prochiral ketone from which it could be made by asymmetric reduction. Indeed a stockpile of the unwanted enantiomer might be available from another process. The combination of two dehydrogenase enzymes with complementary specificity, i.e. Prelog and anti-Prelog, for successive oxidation-reduction reactions constitutes a potentially extremely effective approach for the interconversion of enantiomers via the oxidized (ketone) form.

Medici et al. have used a combined sequential oxidation-reduction to access a range of unsaturated secondary alcohols from their racemates [7] (Scheme 1). Here the *S*-alcohol **2** is oxidized by *B. stereothermophilus* which is displaying Prelog specificity leaving the *R*-enantiomer untouched. The other microorganism, *Y. lipolytica* contains an anti-Prelog dehydrogenase which is therefore able to reduce the ketone **1** to the *R*-alcohol **2**. Thus the combination of the two steps effects a net deracemization of substrate **2**.

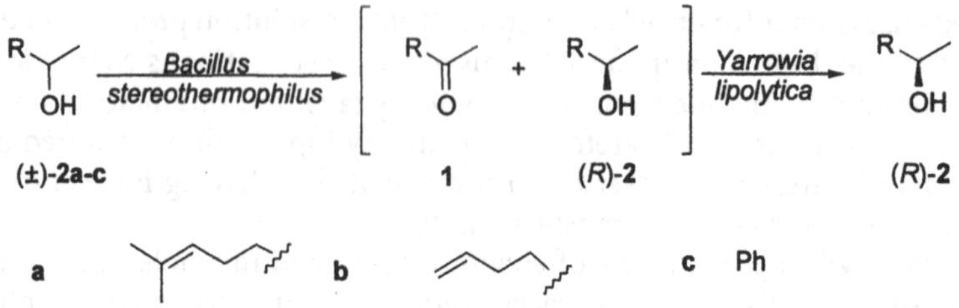

Scheme 1

The oxidation and reduction steps were initially studied in isolation to determine the selectivity and optimal conditions. Then the two steps were combined in different ways (A-C):

A the growth medium containing the ketone **1** and optically active alcohol (*R*)-**2** from the first step was added to spun down cells of *Y. lipolytica* for the second reduction step;
B as in A but the first oxidation step was carried out with resuspended cells in phosphate buffer (pH 7.2);
C suspensions of both microorganisms were grown and mixed and the substrate was then added.

The (*S*)-enantiomer of each substrate alcohol (**2a–c**) was preferentially oxidized in a kinetic resolution leaving optically active (*R*)-alcohol **2**. The ketone **1** was then selectively reduced to the (*R*)-alcohol giving overall yields of 82–100%. The e.e.s for substrates **2a** and **2b** were excellent (100%) using conditions A or B with overall conversions being high (>90%). Use of the mixed culture (conditions C) gave a diminished yield and e.e. for substrate **2a** (82%, 90% e.e.) and **2b** (85%, 40% e.e.). For substrate **2c** the results were somewhat anomalous with conditions B giving the best results (80%, 85% e.e.). The mixed culture C gave no reaction at all which could not be accounted for given that each half-reaction proceeded with complementary enantioselectivity when carried out separately. In addition the effect of using combined cultures for substrates **2a** and **2b** was not nearly as detrimental. Evidently the use of combined cultures in this way and the apparent large differences in efficacy which result is worthy of further study. Subtle effects such as the interaction of cell-signalling mechanisms may be important and research into microbial autoregulators such as the ubiquitous *N*-acylhomoserine lactones which are known to regulate gene expression may improve understanding and future use of mixed microbial populations for biotransformations.

(*R*)-Mandelic acid **3** is a useful chiral synthon for the production of pharmaceuticals such as semi-synthetic penecillins, cephalosporins and antiobesity agents and many methods have been reported for the preparation of the optically pure material. A method to deracemize the racemate which is readily available on a large scale was developed by Ohta et al. using a combination of two biotransformations. The method consists of enantioselective oxidation of (*S*)-

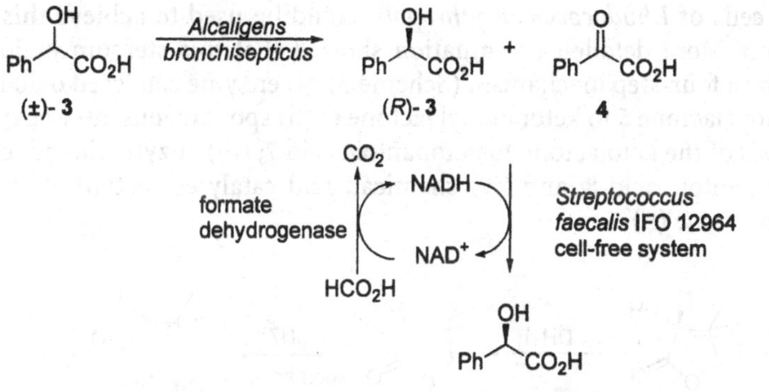

Scheme 2

(+)-mandelic acid **3** by *Alcaligenes bronchisepticus* KU 1201 followed by NADH-dependent asymmetric reduction of the resulting benzoylformic acid **4** to (*R*)-mandelic acid **3** with a cell free extract of *Streptococcus faecalis* IFO 12964 [8] (Scheme 2).

Although a one-pot reaction procedure, it was necessary to carry out each conversion successively. While *A. bronchioseptіcus* oxidized mandelic acid under aerobic conditions with vigorous shaking, the cell free extract of *S. faecalis* was inactivated under such conditions. Optimised conditions involved growing *A. bronchisepticus* for 2 days then addition of 0.5% w/v (+/–)-mandelic acid and shaking for 3 days. The *S. faecalis* cell free extract was added along with NADH, formate dehydrogenase and formic acid and the mixture allowed to stand for 2 days. The overall yield and e.e. of this tandem biotransformation were good but the low substrate concentration would not be useful for preparative purposes without further optimisation. Takahashi et al. [9] have more recently published a similar process using *Pseudomonas polycolor* and *Micrococcus freudenreichii* simultaneously. Although the yield of (*R*)-mandelic acid was lower (70%), intact whole cells were used, precluding the need to add NADH cofactor. In addition the total reaction time was shorter (one day instead of five) and the substrate concentration was nine-fold higher (4.5% w/v).

D-(+)-Pantothenic acid is a water-soluble vitamin and an important constituent of coenzyme A. It is necessary for the metabolism of protein, fat and carbohydrates and for the formation of certain hormones. It is made by condensation of β-alanine with D-(–)-pantoyl lactone **5**. The conventional synthetic process for D-(–)-**5** involves synthesis of the racemate followed by classical resolution and racemization of the unwanted L-(+)- enantiomer. Lanzilotta et al. [10] have used whole cells of *Bassochlamys fulva* to carry out asymmetric reduction on ketopantoyl lactone **6** to give the D-(–)-lactone **5** and Achiwa et al. [11] have used catalytic asymmetric hydrogenation with a rhodium complex in an abiotic version. Shimizu and coworkers [12] have reported two processes using whole-cell biocatalysts for the deracemisation of (D,L)-pantoyl lactone **5**, a more readily available starting material than the ketopantoyl lactone. Initial screening for microorganisms which could convert (L)- to (D)-pantoyl lactone **5** revealed that washed cells of *Rhodococcus erythropolis* could be used to achieve this transformation. More detailed examination showed that this stereoinversion proceeded by a four-step mechanism (Scheme 3): (i) enzyme catalysed oxidation of (L)-pantoyl lactone **5** to ketopantoyl lactone **6**; (ii) spontaneous, non-enzymatic, hydrolysis of the ketolactone to ketopantoic acid **7**; (iii) enzymatic reduction to give (D)-pantoic acid **8**; and (iv) chemical, acid catalysed lactonisation in the work-up to give (D)-**5**.

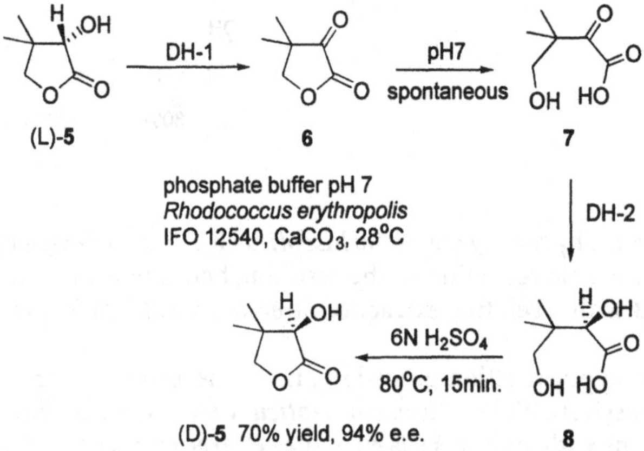

(L)-**5** —DH-1→ **6** —pH7, spontaneous→ **7** —DH-2→ **8** —6N H_2SO_4, 80°C, 15min.→ (D)-**5** 70% yield, 94% e.e.

phosphate buffer pH 7
Rhodococcus erythropolis
IFO 12540, $CaCO_3$, 28°C

Scheme 3

It was necessary to run the reaction in the presence of $CaCO_3$ as solid buffer in order to maintain the optimum pH of 7–8 for the first biooxidation step. When run as a deracemisation reaction the second, reduction, step was initially sluggish with high levels of ketoacid being accumulated as a result of enzyme inhibition by the unreactive (D)-enantiomer. This problem was overcome by supplementing with glucose and $CoCl_2$ after three days of incubation. As an alternative procedure Shimizu's group published an oxidation-reduction using two different micoorganisms in tandem [13] (Scheme 4). The first enantioselec-

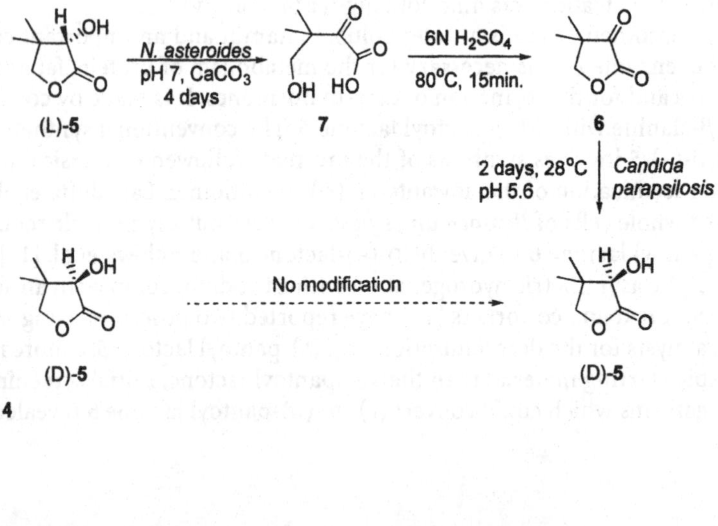

(L)-**5** —*N. asteroides*, pH 7, $CaCO_3$, 4 days→ **7** —6N H_2SO_4, 80°C, 15min.→ **6** —*Candida parapsilosis*, 2 days, 28°C, pH 5.6→ (D)-**5**

(D)-**5** - - - No modification - - -→ (D)-**5**

Scheme 4

tive oxidation step was catalysed by *Nocardia asteroides* AKU 2103 in the presence of $CaCO_3$ to neutralize the ketopantanoic acid formed. It was possible to maximize the dehydrogenase activity in this organism by inducing with 1.2% 1,2-propandiol. In this way the authors report a possible throughput of 80 g l^{-1} of (d,l)-substrate where >90% of the added L-(+)-isomer was converted to ketopantoic acid **7**. After acid treatment to lactonize, the ketolactone was reduced with *Candida parapsilosis* IFO 0784. The throughput here was 72 mg/ml or 72 g l^{-1}.

3 Deracemization Through Stereoinversion Using a Single Microorganism

3.1 Deracemization of Compounds with One Stereocentre

The Baker's yeast asymmetric bioreduction of β-ketoesters to give the (*S*)-hydroxyester is perhaps one of the best known biotransformations and works well for a range of substrates. Azerad and Buisson have used aged cultures of a local *Geotrichum candium* strain for this transformation to give (*R*)-hydroxyesters [14]. These aged cultures, formed by shaking the grown mycelium in buffer for 24 h before adding the substrate, apparently lack the (*S*)-alcohol forming dehydrogenase activity. Reduction of ethyl 3-oxobutanoate **9** gave the (*R*)-hydroxyester **10** in around 50% e.e. (72% conversion) after 7 h. However, continuation of the incubation far beyond the time taken for complete reduction led to the formation of essentially optically pure (96% e.e.) (*R*)-hydroxyester **9** in a 75% isolated yield. It was suspected that this incomplete recovery of material might reflect the presence of a hydrolytic enzyme catalysing the enantioselective hydrolysis of the (*S*)-hydroxyester **10** which would account for the apparent amplification of optical purity of the product. In order to verify this hypothesis, racemic 3-hydroxybutyrate **10** was incubated under the same conditions and the reaction was followed by gas chromatography over 80 h.

The results actually showed a deracemization of the racemic hydroxyester **10** as opposed to enantioselective hydrolysis with formation of optically pure (*R*)-hydroxyester **10** and only 20% loss in mass balance. Small quantities of ethyl 3-oxobutanoate **9** (<5%) were also detected throughout the reaction, leading the authors to suggest a multiple oxidation-reduction system with one dehydrogenase enzyme (DH-2) catalysing the irreversible reduction to the (*R*)-hydroxyester (Scheme 5).

In addition, a sample of (*S*)-hydroxyester of 85% e.e. was converted to the optically pure (*R*)-enantiomer under similar conditions. It is also possible to interconvert short chain 2-methyl-3-oxoalkanoates effecting an overall stereoinversion of the C-2 centre. On prolonged exposure to *G. candidum*, *syn* (2*R*, 3*S*) ethyl 2-methyl-3-hydroxybutanoate **11** was converted, via the 3-oxoester **12**, to the *anti* (2*S*, 3*S*) compound **11** with an e.e. of 97% in around 23 h (Scheme 6).

Evidently equilibration of the C-2 stereocentre occurs via the enol form of the ketester and only the 2(*S*)-3-oxobutanoate **12** is a substrate for the dehydrogenase DH-2 catalysing the irrerversible (*S*)-selective reduction. Because of the

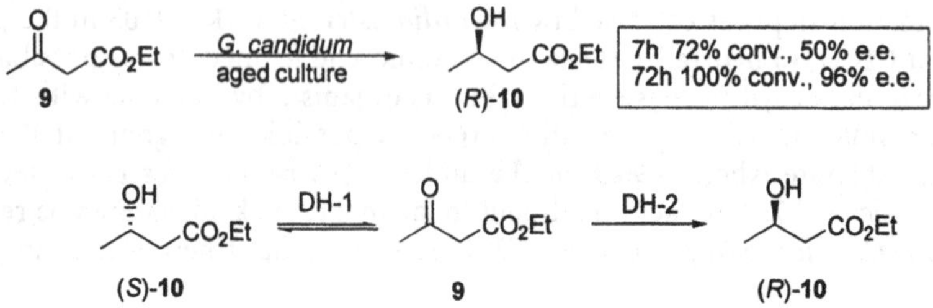

Scheme 5

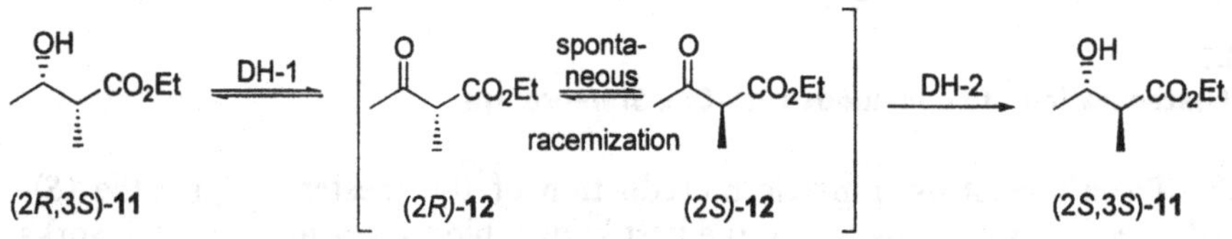

Scheme 6

difficulty of separating two diastereomers which might result from asymmetric reduction using biocatalytic or chemical methods, this stereoinversion is particularly useful. An example of its synthetic utility has been demonstrated in a synthesis of one stereoisomer of the pine sawfly pheromone [15].

Nakamura et al. have used another strain of this fungus, *Geotrichum candidum* IFO 5767, to deracemize various arylethanols to give the (*R*)-alcohol **13** in high yield and enantiomeric excess [16] (Scheme 7).

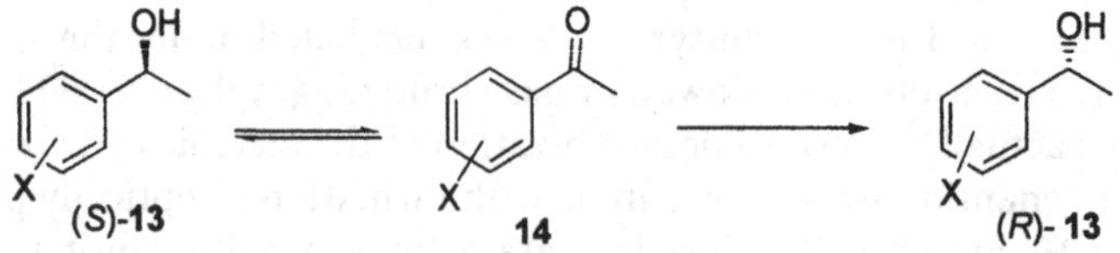

Scheme 7

In a time course study on the conversion of (±)-1-phenylethanol **13** (X=H), formation of acetophenone was observed to a maximum of around 20% during the conversion of (*S*)- to (*R*)- alcohol which occurred over 24 h to give (*R*)-**13** in 96% yield, 99% e.e. The effect of ring substitution on the efficiency of the deracemization was notable. While *para* substituents (Cl, OMe, Me) gave good results, *ortho* derivatives could not be deracemized and the biocatalyst showed little activity towards *meta* substituted compounds. On addition of allyl alcohol, improvements in e.e. were obtained, particularly for the conversion of 1-(*m*-methylphenyl)ethanol (from 21 to 94% e.e.). However these improvements did appear to be at the expense of yield (89% diminished to 55%). The authors suggest that, as noted previously for yeast reductions, the allyl alcohol selectively inhibits an enzyme responsible for reduction of the intermediate ketone **14** to the (*S*)-alcohol. An alternative explanation, given the significantly lower (and close to 50%) yield, is simply selective oxidation of the (*S*)-alcohol to the ketone which was not further processed due to complete inhibition of reductase activity for production of either enantiomer of the alcohol. The yield of the ketone for this transformation was not quoted. As expected, deuterium labelling of the methine position of the model substrate 1-phenylethanol showed a 50% decrease in deuterium content after 24 h of microbial deracemization. Again it was suggested by these workers that at least two enzymes are operating cooperatively within the whole cell system, possibly using different cofactors or different enantiotopic hydrogens (*pro R* or *pro S*) of the same cofactor.

Cell cultures of *Catharanthus roseus* entrapped in calcium alginate have been employed by Takemoto and Achiwa to deracemize pyridyl alcohols such as **15** and **16** [17] (Scheme 8).

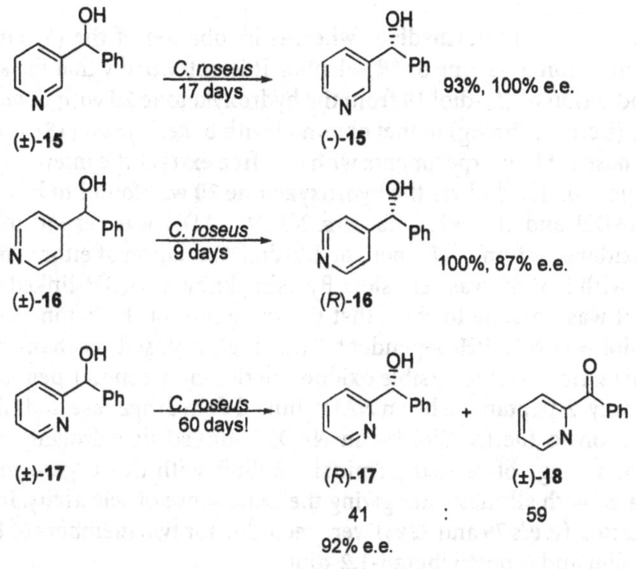

(±)-**15** → *C. roseus*, 17 days → (-)-**15** 93%, 100% e.e.

(±)-**16** → *C. roseus*, 9 days → (*R*)-**16** 100%, 87% e.e.

(±)-**17** → *C. roseus*, 60 days! → (*R*)-**17** + (±)-**18**

41 : 59

92% e.e.

Scheme 8

Reaction times were long but yields and e.e.'s high. (*S*)-(+)-α-phenyl-2-pyridylmethanol **17** has analgesic and anticonvulsant activities and was a target of this research. However, the antipodal (*R*)-alcohol was the product from this biotransformation upon prolonged exposure (60 days) to the *C. roseus* cells. This resulted from a slow selective oxidation of the (*S*)-alcohol giving a 41:59 mixture of (*R*)-alcohol **17** (92% e.e.) and ketone **18**.

Hasegawa and co-workers [18] screened a number of microorganisms for the ability to degrade or deracemize selectively terminal 1,2-diols, useful intermediates for the synthesis of pharmacueticals, agrochemicals and liquid crystals. The yeast *Candida parapsilosis* IFO 0708 was the strain of choice for deracemization of the model substrate, pentan-1,2-diol **19**. This biotransformation produced the (–)-(*S*)-enantiomer (80%, 100% e.e.) and was performed on a relatively large scale, 60 g of product from a 2.6 l fermentation (Scheme 9). In control experiments the optically pure (*R*)-pentan-1,2-diol **19** was completely converted to the (*S*)-enantiomer over 24 h, with 1-hydroxy-2-petanone **20**

OH OH *C. parapsilosis* 24h, pH 6.5 OH OH

(±)-**19** (-)-(*S*)-**19**

80%, 100% e.e.

NAD$^+$ NADPH

OH OH NADH O OH NADP$^+$ OH OH

(+)-(*R*)-**19** **20** (-)-(*S*)-**19**

Scheme 9

being observed as an intermediate, whereas incubation of the (*S*)-enantiomer gave no conversion to ketone or (*R*)-alcohol. It is noteworthy that the selectivity for the production of (*S*)-diol **19** from the hydroxyketone **20** with *C. parapsilosis* is opposite (i.e. anti-Prelog) to that obtained with baker's yeast [15] and glycerol dehydrogenase [16]. In experiments with cell-free extract the interconversion of (*R*) to (*S*)-pentan-1,2-diol via the hydroxyketone **20** was found to be dependent on both NADH and NADPH cofactors. NAD$^+$/NADH was responsible for the selective oxidoreduction of (*R*)-pentan-1,2-diol. Oxidation of either enantiomer of the diol with NADP$^+$ was very slow. By using known NADH-linked reductase inhibitors it was possible to show that the reduction of the ketone to optically pure (*S*)-diol was NADPH-dependent. Thus the proposed mechanism for this biotransformation is (i) reversible oxidoreduction between (*R*)-pentan-1,2-diol and 1-hydroxy-2-pentanone by an NAD$^+$-linked dehydrogenase and (ii) irreversible reduction to the (*S*)-diol by an NADPH-linked dehydrogenase. Deracemization of a range of related terminal 1,2-diols with this organism was also demonstrated with all substrates giving the same sense of selectivity. Incomplete deracemization (e.e.'s 79 and 82%) were recorded for two members of the series, butan-1,2-diol and 4-phenylbutan-1,2-diol.

3.2 Deracemization of Compounds with Two Stereocentres

A more complicated picture arises with substrates containing more than one stereocentre which could be subject to redox stereoinversion of the type described in the previous examples. With two carbinol stereocentres in a symmetrical substrate there exist a maximum of five stereoisomers (the *R,R* and *S,S* enantiomers and *meso* isomer of the diol and two enantiomers of the intermediate α-hydroxyketone) for the dehydrogenase enzyme(s) to discriminate and transform irreversibly to a single enantiomer. Of course for 1,2-diols the intermediate β-hydroxyketone may be spontaneously equilibrating through an enediol intermediate allowing for a dynamic resolution in the reduction step. Two whole cell systems have been discovered, which are able to deracemize a symmetrical 1,3-diol and 1,2-diols respectively.

Matsumura and co-workers [19] investigated the bioreduction of 2,4-pentandione **21** with *Candida boidinii* "methanol yeast" (Scheme 10). The diketone **21** was transformed over 6 h into (2*R*,4*R*)-2,4-pentandiol **22** in 78% yield and 99%

(2*S*,4*S*)-**22** (4*S*)-**24** **23**

double stereoinversion

Scheme 10 **21** (2*R*,4*R*)-**22** (4*R*)-**24**

e.e. along with a small amount of the *meso* isomer **23**. It was necessary to have glucose present in the medium to prevent assimiliation of the substrate by the microorganism. Incubation of the optically pure (2*S*,4*S*)-diol **22** gave a mixture of *meso* diol **23** (39%) and (2*S*,4*S*)-diol **22** with a diminished e.e. (around 68%) showing formation of (*R*,*R*)-diol **22** and double stereoinversion. Biotransformation of the *meso* diol **23** gave a much faster conversion to the (2*R*,4*R*)-diol **22** which was shown to be essentially irreversible. No 4-hydroxy-2-pentanone **24** was observed during the conversions suggesting rapid reduction of the ketone. Thus the (2*S*,4*S*)-diol is presumably transformed via the (4*S*)-β-hydroxyketone **24** to *meso* diol which then undergoes oxidation to (*R*)-4-hydroxy-2-pentanone **24** before irreversible reduction to the (2*R*,4*R*)-diol **22**.

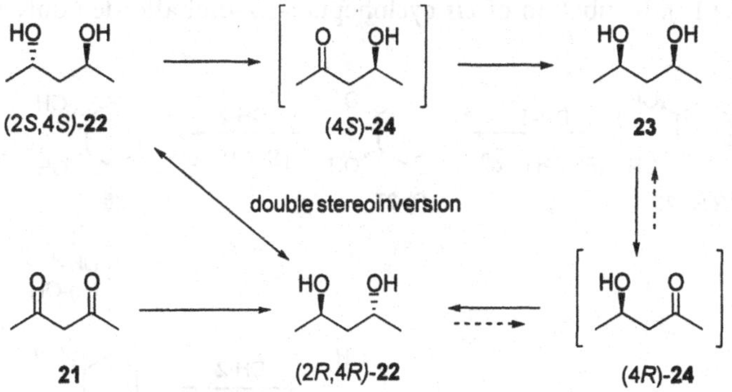

In connection with our own work on the enzyme-catalysed hydrolysis of cyclohexene epoxide with various fungi we made the unexpected observation that the microorganism *Corynesporia casssiicola* DSM 62475 was able to interconvert the (1*R*,2*R*) and (1*S*,2*S*) enantiomers of the product, *trans* cyclohexan-1,2-diol **25**. As the reaction proceeded the (1*R*,2*R*) enantiomer was converted to the (1*S*,2*S*) enantiomer [20]. If the racemic *trans* diol **25** was incubated with the growing fungus over 5 days, optically pure (>99% e.e.) (1*S*,2*S*) diol **25** could be isolated in 85% yield. Similarly biotransformation of *cis (meso)* cyclohexan-1,2-diol **26** yielded the (1*S*,2*S*) diol **25** in 41% (unoptimized) yield (Scheme 11).

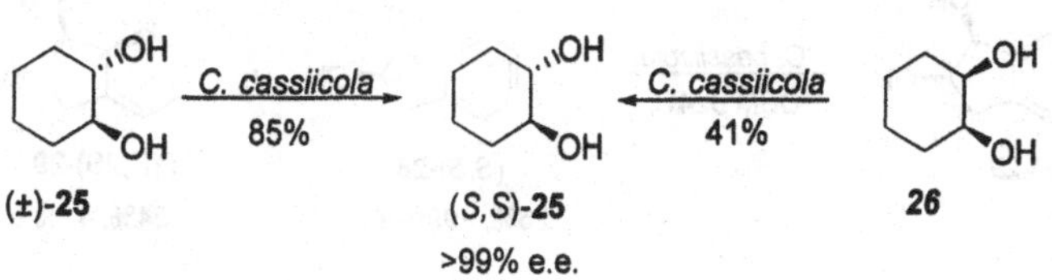

C. cassiicola 85% *C. cassiicola* 41%

(±)-**25** (*S*,*S*)-**25** >99% e.e. ***26***

Scheme 11

Incubation of the (1*R*,2*R*) diol **25** over a shorter time period of 4 days gave a mixture of the *meso cis* diol **26** (26%) and the *trans* diol **25** (*RR*/*SS*: 76:26) showing conversion over to the (1*S*,2*S*) diol. As with the previous examples, no conversion was observed when using the *S*,*S* diol **25** as a substrate. Although no intermediate α-hydroxyketone was observed for this substrate we proposed the operation of at least two dehydrogenase enzymes, DH-1 and DH-2, catalysing the (*R*)-selective oxidation and (*S*)-selective reduction respectively (Scheme 12). Incubation of *cis* cycloheptan-1,2-diol afforded only the (*S*)-2-

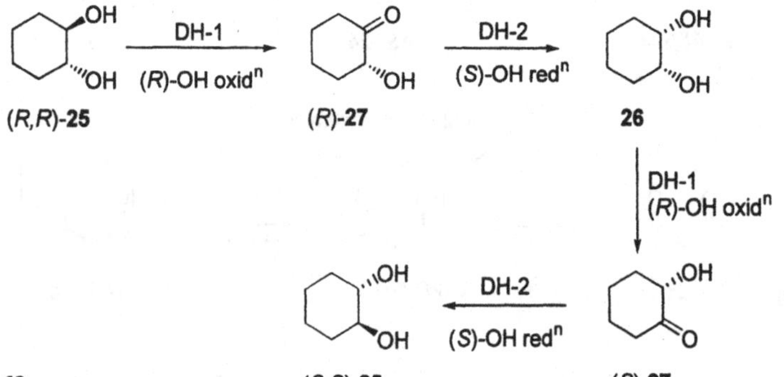

DH-1 (*R*)-OH oxid^n DH-2 (*S*)-OH red^n

(*R*,*R*)-**25** (*R*)-**27** **26** (*S*,*S*)-**25** (*S*)-**27**

Scheme 12

hydroxycycloheptanone (50%, 83%). Evidently this product was not a substrate for DH-2. More recently [21] we have discovered that the non-symmetrical racemic *cis* or *trans* indan-1,2-diols **28** and **29** are excellent substrates for this double stereoinversion and can also be deracemized, although the rate of conversion is slower (around 10–13 days) (Scheme 13). *trans* Indan-1,2-diol **28** gave the (*S*,*S*)-*trans* diol **28** in 82% yield, > 99% e.e. over 10 days. Incubation of the (±)-*cis* diol **29** over a similar time period gave a mixture of (*S*,*S*)-*trans* diol **28** (26%, > 99% e.e.) and (1*R*,2*S*) *cis* diol **29** (54%, 41% e.e.). Similar results were obtained on using indene epoxide as the substrate which initially undergoes non-enzyme catalysed hydrolysis to the racemic *trans* diol **28** under the reaction conditions.

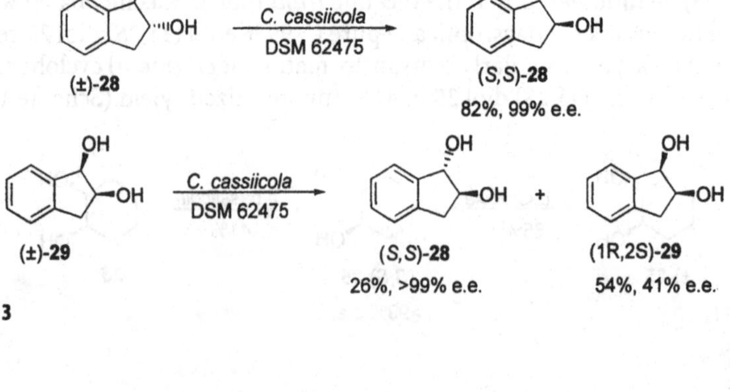

(±)-**28** → *C. cassiicola* DSM 62475 → (*S*,*S*)-**28** 82%, 99% e.e.

(±)-**29** → *C. cassiicola* DSM 62475 → (*S*,*S*)-**28** 26%, >99% e.e. + (1R,2S)-**29** 54%, 41% e.e.

Scheme 13

Closely related substrates such as β-methyl styrene epoxide and dihydronaphthalene epoxide and derived *trans* diols were not substrates for this particular microorganism.

Obviously with the indan-1,2-diol substrates there is no symmetrical *meso* intermediate which makes interpretation of the mechanism more difficult. In both the cyclohexan-1,2-diol and the indan-1,2-diol series the *trans* diols react faster and *cis* diols (both enantiomers for indandiol) are seen as intermediates. The (1*S*,2*R*) *cis* indandiol **29** is faster reacting and on incubation of the racemate only a very small trace of the (*R*,*R*)-*trans* **28** isomer is observed. 2-Hydroxyindan-1-one **30**, an observed intermediate in these biotransformations, undergoes kinetic resolution when incubated as a racemic substrate. The faster reacting enantiomer is reduced to the faster reacting *cis* (1*S*,2*R*)-indan-1,2-diol **29** which is subsequently transformed into both *trans* diols and ultimately the (*S*,*S*)-isomer. Current work is focussing on determining the absolute configuration of the intermediate α-hydroxyketone **30**.

4 Regiospecific Stereoinversions in Sugar Biosynthesis

Sugars are related through many different stereoisomeric forms and nature has provided biocatalysts which are able to mediate their interconversion, precluding the need for de novo synthesis of each isomer of a given structural type. There are many examples of enzymes able to catalyse these transformations which can be broadly classified as isomerases or epimerases. Epimerases, as the name suggests, catalyse stereoinversion reactions such as the interconversion of UDP-glucose **31** and UDP-galactose **32** catalysed by UDP-glucose epimerase, necessary for normal galactose metabolism in animals [22]. This transformation involves selective epimerization of C-4′. Studies involving running the reaction in [^{3}H]H_2O have shown no incorporation of tritium into either nucleoside, suggesting no exchange at C-4 with the solvent. This lack of exchange suggests nicotinamide-coenzyme-dependant catalysis. Enzymes from *E. coli* and yeast have been well studied and contain tightly bound NAD^+. Of several mechanisms which have been proposed for UDP-glucose epimerase it is now generally accepted that stereoinversion occurs through oxidoreduction at the 4′-position. This has raised interesting questions about the apparent non-stereospecificity of the enzyme, which must remove and deliver hydride from opposite faces of the sugar ring (Scheme 14). It is known that oxidoreduction of the nicotinamide cofactor is stereospecific and it was originally proposed that the non-specificity with respect to the substrate occurs through a simple rotation about the saccharide C-1 oxygen and the β-phosphorus of the UDP moiety, thereby allowing the opposite face of the sugar to face the NADH. More recent evidence has shown a requirement for additional rotations about the phosphate backbone of UDP [23]. In addition co-crystallization of structural analogues closely related to the 4′-deoxynucleoside intermediate and X-ray diffraction studies suggest that the active site accomodates various sugar types by simple rearrangements in water molecules rather then large changes in side chain conformations of nucleoside ligands [24].

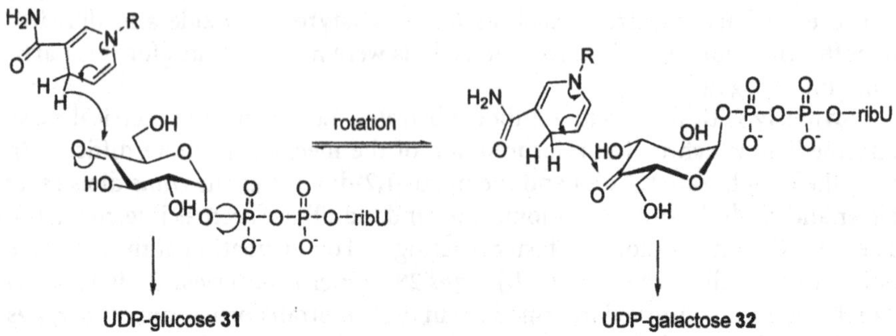

UDP-glucose **31** UDP-galactose **32**

Scheme 14

Obviously the requirement for the pyrimidine diphosphate moiety (the enzyme also accepts TDP-glucose) places limitations on this class of redox enzyme for use in preparatively useful stereoinversions where the aim is not specifically to synthesize nucleosides. The equilibrium for the interconversion favours UDP-glucose. However if this reaction is coupled to a step in which the UDP-galactose **32** is glycosylated using galactosyltransferase the equilibrium can be shifted. This approach has been used very successfully by Wong and others for the large scale preparation of *N*-acetyllactosamine **33** [25, 26] (Scheme 15).

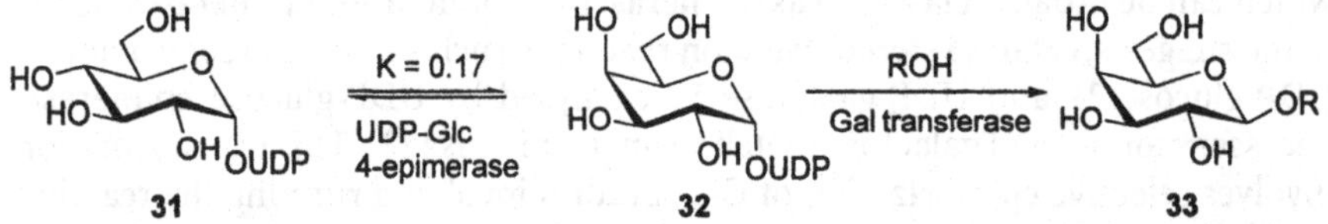

Scheme 15

R = N-acetylglucosamine

Some intriguing examples of stereoinversion are also seen in deoxysugar biosynthesis. The conversion of TDP-glucose **31** to TDP-6-deoxy-L-talose **38** and TDP-L-rhamnose **39** in *E. coli* exemplifies a commonly observed series of transformations in these biosynthetic pathways [27] (Scheme 16). TDP-glucose is converted in several steps to the 4′-keto nucleoside **34**. An enzyme known as TDP-4-keto-L-rhamnose 3,5-epimerase then catalyses two succesive stereoinversions. First the 3′-hydroxyl group undergoes inversion via the stabilized enediol **35** to give **36** prior to the second inversion at C-5. The final reduction of intermediate **37** is catalysed by two different NADPH-dependent 4-reductases with opposite stereospecificity to give TDP-6-deoxy-L-talose **38** or TDP-L-rhamnose respectively **39**.

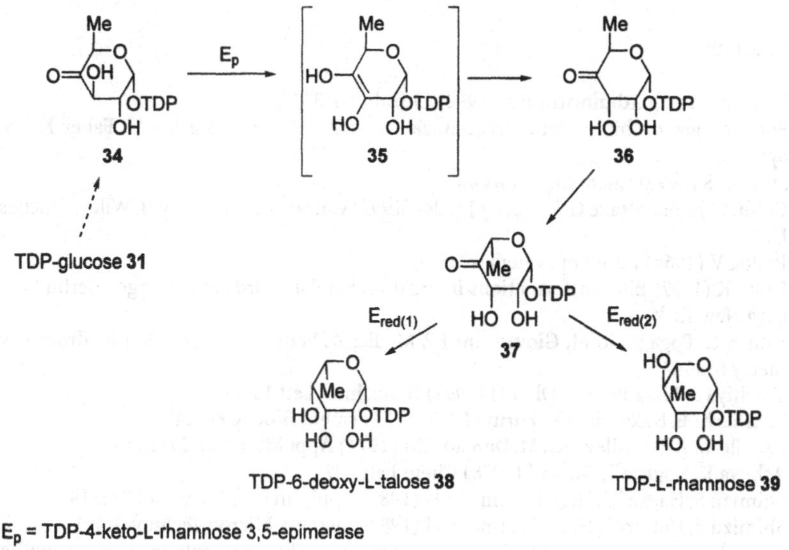

TDP-glucose **31**

TDP-6-deoxy-L-talose **38**

TDP-L-rhamnose **39**

E_p = TDP-4-keto-L-rhamnose 3,5-epimerase

Scheme 16

5 Conclusions

Given that many microorganisms contain several dehydrogenase enzymes with opposite enantiospecificities, one might expect that enantiomer interconversion through redox stereoinversion would be a rather more generally observed phenomenon in microbial reductions. A possible explanation for the apparent paucity of examples might be related to the variance of dehydrogenase activities depending on the relative amounts of redox cofactors available. As demonstrated by Azerad and Buisson [14] for β-hydroxyesters and Nakamura et al. [28] for arylethanols, growth conditions, in particular aeration conditions of the incubation mixture, can have a dramatic effect on the relative activities of dehydrogenase enzymes present in whole cell systems. In performing a biotransformation which utilises microbial redox enzymes it is therefore prudent to pay careful attention to the effect of growth conditions, aeration and reaction time before drawing too many conclusions about the specificity of the "biocatalyst" which may actually correspond to a collection of enzymes with opposite selectivities. The use of whole cell systems and isolated enzymes to carry out stereoinversions and deracemizations constitutes a useful and efficient method for obtaining homochiral intermediates and, in contrast to enzyme kinetic resolution, there is no requirement for prior derivatization or more importantly to discard an unwanted enantiomer. As with the development of any new synthetic method, finding an appropriate biocatalyst for a given substrate may be time consuming but usually rewarding and, as some of the examples above illustrate, re-examination of the whole cell bioreductions already in the literature maybe a good starting point.

6 References

1. Food and Drug Administration (1992) Chirality 4:338
2. For a review on biocatalytic deracemization techniques see: Stecher H, Faber K (1997) Synthesis 1
3. Stinson S (1992) Chem Eng News 46
4. Collins AN, Sheldrake GN, Crosby J (eds) (1997) Chirality in industry II. Wiley, Chichester UK
5. Prelog V (1964) Pure Appl Chem 9:119
6. Faber K (1997) Biotransformations in organic chemistry, 3rd edn. Springer, Berlin Heidelberg New York
7. Fantin G, Fogagnolo M, Giovannini PP, Medici A, Pedrini P (1995) Tetrahedron: Asymmetry 6:3047
8. Tsuchiya S, Miyamoto K, Ohta H (1992) Biotechnol Lett 14:1137
9. Takahashi E, Nakamichi K, Furui M (1995) J Ferment Bioeng 80:247
10. Lanzilotta RP, Bradley DG, McDonald KM (1974) Appl Microbiol 27:130
11. Achiwa K, Kogure T, Ojima I (1978) Chem Lett 297
12. Shimizu S, Hattori S, Hata H, Yamada H (1987) Appl Environ Microbiol 53:519
13. Shimizu S, Hattori S, Hata H, Yamada H (1987) Enzyme Microb Technol 9:411
14. Azerad R, Buisson D (1992) In: Servi S (ed) Microbial reagents in organic synthesis. Kluwar Academic Publishers, Netherlands, p 421
15. Larcheveque M, Sanner C, Azerad R, Buisson D (1988) Tetrahedron 44:6407
16. Nakamura K, Inoue Y, Matsuda T, Ohno A (1995) Tetrahedron Lett 36:6263
17. Takemoto M, Achiwa K (1995) Tetrahedron: Asymmetry 6:2925
18. Hasegawa J, Ogura M, Tsuda S, Maemoto S, Kutsuki H, Ohashi T (1990) Agric Biol Chem 54:1819
19. Matsumura S, Kawai Y, Takahashi Y, Toshima K (1994) Biotechnol Lett 16:485
20. Carnell AJ, Iacazio G, Roberts SM, Willetts AJ (1994) Tetrahedron Lett 35:331
21. Carnell AJ, McKenzie MJ, Page PCB (1997) Manuscript in preparation
22. Walsh C (1979) In: Enzymatic reaction mechanisms. Freeman and Co, San Francsisco, p 347
23. Thoden JB, Frey PA, Holden HM (1996) Biochemistry 35:5137
24. Thoden JB, Hegeman AD, Wesenburg G, Chapeau MC, Frey PA (1997) Biochemistry 36:6294
25. Wong C-H, Haynie SL, Whitesides GM (1982) J Org Chem 47:5416
26. Zervosen A, Elling L (1996) J Am Chem Soc 118:1836
27. Liu H-W, Thorson JS (1994) Annu Rev Microbiol 48:223
28. Nakamura K, Matsuda T, Ohno A (1996) Tetrahedron: Asymmetry 7:3021

Received December 1997

Biotransformations with Peroxidases

Waldemar Adam[1] · Michael Lazarus[2] · Chantu R. Saha-Möller[1] · Oliver Weichold[1] · Ute Hoch[2] · Dietmar Häring[2] · Peter Schreier[2]

[1] Institute of Organic Chemistry, University of Würzburg, Am Hubland, D-97074 Würzburg, Germany. *E-mail: Adam@chemie.uni-wuerzburg.de*

[2] Institute of Pharmacy and Food Chemistry, University of Würzburg, Am Hubland, D-97074 Würzburg, Germany

Enzymes are chiral catalysts and are able to produce optically active molecules from prochiral or racemic substrates by catalytic asymmetric induction. One of the major challenges in organic synthesis is the development of environmentally acceptable chemical processes for the preparation of enantiomerically pure compounds, which are of increasing importance as pharmaceuticals and agrochemicals. Enzymes meet this challenge! For example, a variety of peroxidases effectively catalyze numerous selective oxidations of electron-rich substrates, which include the hydroxylation of arenes, the oxyfunctionalizations of phenols and aromatic amines, the epoxidation and halogenation of olefins, the oxygenation of heteroatoms and the enantioselective reduction of racemic hydroperoxides. In this review, we summarize the important advances achieved in the last few years on peroxidase-catalyzed transformations, with major emphasis on preparative applications.

Keywords: Peroxidase, Biocatalysis, Asymmetric synthesis, Kinetic resolution, Hydroperoxide, Epoxidation, Sulfoxidation, Halogenation, Hydroxylation, Phenol coupling.

1	**Introduction**	74
2	**General Aspects**	75
2.1	Sources and Biological Functions of Peroxidases	75
2.2	Mechanisms of Peroxidase Catalysis	75
3	**Peroxidase-Catalyzed Transformations**	81
3.1	Enantioselective Reduction of Hydroperoxides	81
3.2	Oxidation of Electron-Rich Substrates	87
3.2.1	Oxidation of Aromatic Compounds	87
3.2.1.1	Hydroxylation of Arenes	87
3.2.1.2	Oxidation of Phenols and Aromatic Amines	88
3.2.2	Oxidation of C=C Bonds	91
3.2.2.1	Epoxidation of Olefins	91
3.2.2.2	Halogenation	95
3.2.3	Oxidation of Heteroatoms	98
3.2.3.1	*N*-Oxidation	98
3.2.3.2	Sulfoxidation	99
4	**Conclusions**	103
5	**References**	104

Advances in Biochemical Engineering/
Biotechnology, Vol. 63
Managing Editor: Th. Scheper

List of Symbols and Abbreviations

C*c*P	Cytochrome *c* peroxidase
CPO	Chloroperoxidase
DHF	Dihydroxyfumaric acid
F41T	Threonine mutant of HRP
F41L	Leucine mutant of HRP
HPLC	High performance liquid chromatography
HRP	Horseradish peroxidase
LiP	Ligninperoxidase
LPO	Lactoperoxidase
*m*CPBA	*meta*-Chlorperbenzoic acid
MDGC	Multidimensional gas chromatography
MP-11	Microperoxidase-11
MPO	Myeloperoxidase
PMG	Poly(γ-methyl-L-glutamate)

1 Introduction

Enzymatic transformations are becoming increasingly acceptable in organic synthesis. Although until recently the use of enzymes for synthetic purposes was restricted to some biochemically oriented, specialized research groups, currently many organic chemists are discovering the advantages of utilizing these natural catalysts in asymmetric synthesis. This is reflected in the dramatic increase of the research activities during the last few years on the topic of biotransformations in organic synthesis. Several excellent monographs [1–6] have summarized the recent developments in this current field of chemical research.

The reasons for the increasing acceptance of enzymes as reagents rest on the advantages gained from utilizing them in organic synthesis: Isolated or whole-cell enzymes are efficient catalysts under mild conditions. Since enzymes are chiral materials, optically active molecules may be produced from prochiral or racemic substrates by catalytic asymmetric induction or kinetic resolution. Moreover, these biocatalysts may perform transformations, which are difficult to emulate by transition-metal catalysts, and they are environmentally more acceptable than metal complexes.

A broad spectrum of chemical reactions can be catalyzed by enzymes: Hydrolysis, esterification, isomerization, addition and elimination, alkylation and dealkylation, halogenation and dehalogenation, and oxidation and reduction. The last reactions are catalyzed by redox enzymes, which are classified as oxidoreductases and divided into four categories according to the oxidant they utilize and the reactions they catalyze: 1) dehydrogenases (reductases), 2) oxidases, 3) oxygenases (mono- and dioxygenases), and 4) peroxidases. The latter enzymes have received extensive attention in the last years as biocatalysts for synthetic applications. Peroxidases catalyze the oxidation of aromatic compounds, oxidation of heteroatom compounds, epoxidation, and the enantioselective reduction of racemic hydroperoxides. In this article, a short overview on the recent developments of biotransformations with peroxidases is given and the synthetic applications are highlighted.

2
General Aspects

2.1
Sources and Biological Functions of Peroxidases

Peroxidases (E.C. 1.11.1.7) are ubiquitously found in plants, microorganisms and animals. They are either named after their sources, for example, horseradish peroxidase and lacto- or myeloperoxidase, or akin to their substrates, such as cytochrome *c*, chloro- or lignin peroxidases. Most of the peroxidases studied so far are heme enzymes with ferric protoporphyrin IX (protoheme) as the prosthetic group (Fig. 1). However, the active centers of some peroxidases also contain selenium (glutathione peroxidase) [7], vanadium (bromoperoxidase)

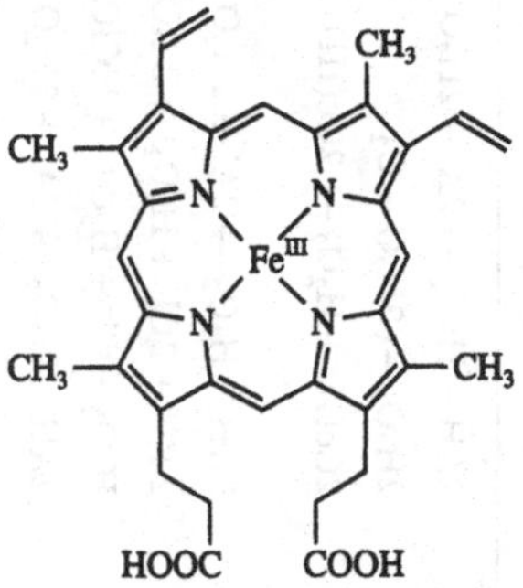

Fig. 1. Ferric protoporphyrin IX as prosthetic group of peroxidases

[8], manganese (manganese peroxidase) [9] and flavin (flavoperoxidase) [10]. A selection of heme enzymes, which exhibits a variety of biological functions, is represented in Table 1. Of these, the peroxidases efficiently catalyze the oxidation of a variety of substances (AH: reducing substrates) by peroxides, such as hydrogen peroxide and alkyl hydroperoxides (Eq. 1).

$$2\,AH + ROOH \rightarrow ROH + A\text{-}A + H_2O \quad (1)$$

As early as 1855, the oxidation of guaiacol by peroxidase with hydrogen peroxide was observed by Schönbein [22]. Furthermore, the introduction of the PZ number (Purpurogallinzahl) by Willstätter and Stoll in 1917 [23] was probably the first attempt to define the catalytic activity for a nonhydrolytic enzyme.

2.2
Mechanisms of Peroxidase Catalysis

Horseradish peroxidase has been intensively studied for the elucidation of the mechanism of peroxidase catalysis [24–28]. Some important mechanistic features of peroxidase-catalyzed reaction are briefly described here.

Table 1. Common peroxidases[a]

Peroxidase	Source	Catalyzed Reaction	Biological Functions
Horseradish Peroxidase [11]	Roots of *Armoracia rusticana*	$2AH + H_2O_2 \rightarrow 2A^{\cdot} + 2H_2O$	Biosynthesis of plant hormones
Catalase [12]	Bovine liver	$2H_2O_2 \rightarrow 2H_2O + O_2$	Detoxification of hydrogen peroxide
Cytochrom *c* Peroxidase [13]	*Saccharomyces cerevisiae*	$2Cc(II) + H_2O_2 \rightarrow 2Cc(III)^{\cdot} + 2H_2O$	Reduction of H_2O_2 and oxidation of Cytochrom *c*
Ligninperoxidase [9, 14]	*Phanerochaete chrysosporium*	$2AH + H_2O_2 \rightarrow 2A^{\cdot} + 2H_2O$	Degradation of lignin
Chloroperoxidase [15-16]	*Caldariomyces fumago*	$2 RH + 2Cl^- + H_2O_2 \rightarrow 2 RCl + 2 H_2O$	Biosynthesis of caldariomycin
Myeloperoxidase [17-19]	Human leukocytes	$H_2O_2 + Cl^- + H_3O^+ \rightarrow HClO + 2H_2O$	Antimicrobial
Lactoperoxidase [20-21]	Bovine milk	$2AH + H_2O_2 \rightarrow 2A^{\cdot} + 2H_2O$	Antimicrobial

[a] Heme as prosthetic group.

The first step of the reaction path involves the addition of H_2O_2 to the Fe^{III} resting state to form an iron-oxo derivative known as *Compound I*, which is formally two oxidation equivalents above the Fe^{III} state (Fig. 2). The well studied *Compound I* contains a $Fe^{IV}=O$ structure and a π cation radical. In the second step, *Compound I* is reduced to *Compound II* with a $Fe^{IV}=O$ structure. The reduction of the π cation radical by a phenol or enol is accompanied by an electron transfer to *Compound I* and a proton transfer to a distal basic group (B), probably His 42 (Fig. 3, step 1). The native state is regenerated on one-electron reduction of *Compound II* by a phenol or an enol. In this process, electron and proton transfers occur to the ferryl group with simultaneous reduction of Fe^{IV} to Fe^{III} (Fig. 3, steps 2–3) and formation of water as the leaving group (Fig. 3, step 4).

The cleavage of the O-O bond of the hydroperoxide is promoted by the push-pull mechanism shown in Fig. 4, in which the native HRP reacts with the unionized form of the hydroperoxide. Thus, the latter is converted into a much better nucleophile upon transfer of its proton to the distal basic group (His 42).

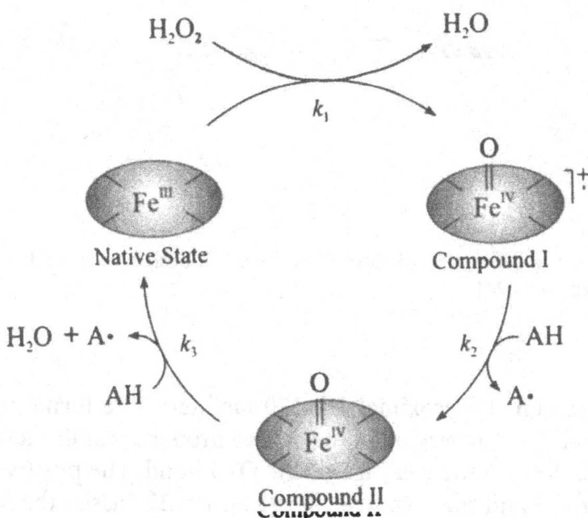

Fig. 2. Catalytic cycle of HRP (from [25])

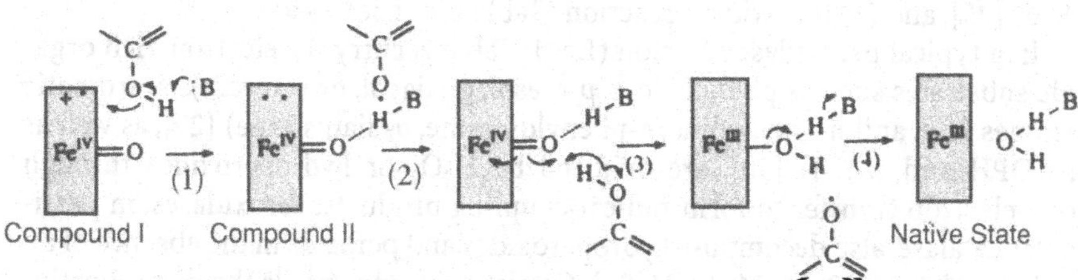

Fig. 3. Mechanism for the generation of Compound II, step (1), and regeneration of the native state, steps (2)-(4), [24]

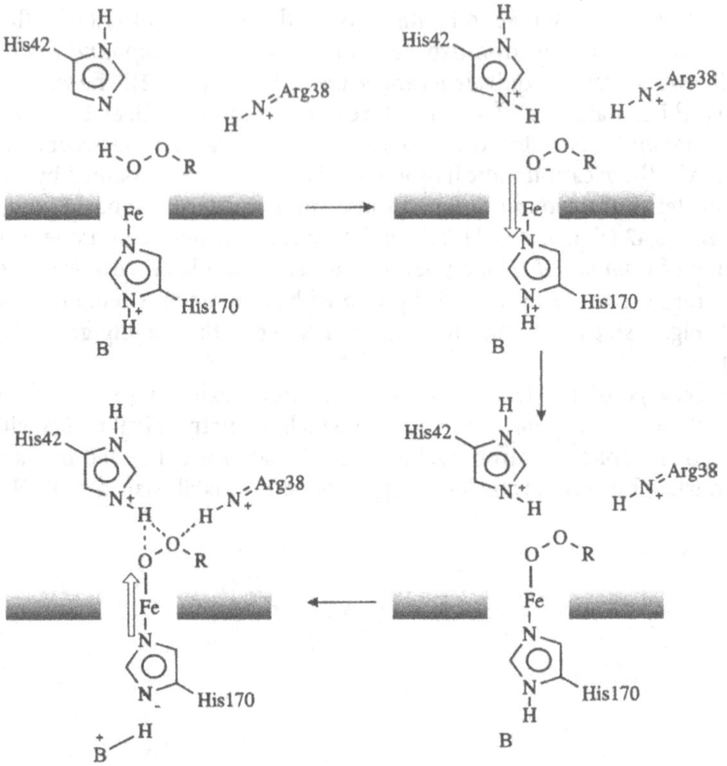

Fig. 4. Mechanism for the generation of Compound I; B is a basic amino acid, which is involved in the catalytic cycle [24]

The positive charge on the proximal His 170 facilitates the formation of the iron-peroxide bond. Thus, inversion of the charge properties at the active site of HRP facilitates the heterolytic cleavage of the O-O bond. The positive charges on His 42 and Arg 38 and the negative charge on His 170 assist the formation of the Fe=O species.

The peroxidase-catalyzed transformations are classified as (1) the peroxidase reaction (Table 2, entries 1–4), (2) the peroxygenase reaction (Table 2, entries 4–6) [25], and (3) the oxidase reaction (Table 2, entries 7–9).

In a typical peroxidase reaction (Eq. 1; Table 2, entry 1), electron-rich organic substrates such as phenols (e.g. *p*-cresol, guaiacol, or resorcin) or aromatic amines (e.g. aniline, benzidine, *o*-phenyldiamine, *o*-dianisidine) [29], as well as NADPH and NADH [30], are oxidized by H_2O_2 or hydroperoxides through one-electron transfer to form radical-coupling products. Peroxidases, in particular catalase, also decompose hydroperoxides and peracids in the absence of an electron donor [12, 16, 24, 25, 31, 32]. Catalase activity entails the disproportionation of H_2O_2 into oxygen and water (Table 2, entry 2). The heme-containing chloroperoxidase and myeloperoxidase are also known to use hydrogen per-

Table 2. Peroxidase-catalyzed reactions

Entry	Reaction		Typical substrates	Peroxidase
1	Electron transfer	$2\,AH + ROOH \rightarrow A\text{-}A + ROH + H_2O$	H_2O_2, ROOH, phenols, aromatic amines	HRP, LiP
2	Disproportionation	$H_2O_2 \rightarrow 2\,H_2O + O_2$	H_2O_2	HRP, CPO
3	Halogenation	$AH + HX + H_2O_2 \rightarrow AX + 2\,H_2O$	Tyrosine, H_2O_2	CPO, LPO
4	Sulfoxidation	 $R1\text{-}S\text{-}R2 + ROOH \rightarrow R1\text{-}S(=O)\text{-}R2 + ROH$	Thioanisole, H_2O_2, ROOH	CPO, HRP
5	Epoxidation	 $R^1\text{-}CH{=}CH\text{-}R^2 + H_2O_2 \rightarrow$ R^1-epoxide-R^2 $+ H_2O$	Alkenes, H_2O_2	CPO, HRP
6	Demethylation	$ROOH + R^1R^2NCH_3 \rightarrow ROH + R^1R^2NH + HCHO$	*N,N*-Dimethylaniline, ROOH	HRP
7	Dehydrogenation	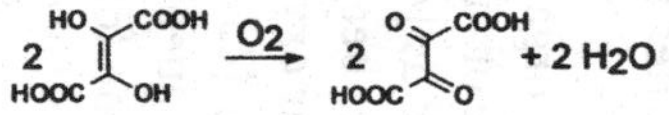 2 HOOC–C(OH)=C(OH)–COOH $\xrightarrow{O_2}$ 2 HOOC–C(=O)–C(=O)–COOH $+ 2\,H_2O$	Dihydroxyfumaric acid	HRP
8	Hydroxylation	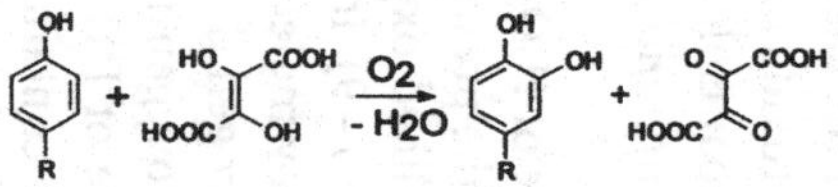 4-R-phenol + HOOC–C(OH)=C(OH)–COOH $\xrightarrow[-H_2O]{O_2}$ 4-R-catechol + HOOC–C(=O)–C(=O)–COOH	L-Tyrosine, adrenaline	HRP
9	α-Oxidation	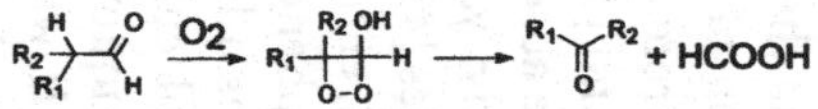 $R_2R_1CH\text{-}CHO \xrightarrow{O_2}$ R_1R_2C–C(OH)(H) (dioxetane, O–O) $\rightarrow R_1\text{-}C(=O)\text{-}R_2 + HCOOH$	Aldehydes	HRP

oxide to oxidize halide ions (Cl^-, Br^- or I^-, not F^-) to reactive halogenating species (Table 2, entry 3). In the halogenation, the halide ion may interact directly with the oxo ligand of *Compound I* to produce a ferric hypochlorite adduct [16, 25, 27]. It is still controversial whether the chlorinating agent is an enzyme-bound (Fe-O-Cl) species or the catalytically generated hypochlorous acid. A variety of substrates, e.g. phenols, anilines, pyrazoles, pyridines, and alkenes may be halogenated [3].

Peroxidases, in particular CPO, may also catalyze P-450-like monooxygenation reactions, which include thioether sulfoxidation, olefin epoxidation, and *N*-dealkylation of aromatic alkylamines (Table 2, entries 4–6) [16, 25, 27]. The oxygen atom that is incorporated into the product in the CPO-catalyzed oxyfunctionalizations derives primarily or exclusively from the peroxide. The two-electron process proceeds with direct oxygen-atom transfer from *Compound I* to the substrate [16, 25, 27]. This type of peroxidase-catalyzed oxygen transfer is called the peroxygenase reaction. The HRP enzyme catalyzes some oxygen-incorporating reactions in the presence of H_2O_2, wherein the source of the incorporated oxygen atom is not H_2O_2. The oxygen-atom incorporation requires a cosubstrate (co-oxidation), which by one-electron transfer is converted to a carbon radical (peroxidase reaction). The latter reacts with molecular oxygen to a peroxy radical ($ROO^\bullet$), which is responsible for the formation of the oxidation product [24, 25, 27, 33]. An exception to this reaction pathway has been reported in the case of HRP oxygenations of thioanisoles. Evidence indicates that the peroxide oxygen-atom is incorporated during the HRP-catalyzed oxidation of thioanisoles and that the actual oxygen atom transfer to sulfur occurs from *Compound II,* after an initial one-electron oxidation of the sulfur compound [34, 35].

In the absence of H_2O_2 or organic hydroperoxides, the peroxidases show oxidase activity. In the presence of molecular oxygen, indole-3-acetic acid [36–40], aldehydes [41], dihydroxy acids [42] or NADH [30] are oxidized by peroxidases (Table 2, entries 7–9). It has been reported that HRP catalyzes the hydroxylation of phenols by molecular oxygen in the presence of dihydroxyfumaric acid (DHF) as a hydrogen donor [43–46]. It has been shown [47] that DHF reacts with molecular oxygen in aqueous solution to produce superoxide ($O_2^{\bullet-}$) and the semiquinone of DHF; the former reacts directly with HRP to form *Compound III,* which contains an Fe(III)-$O_2H^\bullet$ structure. Hydroxyl radicals generated from *Compound III* in the presence of DHF hydroxylate phenols in a non-catalytic step (Table 2, entry 8) [48, 49]. The peroxidase-catalyzed aerobic oxidation of appropriate substrates (Table 2, entry 9) leads to carbonyl products in the electronically excited triplet state, which are postulated to be derived from dioxetane intermediates [50–52]. It has been shown that the enol form of the carbonyl compound serves as the actual substrate for peroxidase [53]. The electronically excited species transfers its energy to suitable acceptors and, therefore, sensitizes *photochemistry in the dark*, a new field of bioorganic photochemistry developed primarily by the late Prof. G. Cilento [54, 55].

3
Peroxidase-Catalyzed Transformations

The heme peroxidases oxidize a variety of structurally diverse substrates by using hydrogen peroxide or hydroperoxides as oxidants, whereby the oxidants are reduced to water or alcohol. In the following, the present state of such peroxidase-catalyzed reactions in organic synthesis is described, which unquestionably is important for preparative work.

3.1
Enantioselective Reduction of Hydroperoxides

Hydroperoxides play an important role as oxidants in organic synthesis [56–58]. Although several methods are available for the preparation of racemic hydroperoxides, no convenient method of a broad scope was until recently [59] known for the synthesis of optically active hydroperoxides. Such peroxides have potential as oxidants in the asymmetric oxidation of organic substrates, currently a subject of intensive investigations in synthetic organic chemistry [60, 61]. The application of lipoxygenase [62–65] and lipases [66, 67] facilitated the preparation of optically active hydroperoxides by enzymes for the first time.

In the oxidation of aryl methyl sulfides catalyzed by chloroperoxidase from *Caldariomyces fumago* with racemic 1-phenylethyl hydroperoxide instead of H_2O_2 as oxygen donor, it was found that (*R*)-sulfoxides, the (*S*)-hydroperoxides and the corresponding (*R*)-alcohol are produced in moderate to good enantiomeric excesses by double stereodifferentiation of the substrate and oxidant (Eq. 2, Table 3) [68].

X S + OOH R1 R2 rac —CPO→ X S O + OH R1 R2 + OOH R1 R2 (2)

However, this reaction is not suitable for the kinetic resolution of hydroperoxides on the preparative scale, because to obtain the hydroperoxides in high enantioselectivity, conversions significantly higher than 50% are required (Table 3, entry 4).

OOH R1 R2 rac —HRP, OH OCH3→ OH R1 R2 + OOH R1 R2 (3)

In contrast, the HRP-catalyzed kinetic resolution of racemic secondary hydroperoxides in the presence of guaiacol afford the hydroperoxides and their alcohols in high enantiomeric excesses (Eq. 3) [69]. In the case of the aryl alkyl-substituted hydroperoxides and cyclic derivatives (Table 4, entries 1–3, 6–10), HRP preferentially accepts the (*R*)-enantiomers as substrates with concurrent formation of the (*R*)-alcohols; the (*S*)-hydroperoxides are left behind. Further-

Table 3. CPO-catalyzed oxidations of aryl methyl sulfides with racemic hydroperoxides

Entry	Substrate	Oxidant	Sulfide conv. [%]	Peroxide conv. [%]	(*R*)-Sulfoxide ee [%]	(*S*)-Peroxide ee [%]	(*R*)-Alcohol ee [%]	E[a]
1	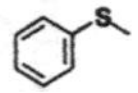	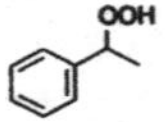	34	46	86	62	71	11
2	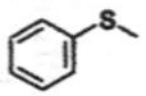	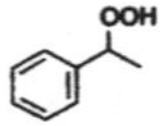	nd[b]	64	70	89	50	8.4
3	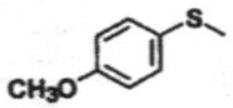	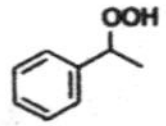	21	38	76	24	39	2.9
4	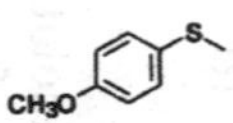	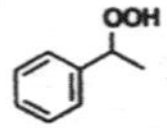	66	71	61	91	38	6.3
5	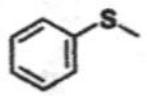	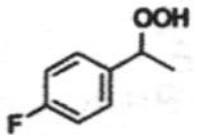	nd[b]	45	58	56	68	9.1
6	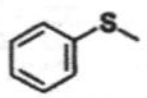	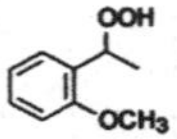	nd[b]	nd[b]	13	nd[b]	17	nc[c]
7	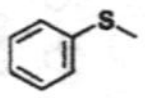	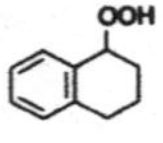	50	nd[b]	0	0	0	1.0
8	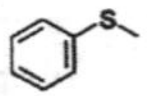	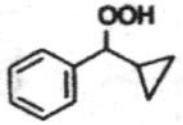	50	nd[b]	0	nd[b]	nd[b]	nc[c]

[a] For the resolution of the hydroperoxide.
[b] Not determined.
[c] Not calculable.

Table 4. Enantioselectivities of the HRP-catalyzed kinetic resolution of hydroperoxides in the presence of guaiacol

Entry	Hydro-peroxide	Hydro-peroxide ee [%][a]	Alcohol ee [%][a]	E	Ref.
1	OOH	>99 (*S*)	>99 (*R*)	>200	[59, 70–72]
2	OOH, Cl	>95 (*S*)	>95 (*R*)	>146	[70–72]
3	OOH	93 (*S*)	95 (*R*)	133	[70–72]
4	OOH	15 (*R*)	14 (*S*)	1.5	[70]
5	OOH	36 (*R*)	47 (*S*)	3.9	[70]
6	OOH	44 (*S*)	36 (*R*)	3.2	[70]
7	OOH	>99 (*S*)	91 (*R*)	111	[70, 72]
8	OOH	95 (*S*)	97 (*R*)	>200	[70, 72]
9	OOH, OH	>95 (*S*)	>95 (*R*)	>146	[70, 72]
10	EtO, O, O, OOH	>99 (*S*)	89 (*R*)	90	[73]
11	OOH, OMe, O	97 (*R*)	97 (*S*)	>200	[74]
12	OOH, OMe, O	79 (*R*)	64 (*S*)	10.7	[74]

Table 4 (continued)

Entry	Hydro-peroxide	Hydro-peroxide ee [%][a]	Alcohol ee [%][a]	E	Ref.
13	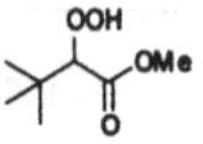	–[b]	–	–	[74]
14	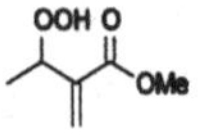	97 (*S*)	>99 (*R*)	>200	[59, 70–72]
15	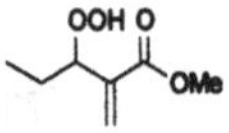	>99 (*S*)	>99 (*R*)	>200	[59, 72]
16	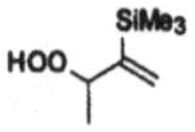	58 (*S*)	42 (*R*)	4.2	[75]
17	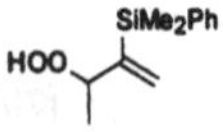	75 (*S*)	73 (*R*)	14.2	[75]
18	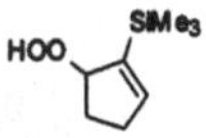	74 (*S*)	42 (*R*)	5.1	[75]
19	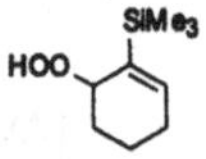	–[b]	–	–	[75]

[a] The enantiomeric excess was established by MDGC or HPLC.
[b] No conversion of the hydroperoxide with HRP.

more, the enzyme also reduces functionalized hydroperoxides, e.g. the sterically non-demanding *n*-alkyl derivatives of α- and β-hydroperoxy esters (Table 4, entries 11–15) in excellent enantioselectivities (ee >97%). Branched alkyl groups at the chirality center of the hydroperoxides diminish the reactivity and selectivity of the enzyme dramatically (Table 4, entries 4, 5, 12 and 13). Also the sterically encumbered silyl-substituted allyl hydroperoxides are not or are only reluctantly converted by HRP (Table 4, entries 16–19).

The HRP also exhibits good to excellent enantioselectivities in the reduction of *threo*- and *erythro*-diastereomeric, hydroxy-functionalized hydroperoxides with preference for the (*R*)-configured enantiomers (Table 5). Very recently, diasteromeric α,β-hydroxy hydroperoxides (entries 6 and 15) were also resolved by HRP with high ee values. Moreover, the *threo*- and *erythro*-3-hydroperoxy-2-methyl-4-pentenoates are selectively reduced by HRP to the corresponding 3-hydroxy-2-methyl-4-pentenoates (entries 8, 9, 17 and 18). Importantly for

Table 5. Enantioselectivities of the HRP-catalyzed kinetic resolution of hydroperoxides in the presence of guaiacol

Entry	Hydro-peroxide	Hydro-peroxide ee [%][a]	Alcohol ee [%][a]	E	Ref.
	threo				
1	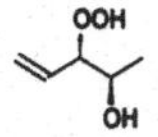 OOH, OH	63 (*S*,*S*)	67 (*R*,*R*)	9.5	[76–77]
2	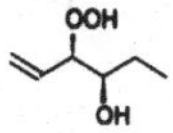 OOH, OH	>99 (*S*,*S*)	>99 (*R*,*R*)	>200	[76–77]
3	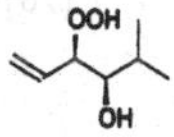 OOH, OH	>99 (*S*,*S*)	>99 (*R*,*R*)	>200	[76–77]
4	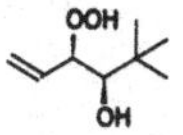 OOH, OH	91 (*S*,*S*)	84 (*R*,*R*)	36	[76–77]
5	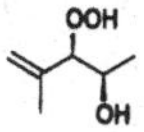 OOH, OH	14 (*S*,*S*)	18 (*R*,*R*)	1.6	[76–77]
6[b]	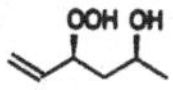 OOH OH	98	98	>200	[78]
7	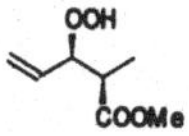 OOH, COOMe	60	88	29	[79]
8	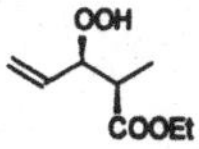 OOH, COOEt	81	[c]	nc[d]	[79]
9	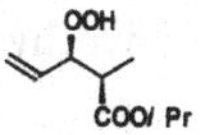 OOH, COO*i* Pr	93	>95	133	[79]
	erythro				
10	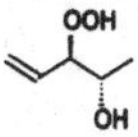 OOH, OH	>99 (*S*,*R*)	84 (*R*,*S*)	60	[76–77]
11	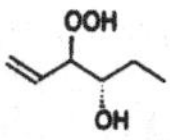 OOH, OH	>99 (*S*,*R*)	89 (*R*,*S*)	90	[76–77]
12	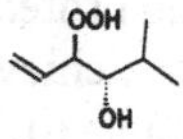 OOH, OH	>99 (*S*,*R*)	>99 (*R*,*S*)	>200	[76–77]

Table 5 (continued)

Entry	Hydro-peroxide	Hydro-peroxide ee [%][a]	Alcohol ee [%][a]	E	Ref.
13		>99 (*S*,*R*)	89 (*R*,*S*)	90	[76–77]
14		62 (*S*,*R*)	73 (*R*,*S*)	12	[76–77]
15[e]		98	84	52	[78]
16		53	86	23	[79]
17		76	81	22	[79]
18		88	93	81	[79]

[a] The enantiomeric excess was established by MDGC.
[b] Like (l) diastereomer.
[c] The alcohol was not detected.
[d] Not calculable.
[e] Unlike (u) diastereomer.

practical purposes, the enzyme-catalyzed resolution of hydroperoxides may be performed on the preparative scale to provide the optically active hydroperoxides necessary for asymmetric oxidation.

Recently, the kinetic resolution of alkyl aryl hydroperoxides with the semisynthetic peroxidase selenosubtilisin has been reported for the first time (Eq. 4) [80–81].

OOH R1 R2 *rac* —Selenosubtilisin→ OH R1 R2 + OOH R1 R2 (4)

By selective modification of the serine 221 at the active site of the protease subtilisin *Carlsberg* to its seleno derivative, peroxidase activity was achieved and utilized for the enantioselective reduction of hydroperoxides (Table 6). In con-

Table 6. Enantioselectivities of the selenosubtilisin-catalyzed kinetic resolution of hydroperoxides [80–81]

Entry	Hydroperoxide	Hydroperoxide ee [%][a]	Alcohol ee [%][a]	E
1	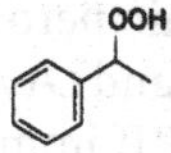	52 (*R*)	60 (*S*)	6.6
2	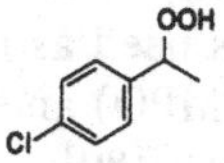	48 (*R*)	56 (*S*)	5.6
3	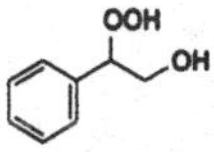	98 (*S*)	98 (*R*)	>200
4	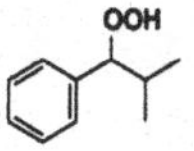	40 (*S*)	42 (*R*)	3.6
5	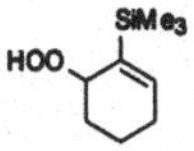	96[b]	80[b]	120

[a] The enantiomeric excess was established by MDGC.
[b] Configuration not determined.

trast to the natural enzyme HRP, the semisynthetic peroxidase selenosubtilisin exhibits selectivity for the (*S*)-hydroperoxides and its activity is not restricted to sterically unhindered substrates. The kinetic parameters indicated selenosubtilisin as the first semisynthetic enzyme with catalytic efficiency comparable to native enzymes.

In the oxidative lipid metabolism, an intermediary α-hydroperoxy acid is formed by α-oxidation of the corresponding fatty acid [82, 83]. Presumably, peroxidase-catalyzed reduction of the hydroperoxide leads to enantiomerically pure (*R*)-2-hydroxy acids [84].

3.2 Oxidation of Electron-Rich Substrates

3.2.1 *Oxidation of Aromatic Compounds*

3.2.1.1 Hydroxylation of Arenes

Mason and coworkers [44, 45] have for the first time observed that HRP catalyzes the hydroxylation of aromatic compounds by molecular oxygen in the presence of dihydroxyfumaric acid (DHF) as cofactor. The preparative potential of this biotransformation may offer attractive opportunities for the synthesis of valuable fine chemicals. For example, L-3,4-dihydroxyphenylalanine was produced from L-tyrosine, D-(–)-3,4-dihydroxyphenylglycine from D-(–)-*p*-hydroxyglycine, L-epinephrine (adrenaline) from L-(–)-phenylephrine, and catechol from phenol (Fig. 5) [85–86]. Furthermore, the lignin peroxidase (LiP) [87] catalyzes the hydroxylation of phenol, *p*- and *m*-cresol, and L-tyrosine with molecular oxygen in the presence of DHF. The isolated HRP, immobilized on poly(γ-methyl-L-glutamate) PMG, even catalyzes the hydroxylation of benzene with H_2O_2 or *m*CPBA as oxygen sources, when benzene is used as the solvent [88]. Incubation of phenylalanine with myeloperoxidase (MPO) and hydrogen peroxide gave *p*-, *m*- and *o*-tyrosine as hydroxylated products [89].

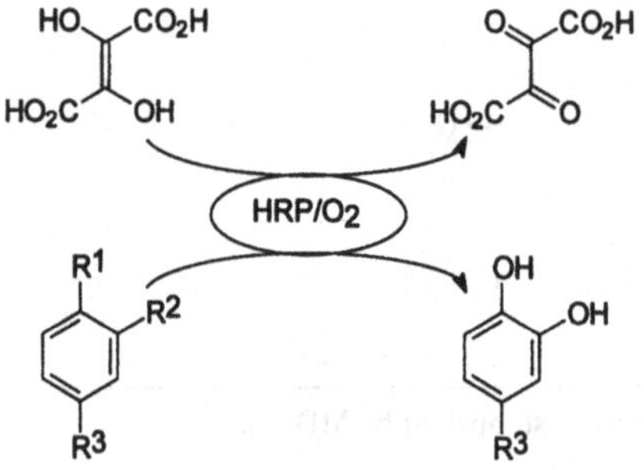

a: R1 = OH, R2 = H, R3 = $CH_2CH(NH_2)COOH$
b: R1 = OH, R2 = H, R3 = $CH(NH_2)COOH$
c: R1 = H, R2 = OH, R3 = $CH(OH)CH_2NHCH_3$
d: R1 = OH, R2 = H, R3 = H

Fig. 5. Peroxidase-catalyzed hydroxylation of phenols [85–86]

3.2.1.2
Oxidation of Phenols and Aromatic Amines

Phenols (*p*-cresol, guaiacol, pyrogallol, catechol) and aromatic amines (aniline, *p*-tolidine, *o*-phenyldiamine, *o*-dianisidine) are typical substrates for peroxidases [90–109]. These compounds are oxidized by hydrogen peroxide or hydroperoxides under peroxidase catalysis to generate radicals, which after diffusion from the active center of the enzyme react with further aromatic substrates to form dimeric, oligomeric or polymeric products.

In recent years, numerous applications of such peroxidase-catalyzed oxidative coupling of phenols and aromatic amines have been reported (Table 7). These peroxidase-catalyzed biotransformations lead to modified natural products with high biological activities [110–118]. Several examples have also been described for the oxidative coupling of phenols with peroxidases and other oxidative enzymes from a variety of fungal and plant sources as whole cell systems

Table 7. Peroxidase-catalyzed coupling of phenols and aromatic amines

Entry	Substrates	Products	Application
1	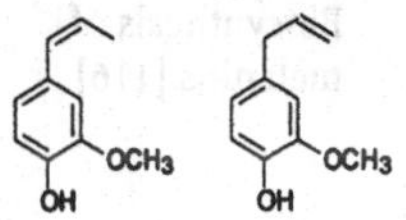 OCH_3, OH; OCH_3, OH	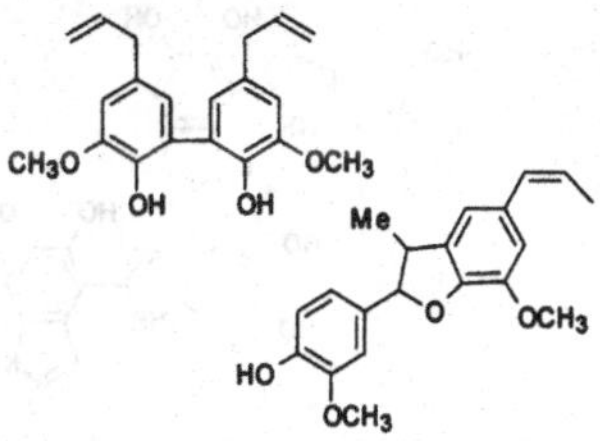 CH_3O, OH, OH, OCH_3; Me, O, OCH_3, HO, OCH_3	Synthesis of *Lythraceae* alkaloids [110][a]
2	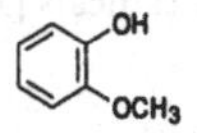 OH, OCH_3	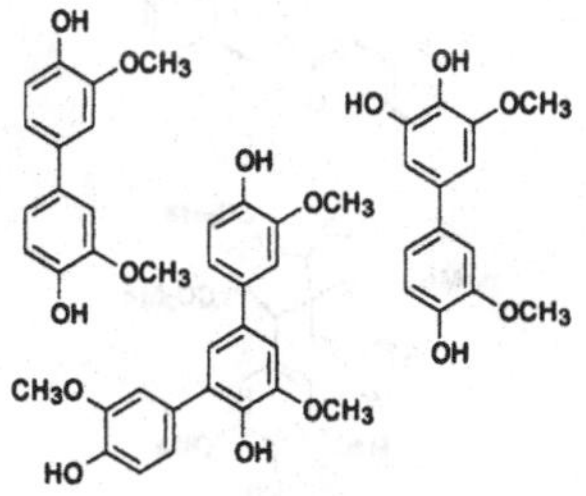 OH, OCH_3, OCH_3, OH; OH, HO, OCH_3, OCH_3, OH; OH, OCH_3, CH_3O, HO, OCH_3, OH	Antimicrobial compounds [111]
3	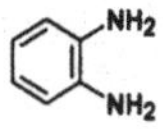 NH_2, NH_2	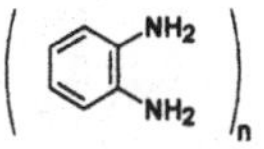 $(NH_2, NH_2)_n$	Polymerization [112]
4	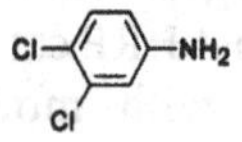 Cl, Cl, NH_2	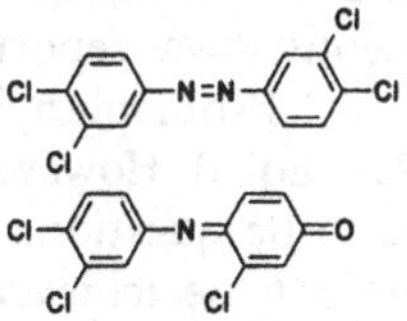 Cl, Cl, N=N, Cl, Cl; Cl, Cl, N=, Cl, =O	Biological decontamination [113]
5	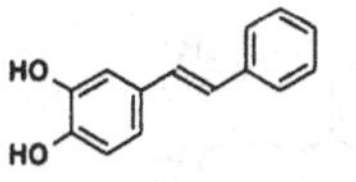 HO, HO	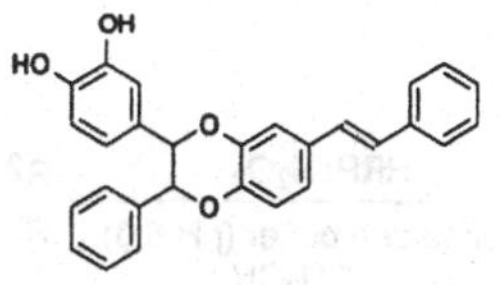 OH, HO, O, O	Stilbene oxidation, phytoalexin activity [114][a]
6	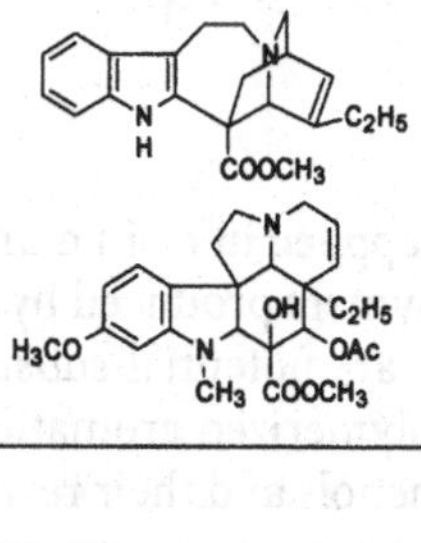 N, H, N, C_2H_5, $COOCH_3$; N, OH, C_2H_5, H_3CO, N, OAc, CH_3, $COOCH_3$	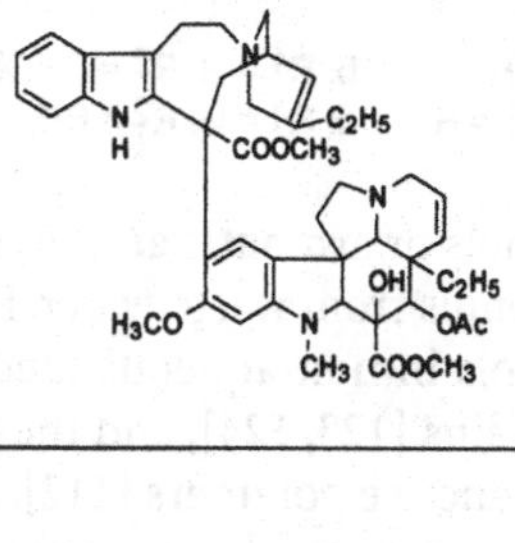 N, H, N, C_2H_5, $COOCH_3$, N, OH, C_2H_5, H_3CO, N, OAc, CH_3, $COOCH_3$	Cancer chemotherapy [115][a]

Table 7 (continued)

Entry	Substrates	Products	Application
7	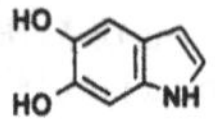	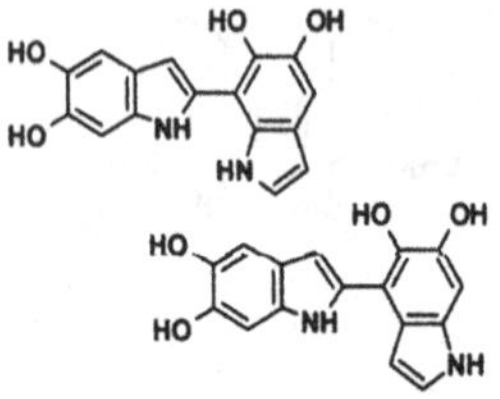	Biosynthesis of melanins [116]
8	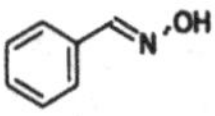	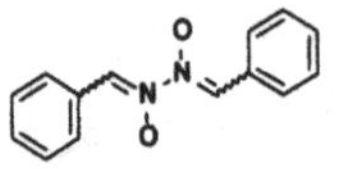	Iminoxy radicals [117]
9	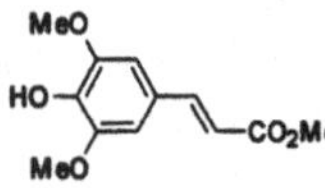	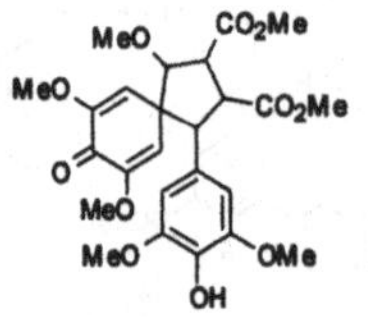	Diasteromeric spiro compounds [118]

[a] Configuration not determined.

[119, 120]. This biocatalytic method is therefore quite promising for the synthesis of complex molecules. Very recently, it was reported [121] that HRP catalyzes the oxidation of 2-naphthols to 1,1′-binaphthyl-2,2′-diols with moderate enantiomeric excess (ee 38–64%) (Eq. 5). However, in view of the analytical techniques used, these data have to be questioned [167]. As shown recently, atrop-selective biaryl coupling can only be achieved by means of dirigent protein as chiral auxiliary [122].

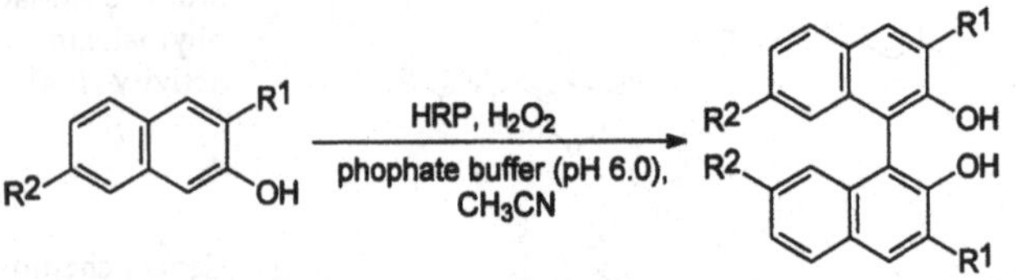

HRP, H_2O_2 / phophate buffer (pH 6.0), CH_3CN (5)

a: $R^1 = R^2 = H$ b: $R^1 = H$, $R^2 = COOMe$
c: $R^1 = Br$, $R^2 = H$ d: $R^1 = H$, $R^2 = CH_3$

The polymerization of phenols or aromatic amines is applied in resin manufacture and the removal of phenols from waste water. Polymers produced by HRP-catalyzed coupling of phenols in non-aqueous media are potential substitutes for phenol-formaldehyde resins [123, 124], and the polymerized aromatic amines find applications as conductive polymers [112]. Phenols and their resins are pollutants in aqueous effluents derived from coal conversion, paper-making, production of semiconductor chips, and the manufacture of resins and plastics. Their transformation by peroxidase and hydrogen peroxide constitutes a convenient, mild and environmentally acceptable detoxification process [125–127].

3.2.2
Oxidation of C=C Bonds

Peroxidases, in particular haloperoxidases, catalyze the oxidation of olefins in the presence or absence of halide ions. In the latter case, epoxides are produced directly (Fig. 6, route A), while in the former process halohydrins are formed (Fig. 6, route B), which may be transformed to epoxides either enzymatically by halohydrin epoxidase [128] or chemically by base treatment.

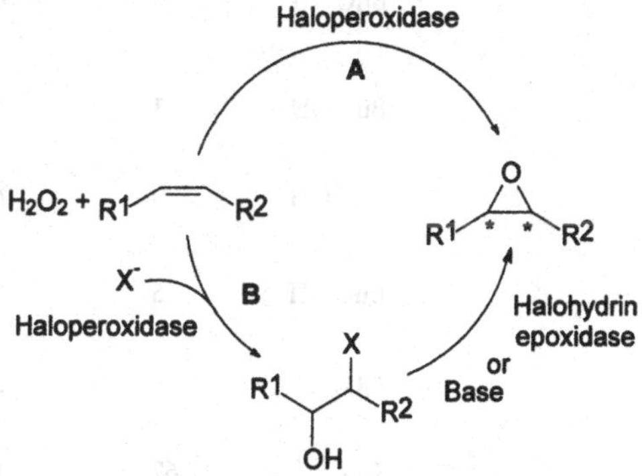

Fig. 6. Direct and indirect epoxidation catalyzed by haloperoxidases

3.2.2.1
Epoxidation of Olefins

Optically active epoxides are important building blocks in asymmetric synthesis of natural products and biologically active compounds. Therefore, enantioselective epoxidation of olefins has been a subject of intensive research in the last years. The Sharpless [56] and Jacobsen [129] epoxidations are, to date, the most efficient metal-catalyzed asymmetric oxidation of olefins with broad synthetic scope. Oxidative enzymes have also been successfully utilized for the synthesis of optically active epoxides. Among the peroxidases, only CPO accepts a broad spectrum of olefinic substrates for enantioselective epoxidation (Eq. 6), as shown in Table 8.

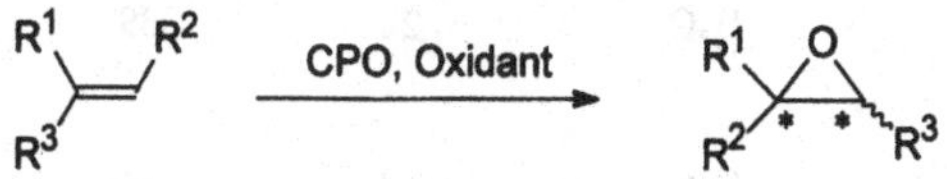

CPO, Oxidant (6)

The first CPO-catalyzed epoxidation was reported by McCarthy and White in 1983 [137]. Since this discovery, several research groups have intensively studied the substrate selectivity of this enzyme for the enantioselective epoxidation. As shown in Table 8, styrene and its ring-substituted derivatives are epoxidized by CPO in moderate enantiomeric excesses (entries 1–8). In contrast, for the *cis*-*β*-methylstyrene (entry 9) and tetrahydronaphthalene (entry 11) high enantioselectivities were observed, while the *trans*-*β*-methylstyrene was not converted by CPO (entry 10).

Table 8. Chloroperoxidase-catalyzed epoxidation of olefins

Entry	Substrate	Oxidant	Yield [%]	ee [%]	Ref.
1		*t*BuOOH	23	49	[130]
2		H_2O_2	89	49	[131]
3	Cl	*t*BuOOH	35	66	[130]
4	Cl	*t*BuOOH	34	62	[130]
5	Cl	*t*BuOOH	3	64	[130]
6	Br	*t*BuOOH	30	68	[130]
7	O_2N	*t*BuOOH	5	28	[130]
8		H_2O_2	55	89	[131]
9		H_2O_2	67	96	[132]
10		H_2O_2	–[a]		[133]
11		H_2O_2	85	97	[132]
12		H_2O_2	1	81	[131]
13		H_2O_2	7	37	[131]
14		H_2O_2	41	70	[131]
15	PhO	H_2O_2	1	46	[131]
16	PhO	H_2O_2	22	89	[131]
17	EtO_2C	H_2O_2	12	24	[131]
18	EtO_2C	H_2O_2	34	94	[131]
19	HO	*t*BuOOH	–[b]		[134]
20	MeO	*t*BuOOH	–[c]	76	[134]

Table 8 (continued)

Entry	Substrate	Oxidant	Yield [%]	ee [%]	Ref.
21	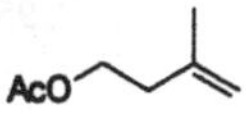 AcO	*t*BuOOH	–[c]	44	[134]
22	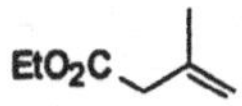 EtO_2C	*t*BuOOH	–[c]	93	[134]
23	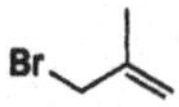 Br	*t*BuOOH	61	62	[135]
24	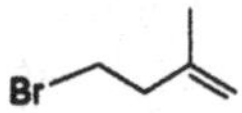 Br	*t*BuOOH	93	88	[135]
25	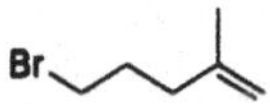 Br	*t*BuOOH	89	95	[135]
26	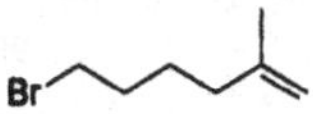 Br	*t*BuOOH	33	87	[135]
27	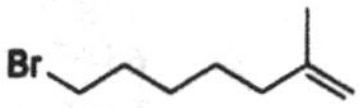 Br	*t*BuOOH	42	50	[135]
28		H_2O_2	2	10	[131]
29	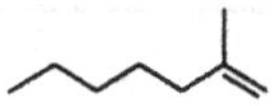	H_2O_2	23	95	[129]
30	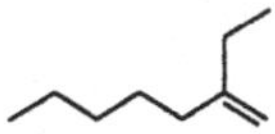	H_2O_2	4	nd[d]	[131]
31	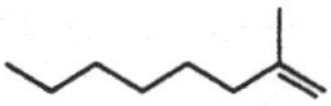	H_2O_2	nd[d]	74	[132]
32	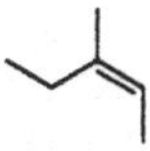	H_2O_2	nd[d]	81	[132]
33	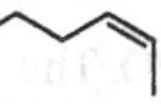	H_2O_2	12	97	[132]
34	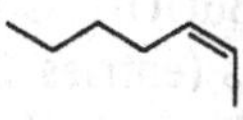	H_2O_2	78	96	[132]
35	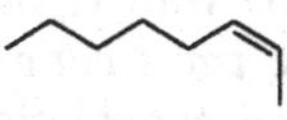	H_2O_2	82	92	[132]
36	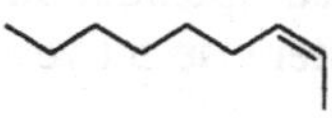	H_2O_2	–[b]		[132]
37	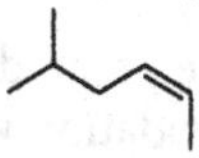	H_2O_2	33	94	[132]
38	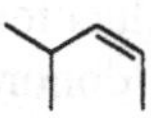	H_2O_2	nd[d]	66	[132]
39	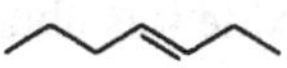	H_2O_2	40	95	[132]

Table 8 (continued)

Entry	Substrate	Oxidant	Yield [%]	ee [%]	Ref.
40		H_2O_2	20	95	[132]
41		H_2O_2	45	-[c]	[133]
42		H_2O_2	3[e]	-[c]	[133]
43	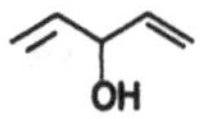 OH	H_2O_2	50	65 (de 98)	[133]
44	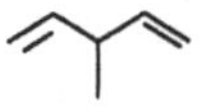	H_2O_2	30	63 (de 90)	[133]
45	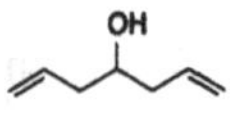 OH	H_2O_2	15	65 (de 6.2)	[133]
46	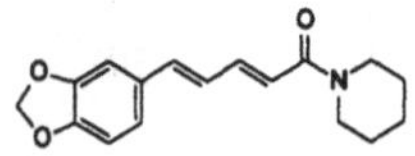 O O N O	H_2O_2	75[f]	0	[136]

[a] No epoxide, but 25% aldehyde and 5% acid were observed.
[b] No reaction.
[c] Not given.
[d] Not determined.
[e] 30% of α,β-unsaturated aldehyde was observed.
[f] HRP was used instead of CPO.

Alkyl-substituted olefins are also epoxidized by CPO. The oxidation of functionalized terminal olefins by CPO catalysis with *t*BuOOH as oxygen donor affords the corresponding epoxides in good ee values (entries 20, 22, 24–26). While *cis*-substituted olefins were epoxidized by CPO with H_2O_2 in good to excellent enantiomeric excesses (entries 32–35), the *trans* compounds were poorly converted. The enantioselective CPO-catalyzed epoxidation has been intensively investigated, although examples of diastereoselective oxidations (entries 43–45) are rare.

In a comparative study [130] with *p*-chlorostyrene as model substrate, the influence of the oxygen source in CPO-catalyzed epoxidation was investigated. The time profile for the substrate conversion and product formation are shown in Fig. 7 for *t*BuOOH (a) and H_2O_2 (b). Both oxygen donors exhibit the same selectivities (ee 66–67%); but with *t*BuOOH the epoxide was obtained in 35% yield and H_2O_2 afforded only 11% of the product.

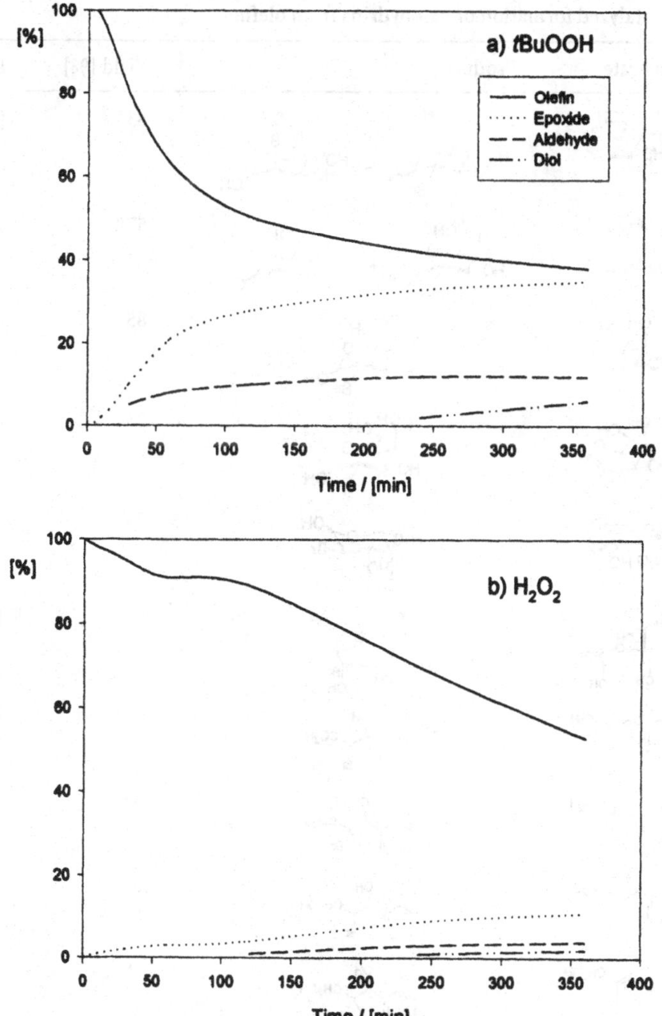

Fig. 7. Comparison of *t*BuOOH (a) and H_2O_2 (b) in the CPO-catalyzed epoxidation of *p*-chlorostyrene [130]

3.2.2.2
Halogenation

Haloperoxidases act as halide-transfer reagents in the presence of halide ions and hydrogen peroxide. In the first step, the halide ion is oxidized to a halonium-ion carrier, from which the positive halogen species is then transferred to the double bond. In an aqueous medium, the intermediary carbocation is trapped and racemic halohydrins are formed (Eq. 7). Selective examples of CPO-catalyzed formation of halohydrins are given in Table 9. In CPO-catalyzed reaction,

Table 9. CPO-catalyzed formation of halohydrins from olefins

Entry	Substrate	Products	Yield [%]	Ref.
1	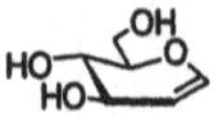	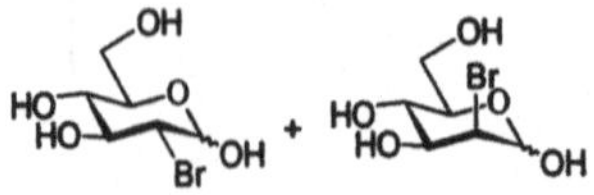	83[a]	[138]
2	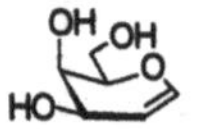	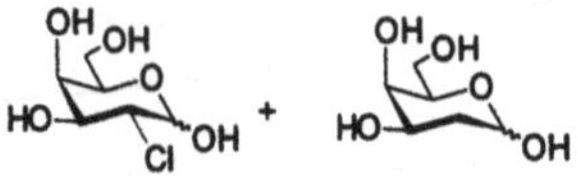	57[b]	[138]
3	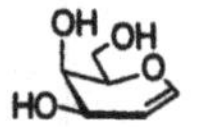	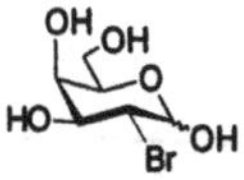	85	[138]
4	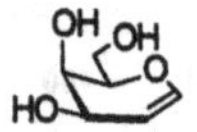	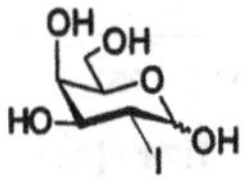	63	[138]
5	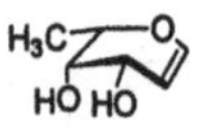	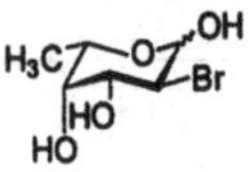	79	[138]
6	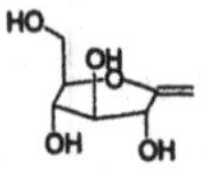	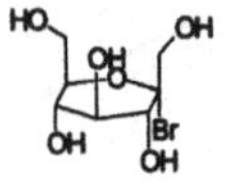	–[c]	[139]
7	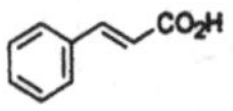	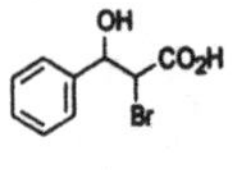	–[c]	[140]
8	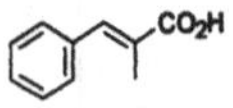	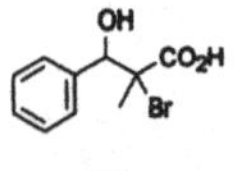	–[c]	[140]
9	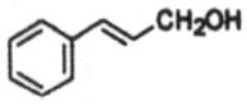	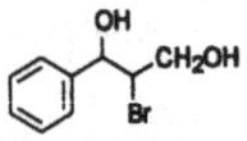	–[c]	[140]
10	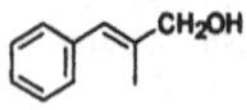	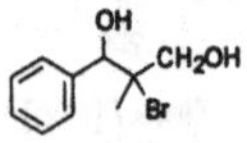	-[c]	[140]
11	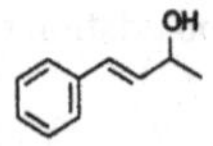	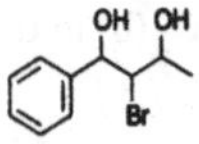	-[c]	[140]

[a] Product ratio 50:50.
[b] 15% of Chlorhydrin were formed.
[c] Not given.

halohydrins are formed in high regioselectivity; however, the diastereoselectivity is generally very low.

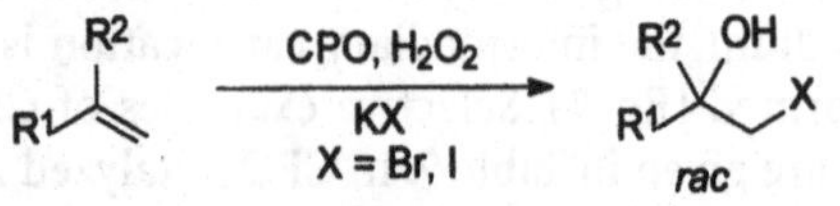

$$R^1(R^2)C{=}CH_2 \xrightarrow[\substack{KX \\ X = Br, I}]{CPO, H_2O_2} R^1R^2C(OH)CH_2X \ (rac) \tag{7}$$

CPO has been applied almost exclusively as biocatalyst for the hydroxyhalogenation. Recently, lactoperoxidase-catalyzed bromination of laurediols was reported [141, 142]. In this reaction, cyclic ethers were unspecifically formed by intramolecular trapping of the carbocation with hydroxy groups (Eq. 8).

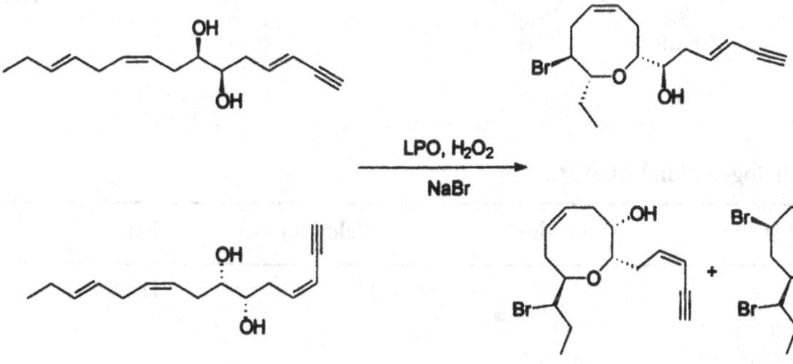

OH OH → (LPO, H_2O_2, NaBr) Br O OH (8)

OH OH → OH Br O + Br O Br O Br

Peroxidases catalyze not only the formation of halohydrines, but also the halogenation of 1,3-dicarbonyl compounds. In CPO-catalyzed halogenation of enolizable substrates, the halonium ion is trapped by the enolate to afford the corresponding mono- and dihalogenated products (Eq. 9, Table 10).

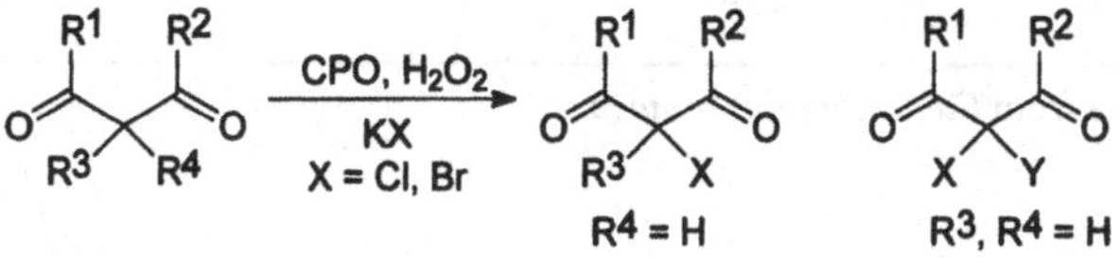

R1 R2 O O R3 R4 → (CPO, H_2O_2, KX, X = Cl, Br) R1 R2 O O R3 X, R4 = H; R1 R2 O O X Y, R3, R4 = H (9)

Table 10. CPO-catalyzed a halogenation of 1,3-dicarbonyl compounds

Entry	Substrates	Products	Ref.
1	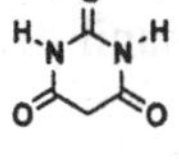	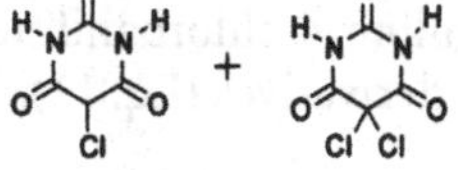 +	[143–144]
2	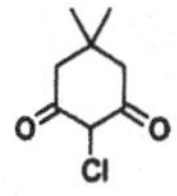	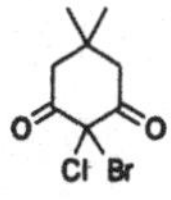	[145–146]
3	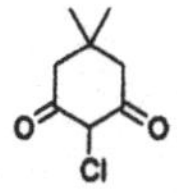	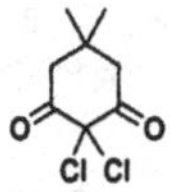	[147–148]

[a] Chloroperoxidase from *Caldariomyces fumago*.

Another example of halogen-atom transfer by CPO is the vinylic halogenation of cyclic enaminones and enamines (Eq. 10, Table 11).

CPO, H_2O_2 / KX / X = Cl, Br (10)

Table 11. Vinylic halogenation by CPO[a]

Entry	Substrate	Product	Yield [%]	Ref.
1		Br	11	[149]
2		Cl	7	[150]
3		Cl	68	[151]

[a] Chloroperoxidase from *Caldariomyces fumago.*

3.2.3 Oxidation of Heteroatoms

3.2.3.1 N-Oxidation

Peroxidases, in particular CPO, catalyze the oxidation of aryl amines, e.g. *p*-toluidine, 3,4-dichloroaniline, 4-chloroaniline and 4-nitroaniline, by H_2O_2 to the corresponding nitroso derivatives (Eq. 11).

CPO/H_2O_2 (11)

a: R1 = H, R2 = CH_3 b: R1 = H, R2 = Cl
c: R1 = Cl, R2 = Cl d: R1 = H, R2 = NO_2

It was postulated [152, 153] that the aryl amine is oxidized by direct oxygen transfer from *Compound I* to the substrate. In contrast, for the oxidation of alkaloids, e.g. morphine, codeine and thebaine (Eq. 12), to the corresponding *N*-oxides by hydrogen peroxide in the presence of HRP or crude enzyme preparation from poppy seedlings, a radical mechanism was proposed [154].

R1O, O, NCH3, R2O — HRP, H_2O_2 → R1O, O, $\overset{+}{N}CH_3$, O^-, R2O (12)

Morphine ($R^1 = R^2 = H$)
Codeine ($R^1 = CH_3$, $R^2 = H$)
Thebaine ($R^1 = R^2 = CH_3$)

Furthermore, the preparation of the antibiotic Pyrrolnitrine by oxidation of the amino-functionalized precursor with the chloroperoxidase from *Pseudomonas pyrrocinia* was reported (Eq. 13) [155]. The mechanism of this peroxidase-catalyzed *N*-oxidation has not been elucidated.

Cl, NH_2, N–H, Cl — CPO, H_2O_2 → Cl, NO_2, N–H, Cl (13)

3.2.3.2
Sulfoxidation

Enantiomerically pure sulfoxides play an important role in asymmetric synthesis either as chiral building blocks or stereodirecting groups [156]. In the last years, metal- and enzyme-catalyzed asymmetric sulfoxidations have been developed for the preparation of optically active sulfoxides. Among the metal-catalyzed processes, the Kagan sulfoxidation [157] is the most efficient, in which the sulfide is enantioselectively oxidized by $Ti(OiPr)_4$/*t*BuOOH in the presence of tartrate as chirality source. However, only alkyl aryl sulfides may be oxidized by this system in high enantiomeric excesses, and poor enantioselectivities were observed for dialkyl sulfides.

Enzymes, in particular peroxidases, catalyze efficiently the enantioselective oxidation of alkyl aryl sulfides and also dialkyl sulfides, provided that the alkyl substituents are sterically differentiable by the enzyme. The peroxidases HRP, CPO, MP-11, and the mutants of HRP, e.g. F41L and F41T, were successfully used as biocatalysts for the asymmetric sulfoxidation (Eq. 14). A selection of sulfides,

$$R^1{-}S{-}R^2 \xrightarrow[\text{Oxidant}]{\text{Peroxidase}} R^1{-}\overset{O}{\overset{\|}{S^*}}{-}R^2 \qquad (14)$$

Table 12. Peroxidase-catalyzed enantioselective sulfoxidation

Entry	Substrate	Enzyme	Oxidant	Yield [%]	ee [%]	Ref.
1	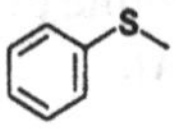	HRP	H_2O_2	95	46	[158]
2		HRP	H_2O_2	–[a]	77	[159]
3		F41L	H_2O_2	–[a]	97	[159]
4		F41T	H_2O_2	–[a]	10	[159]
5		CPO	H_2O_2	90	99	[160]
6		CPO	*t*BuOOH	90	80	[161]
7		MP-11	H_2O_2	45	20	[162]
8	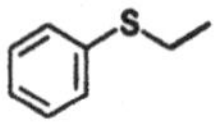	HRP	H_2O_2	–[a]	35	[163]
9		F41L	H_2O_2	–[a]	94	[163]
10		F41T	H_2O_2	–[a]	10	[163]
11		MP-11	H_2O_2	35	16	[162]
12	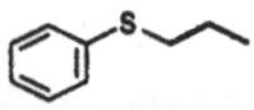	HRP	H_2O_2	–[a]	12	[159]
13		F41L	H_2O_2	–[a]	94	[159]
14		F41T	H_2O_2	–[a]	44	[159]
15	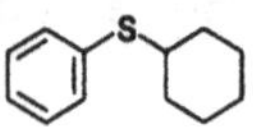	HRP	H_2O_2	–[a]	7	[159]
16		F41L	H_2O_2	–[a]	94	[159]
17		F41T	H_2O_2	–[a]	5	[159]
18	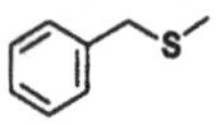	HRP	H_2O_2	55	0	[158]
19		CPO	H_2O_2	100	90	[161]
20		CPO	*t*BuOOH	51	91	[164–165]
21	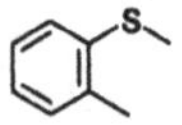	HRP	H_2O_2	50	0	[158]
22		CPO	H_2O_2	27	33	[161]
23		CPO	*t*BuOOH	56	43	[161]
24	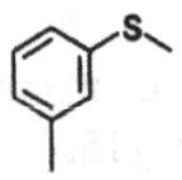	MP-11	H_2O_2	40	25	[162]
25	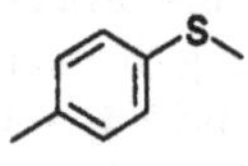	HRP	H_2O_2	47	68	[158]
26		HRP	*t*BuOOH	30	0	[164]
27		HRP	CumylOOH	50	0	[164]
28		F41L	H_2O_2	–[a]	97	[163]
29		CPO	H_2O_2	92	99	[160]
30		CPO	*t*BuOOH	80	70	[161]
31		CPO	H_2O_2[b]	48	35	[164]
32		CPO	*t*BuOOH[b]	60	86	[164-165]
33		CPO	CumylOOH[b]	52	5	[164]
34		CPO	PhIO[b]	49	0	[164]
35		CPO	Ph_3COOH[b]	40	0	[164]
36		CPO	*m*CPBA[b]	30	0	[164]
37		MP-11	H_2O_2	50	24	[162]
38	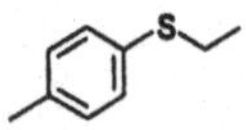	CPO	H_2O_2	50	68	[161]
39		CPO	*t*BuOOH	50	68	[161]

Table 12 (continued)

Entry	Substrate	Enzyme	Oxidant	Yield [%]	ee [%]	Ref.
40	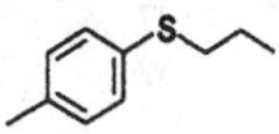	CPO	H_2O_2	53	5	[161]
41		CPO	*t*BuOOH	30	5	
42	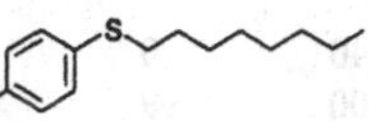	CPO	*t*BuOOH	13	0	[164]
43	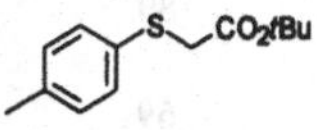	CPO	*t*BuOOH	2	0	[164]
44	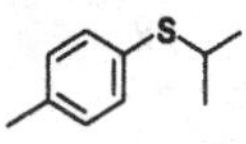	CPO	*t*BuOOH	24	0	[164]
45	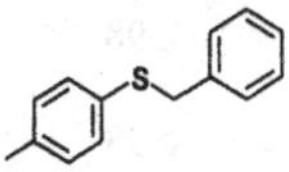	CPO	*t*BuOOH	0	0	[164]
46	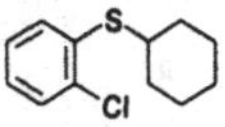	HRP	H_2O_2	50	0	[158]
47		CPO	H_2O_2	33	85	[161]
48		CPO	*t*BuOOH	17	45	[161]
49	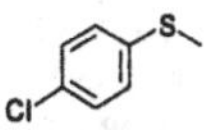	HRP	H_2O_2	– [a]	68	[163]
50		F41L	H_2O_2	– [a]	88	[163]
51		CPO	H_2O_2	77	90	[161]
52		CPO	*t*BuOOH	44	85	[164-165]
53		MP-11	H_2O_2	38	24	[162]
54	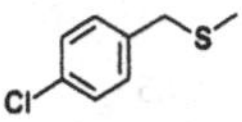	HRP	H_2O_2	– [a]	nd	[163]
55		F41L	H_2O_2	– [a]	nd	[163]
56	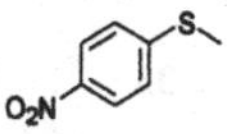	HRP	H_2O_2	– [a]	66	[163]
57		F41L	H_2O_2	– [a]	72	[163]
58		CPO	H_2O_2	10	80	[161]
59		CPO	*t*BuOOH	16	80	[161]
60	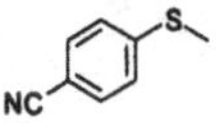	HRP	H_2O_2	– [a]	64	[163]
61		F41L	H_2O_2	– [a]	90	[163]
62	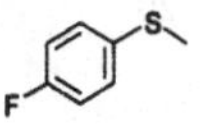	CPO	H_2O_2	100	97	[161]
63		CPO	*t*BuOOH	90	70	[161]
64	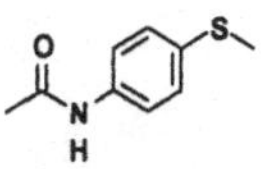	HRP	H_2O_2	– [a]	78	[163]
65		F41L	H_2O_2	– [a]	94	[163]
66		CPO	H_2O_2	86	67	[161]
67		CPO	*t*BuOOH	86	70	[161]
68	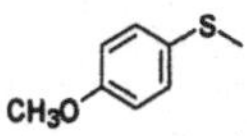	HRP	H_2O_2	84	30	[158]
69		F41L	H_2O_2	– [a]	97	[163]
70	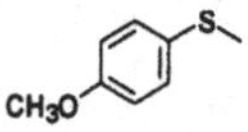	CPO	H_2O_2	66	100	[160]
71		CPO	*t*BuOOH	71	92	[164–165]
72		HRP	*t*BuOOH	24	0	[164]
73		HRP	H_2O_2	22	0	[164]

Table 12 (continued)

Entry	Substrate	Enzyme	Oxidant	Yield [%]	ee [%]	Ref.
74	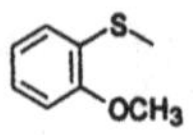	CPO	H_2O_2	24	27	[161]
75		CPO	*t*BuOOH	30	37	[161]
76	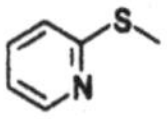	HRP	H_2O_2	40	0	[158]
77		CPO	H_2O_2	100	99	[161]
78		CPO	*t*BuOOH	61	89	[161]
79	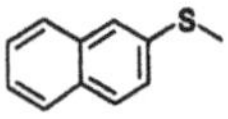	HRP	H_2O_2	–[a]	69	[163]
80		F41L	H_2O_2	–[a]	99	[163]
81		CPO	*t*BuOOH	–[a]	–	[164–165]
82	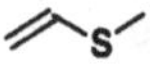	CPO	H_2O_2	≥98[c]	≥98	[166]
83	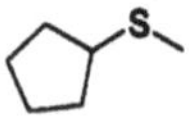	CPO	H_2O_2	≥98[c]	≥98	[166]
84	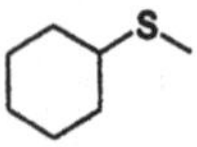	CPO	H_2O_2	85[c]	85	[166]
85		CPO	*t*BuOOH	54	38	[164–165]
86		CPO	H_2O_2	75[c]	≥98	[166]
87	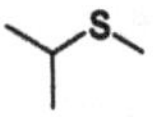	CPO	H_2O_2	≥98[c]	≥98	[166]
88	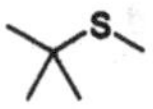	CPO	H_2O_2	80[c]	85	[166]
89	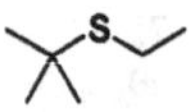	CPO	H_2O_2	30[c]	35	[166]
90	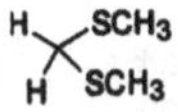	CPO	H_2O_2	75[c]	≥98	[166]
91	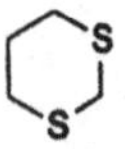	CPO	*t*BuOOH	63	20	[164]

[a] Not given.
[b] Performed at 4 °C in aqueous citrate buffer.
[c] Percent conversion.

which have been oxidized enantioselectively by peroxidases with H_2O_2 or hydroperoxides, is shown in Table 12.

Among peroxidases, CPO is the most efficient and selective biocatalyst for such sulfoxidation. This enzyme oxidizes alkyl aryl, dialkyl and alkyl vinyl sulfides to the corresponding sulfoxides in up to 99% enantiomeric excess. In contrast, poor to moderate selectivities were observed for the HRP and MP-11 enzymes. The latter peroxidase is a heme peptide obtained by combining the cytochrome *c* with proteolytic enzymes. To enhance the selectivity of HRP, its mutants F41L and F41T were prepared by replacement of the phenylalanine in the native enzyme by leucine or threonine. Indeed, the leucine mutant F41L exhibited higher selectivity than the native HRP in sulfoxidations (Table 12, entries 3, 9, 16, 50, 57, 61, 65, 69 and 80); the threonine mutant F41T was in all cases less selective.

As shown in Table 12, H_2O_2 and *t*BuOOH have been used frequently as oxygen donors in peroxidase-catalyzed sulfoxidations. Other achiral oxidants, e.g. iodosobenzene and peracids, are not accepted by enzymes and, therefore, only racemic sulfoxides were found (c.f. entries 34–36). Interestingly, racemic hydroperoxides oxidize sulfides to sulfoxides enantioselectively under CPO catalysis [68]. In this reaction, not only the sulfoxides but also the hydroperoxide and the corresponding alcohol were produced in optically active form by enzyme-catalyzed kinetic resolution (cf. Eq. 3 and Table 3 in Sect. 3.1).

4
Conclusions

Peroxidases have been used very frequently during the last ten years as biocatalysts in asymmetric synthesis. The transformation of a broad spectrum of substrates by these enzymes leads to valuable compounds for the asymmetric synthesis of natural products and biologically active molecules. Peroxidases catalyze regioselective hydroxylation of phenols and halogenation of olefins. Furthermore, they catalyze the epoxidation of olefins and the sulfoxidation of alkyl aryl sulfides in high enantioselectivities, as well as the asymmetric reduction of racemic hydroperoxides. The less selective oxidative coupling of various phenols and aromatic amines by peroxidases provides a convenient access to dimeric, oligomeric and polymeric products for industrial applications.

Most of the peroxidases described in this review are commercially available; thus, the transformations may be performed on a preparative scale, unless the substrates are poorly accepted by the enzyme. For the future, biotechnology may offer new peroxidases and mutants of natural enzymes, as well as microbiological cell sytems with peroxidase activity, which will enable one to perform new selective chemical transformations with these biocatalysts.

Acknowledgements. The financial support of our current studies by the *Deutsche Forschungsgemeinschaft* (Sonderforschungsbereich 347 "Selektive Reaktionen Metall-aktivierter Moleküle"), the *Bayerische Forschungsstiftung* (Bayerischer Forschungsverbund Katalyse-FORKAT) and the *Fonds der Chemischen Industrie* is gratefully acknowledged.

5
References

1. Faber K (1997) Biotransformations in organic chemistry, 3rd edn. Springer, Berlin Heidelberg New York
2. Schreier P (1997) Enzymes and flavour biotechnology. In: Berger (ed) RG Biotechnology of aroma compounds. Springer Berlin, Heidelberg New York, p 51
3. Drauz K, Waldmann H (1995) Enzmye catalysis in organic synthesis. VCH, Weinheim
4. Wong CH, Whitesides GM (1994) Enzymes in synthetic organic chemistry. Pergamon, New York
5. Poppe L, Novák L (1992) Selective biocatalysis. A synthetic approach. VCH, Weinheim
6. Roberts SM, Wiggins K, Casy G (1992) Preparative biotransformations. Whole cell and isolated enzymes in organic synthesis. Wiley, Chichester
7. Flohé L (1979) CIBA Foundation Symposium 65:95
8. de Boer E, van Kooyk Y, Tromp MGM, Plat H, Wever R (1986) Biochim Biophys Acta 869:48
9. Kuwahara M, Glenn JK, Morgan MA, Gold MH (1984) FEBS Lett 169:247
10. Dolin MI (1957) J Biol Chem 225:557
11. Bach A, Chodat R (1903) Chem Ber 36:600
12. Schonbaum GR, Chance B (1976) Catalase. In: Boyer PD (ed) The enzymes. Academic Press, New York, p 363
13. Bosshard HR, Anni H, Yonetani T (1991) Yeast cytochrome *c* peroxidase. In: Everse J, Everse KE, Grisham MB (eds) Peroxidases in chemistry and biology. CRC Press, Boca Raton, p 51
14. Gold MH, Wariishi H, Valli K (1989) Extracellular peroxidases involved in lignin degradation by the white rot basidiomycete *Phanerochaete chrysosporium*. In: Whitaker JR, Sonnet PE (eds) Biocatalysis in agricultural biotechnology. ACS symposium series 389. American Chemical Society, Washington, D.C., p 127
15. Thomas JA, Morris DR, Hager LP (1970) J Biol Chem 245:3135
16. Griffin BW (1991) Chloroperoxidase: A review. In: Everse J, Everse KE, Grisham MB (eds) Peroxidases in chemistry and biology. CRC Press, Boca Raton, p 85
17. Harrison JE, Schultz J (1976) J Biol Chem 251:1371
18. Dugad LB, La Mar GN, Lee HC, Ikeda-Saito M, Booth KS, Caughey WS (1990) J Biol Chem 265:7173
19. Zeng J, Fenna RE (1992) J Mol Biol 226:185
20. Paul KG, Ohlsson PI (1978) Acta Chem Scand B32:395
21. Pruitt KM, Tenovuo JO (1985) The lactoperoxidase system. Chemistry and biological significance. Marcel Dekker, New York
22. Schönbein CF (1856) Verh Nat Ges Basel 1:467
23. Willstätter R, Stoll A (1919) Ann Chem 416:21
24. Dunford HB (1991) Horseradish peroxidase: structure and kinetic properties. In: Everse J, Everse KE, Grisham MB (eds) Peroxidases in chemistry and biology. CRC Press, Boca Raton, p 1
25. Dawson JH (1988) Science 240:433
26. Anni H, Yonetani T (1992) Mechanism of action of peroxidases. In: Sigel H, Sigel (eds) A Metal ions in biological systems: degradation of environmental pollutants by microorganisms and their metalloenzymes. Marcel Dekker, New York, p 219
27. Oritz de Montellano PR (1992) Annu Rev Pharmacol Toxicol 32:89
28. van Deurzen MPJ, van Rantwijk F, Sheldon RA (1997) Tetrahedron 53:13,183
29. Whitaker JR (1972) Principles of enzymology for the food sciences. Marcel Dekker, New York, p 591
30. Yokota K, Yamazaki I (1977) Biochem 16:1913
31. Deisseroth A, Dounce AL (1970) Physiol Rev 50:319
32. Chance B, Sies H, Boveris A (1979) Physiol Rev 59:527
33. Oritz de Montellano PR, Grab LA (1987) Biochem 26:5310
34. Kobayashi S, Nakano M, Kimura T, Schaap AP (1987) Biochem 26:5019

35. Kobayashi S, Nakano M, Goto T, Kimura T, Schaap AP (1986) Biochem Biophys Res Commun 135:166
36. Ricard J, Job D (1974) Eur J Biochem 44:359
37. Smith AM, Morrison WL, Milham PJ (1982) Biochem 21:4414
38. Nakajima R, Yamazaki I (1979) J Biol Chem 254:872
39. Grambow HJ (1982) Z Naturforsch Teil 37c:884
40. Mottley C, Mason RP (1986) J Biol Chem 261:16,860
41. Campa A, Nassi L, Cilento G (1984) Photochem Photobiol 40:127
42. Halliwell B, de Rycker J (1978) Photochem Photobiol 28:757
43. Mason HS, Onopryenko I, Buhler DR (1957) Biochim Biophys Acta 24:225
44. Mason HS (1957) Proc Int Symp Enzyme Chem 2:224
45. Buhler DR, Mason HS (1961) Arch Biochem Biophys 92:424
46. Daly JW, Jerina DM (1970) Biophys Biochim Acta 208:340
47. Nilsson R, Pick FM, Bray RC (1969) Biochim Biophys Acta 192:145
48. Courteix A, Bergel A (1995) Enzyme Microb Technol 17:1087
49. Courteix A, Bergel A (1995) Enzyme Microb Technol 17:1094
50. Cilento G (1984) Pure Appl Chem 56:1179
51. Cilento G (1988) Stud Org Chem 33:435
52. Cilento G, Adam W (1995) Free Rad Biol Med 19:103
53. Adam W, Baader WJ, Cilento G (1986) Biochim Biophys Acta 881:330
54. Adam W, Cilento G (1982) Chemical and biological generation of electronically excited states. Academic Press, New York
55. Adam W, Cilento G (1983) Angew Chem Int Ed Engl 22:529
56. Johnson RA, Sharpless KB (1993) Catalytic asymmetric epoxidation of allylic alcohols. In: Ojima I (ed) Catalytic asymmetric synthesis. VCH, Weinheim, p 103
57. Adam W, Richter MJ (1994) Acc Chem Res 27:57
58. Shum WPS, Saxton RJ, Zajacek JG (1997) US Patent 5,663,384
59. Adam W, Hoch U, Saha-Möller CR, Schreier P (1993) Angew Chem Int Ed Engl 32:1737
60. Ojima I (1993) Catalytic asymmetric synthesis. VCH, Weinheim
61. Noyori R (1994) Asymmetric catalysis in organic synthesis. Wiley Interscience Press, New York
62. Datcheva VK, Kiss K, Solomon L, Kyler KS (1991) J Am Chem Soc 113:270
63. Scheller G, Jäger E, Hoffmann B, Schmitt M, Schreier P (1995) J Agric Food Chem 43:1768
64. Dussault P, Sahli A (1990) Tetrahedron Lett 31:5117
65. Dussault P, Lee IQ, Kreifels SJ (1991) J Org Chem 56:4087
66. Baba N, Mimura M, Hiratake J, Uchida K, Oda J (1988) Agric Biol Chem 52:2685
67. Baba N, Tateno K, Iwasa J, Oda J (1990) Agric Biol Chem 54:3349
68. Fu H, Kondo H, Ichkawa Y, Look GC, Wong CH (1992) J Org Chem 57:7265
69. Hoch U, Adam W, Fell R, Saha-Möller CR, Schreier P (1997) J Mol Catal A: Chemical 117:321
70. Adam W, Hoch U, Lazarus M, Saha-Möller CR, Schreier P (1995) J Am Chem Soc 117:11,898
71. Höft E, Hamann HJ, Kunath A, Adam W, Hoch U, Saha-Möller CR, Schreier P (1995) Tetrahedron: Asymmetry 6:603
72. Hoch U, Scheller G, Schmitt M, Schreier P, Adam W, Saha-Möller CR (1995) Enzymes in synthetic organic chemistry: selective oxidoreductions catalyzed by the metalloenzymes lipoxygenase and peroxidase. In: Werner H, Sundermeyer J (eds) Stereoselective reactions of metal-activated molecules, 2nd Symposium. Vieweg, Braunschweig, p 33
73. Adam W, Korb MN (1997) Tetrahedron: Asymmetry 8:1131
74. Adam, W, Fell RT, Hoch U, Saha-Möller CR, Schreier P (1995) Tetrahedron: Asymmetry 6:1047
75. Adam W, Mock-Knoblauch C, Saha-Möller CR (1997) Tetrahedron: Asymmetry 8:1947
76. Adam W, Hoch U, Humpf HU, Saha-Möller CR, Schreier P (1996) Chem Commun 2701
77. Hoch U, Humpf HU, Schreier P, Saha-Möller CR, Adam W (1997) Chirality 9:69
78. Schmidt K (1997) Diploma thesis, University of Würzburg

79. Weichold O (1996) Diploma thesis, University of Würzburg
80. Häring D, Herderich M, Schüler E, Withopf B, Schreier P (1997) Tetrahedron: Asymmetry 8:853
81. Schüler E, Häring D, Boss B, Herderich M, Schreier P, Adam W, Mock-Knoblauch C, Renz M, Saha-Möller CR, Weichold O (1998) The potential of selenium-containing peroxidases in asymmetric catalysis: glutathione peroxidase and seleno subtilisin. In: Werner H, Schreier P (eds) Selective reactions of metal-activated molecules. Proceedings of the 3rd international symposium SFB 347. Vieweg, Braunschweig, p 35
82. Shine WE, Stumpf PK (1974) Arch Biochem Biophys 162:147
83. Kajiwara T, Matsui K, Akakabe Y (1996) Biogeneration of flavor compounds via edible seaweeds. In: Takeoka GR, Teranishi R, Williams PJ, Kobayashi A (eds) Biotechnology for improved foods and flavours. ACS symposium series 637. American Chemical Society, Washington, D.C., p 146
84. Adam W, Lazarus M, Saha-Möller CR, Schreier P (1996) Tetrahedron: Asymmetry 7:2287
85. Klibanov AM, Berman Z, Alberti BN (1981) J Am Chem Soc 103:6263
86. Dordick JS, Klibanov AM, Marletta MA (1986) Biochem 25:2946
87. Schmall MW, Gorman LS, Dordick JS (1989) Biochim Biophys Acta 999:267
88. Akasaka R, Mashino T, Hirobe M (1995) Bioorg Med Chem Lett 5:1861
89. Fujimoto S, Ishimitsu S, Hirayama S, Kawakami N, Ohara A (1991) Chem Pharm Bull 39:1598
90. Booth H, Saunders BC (1956) J Chem Soc 940
91. Hughes GMK, Saunders BC (1954) J Chem Soc 4630
92. Scott AI (1965) Quart Rev 19:1
93. Pugh CEM, Raper HS (1927) Biochem J 21:1370
94. Willstätter R, Heiss H (1923) Ann Chem 433:17
95. Hathway DE (1957) Biochem J 67:445
96. Saunders BC, Holmes-Siedle AG, Stark BP (1964) Peroxidase. Butterworths, London
97. Sizer IW (1953) Adv Enzymol 14:2089
98. Gross AJ, Sizer IW (1959) J Biol Chem 234:1611
99. Harkin JM (1960) Experienta 16:80
100. Freudenberg K, Harkin JM, Reichert M, Fukuzumi T (1958) Ber 91:581
101. Erdtman H (1933) Biochem Z 258:177
102. Westerfeld WW, Lowe C (1942) J Biol Chem 145:463
103. Dunford HB, Stillman JS (1976) Coord Chem Rev 19:187
104. Dunford HB, Adeniran AJ (1986) Arch Biochem Biophys 251:536
105. Frew JE, Jones P (1984) Structure and functional properties of peroxidases and catalases. In: Sykes G (ed) Advances in inorganic and bioinorganic mechanisms. Academic Press, New York, p 175
106. Sakurada J, Sekiguchi R, Sata K, Hosoya T (1990) Biochem 29:4093
107. Casella L, Poli S, Gullotti M, Selvaggini C, Beringhelli T, Marchesini A (1994) Biochem 33:6377
108. Pietkäinen P, Adlercreutz P (1990) Appl Microbiol Biotechnol 33:455
109. Sawahata T, Neal RA (1982) Biochem Biophys Res Commun 109:988
110. Krawczyk AR, Lipkowska E, Wróbel JT (1991) Collect Czech Chem Commun 56:1147
111. Kobayashi A, Koguchi Y, Kanzaki H, Kajiyama SI, Kawazu K (1994) Biosci Biotech Biochem 58:133
112. Kobayashi S, Kaneko I, Uyama H (1992) Chem Lett 393
113. Pieper DH, Winkler R, Sandermann H (1992) Angew Chem Int Ed Engl 31:68
114. Donelly DMX, Murphy FG, Polonski J, Prangé T (1987) J Chem Soc Perkin Trans I 2719
115. Goodbody AE, Endo T, Vukovic J, Kutney JP, Choi LSL, Misawa M (1988) Planta Med 136
116. d'Ischia M, Napolitano A, Tsiakas K, Prota G (1990) Tetrahedron 46:5789
117. Fukunishi K, Kitada K, Naito I (1991) Synthesis 237
118. Setälä H, Pajunen A, Kilpeläinen I, Brunow G (1994) J Chem Soc Perkin Trans I 1163
119. Brown BR, Bocks SM (1963) Some new enzymic reactions of phenols. In: Pridham JB (ed) Enzyme chemistry of phenolic compounds. Pergamon Press, Oxford, p 129

120. Holland HL (1992) Organic synthesis with oxidative enzymes. VCH, New York, p 341
121. Sridhar M, Vadivel SK, Bhalerao UT (1997) Tetrahedron Lett 38:5695
122. Davin LB, Wang HB, Crowell AL, Bedgar DL, Martin DM, Sarkanen S, Lewis NG (1997) Science 275:362
123. Dordick JS, Marletta MA, Klibanov AM (1987) Biotechnol Bioeng 30:31
124. Ryu K, Stafford DR, Dordick JS (1989) Peroxidase-catalyzed polymerization of phenols. In: Whitaker JR, Sonnet PE, (eds) Biocatalysis in agricultural biotechnology. American Chemical Society, Washington, D.C., p 141
125. Klibanov AM, Alberti BN, Morris ED, Felshin LM (1980) J App Biochem 2:414
126. Klibanov AM, Tu TM, Scott KP (1983) Science 221:259
127. Pokora AR, Stolfo JJ (1992) US Pat 5,110,740
128. Besse P, Veschambre H (1994) Tetrahedron 50:8885
129. Jacobsen EN (1993) Asymmetric catalytic epoxidation of unfunctionalized olefins. In: Ojima I (ed) Catalytic asymmetric synthesis. VCH, Weinheim, p 159
130. Colonna S, Gaggero N, Casella L, Carrea G, Pasta P (1993) Tetrahedron: Asymmetry 4:1325.
131. Dexter AF, Lakner FJ, Campbell RA, Hager LP (1995) J Am Chem Soc 117:6412
132. Allain EJ, Hager LP, Deng L, Jacobsen EN (1993) J Am Chem Soc 115:4415
133. Zaks A, Dodds DR (1995) J Am Chem Soc 117:10,419
134. Lakner FJ, Hager LP (1996) J Org Chem 61:3923
135. Lakner FJ, Cain KP, Hager LP (1997) J Am Chem Soc 119:443
136. Rao AB, Rao MV (1994) Tetrahedron Lett 35:279
137. McCarthy MB, White RE (1983) J Biol Chem 258:9153
138. Liu KKC, Wong CH (1992) J Org Chem 57:3748
139. Fang JM, Lin CH, Bradshaw CW, Wong CH (1995) J Chem Soc Perkin Trans I 967
140. Coughlin P, Roberts S, Rush C, Willetts A (1993) Biotech Lett 15:907
141. Fukuzawa A, Aye M, Murai A (1990) Chem Lett 1579
142. Fukuzawa A, Takasugi Y, Murai A, Nakamura M, Tamura M (1992) Tetrahedron Lett 33:2017
143. Franssen MCR, van der Plas HC (1987) Bioorg Chem 15:59
144. van der Plas HC, Franssen MCR, Jansma JD, van Boven HG (1987) Eur Congr Biotechnol 2:202
145. Franssen MCR, Weijnen JGJ, Vincken JP, Laane C, van der Plas HC (1988) Biocatalysis 1:205
146. Renganathan V, Miki K, Gold MH (1987) Biochemistry 26:5127
147. DeBoer E, Plat H, Wever R (1987) Algal vanadium(V)-bromoperoxidase. A halogenating enzyme retaining full activity in apolar solvent systems. In: Laane C, Tramper J, Lilly MD (eds) Biocatalysis in organic media. Elsevier, Amsterdam, p 317
148. Franssen MCR, Weijnen JGJ, Vincken JP, Laane C, van der Plas HC (1987) Haloperoxidase in reversed micelles: use in organic synthesis and optimisation of the system. In: Laane C, Tramper J, Lilly MD (eds) Biocatalysis in organic media. Elsevier, Amsterdam, p 289
149. Itoh N, Izumi Y, Yamada H (1987) Biochemistry 26:282
150. Itahara T, Ide N (1987) Chem Lett 2311
151. Franssen MCR, van Boven HG, van der Plas HC (1987) J Heterocyclic Chem 24:1313
152. Corbett MD, Corbett BR (1985) Biochem Arch 1:115
153. Doerge DR, Corbett MD (1991) Chem Res Toxicol 4:556
154. Vagujfalvi D, Petz-Stifter M (1982) Phytochem 21:1533
155. Kirner S, van Pée KH (1994) Angew Chem Int Ed Engl 33:352
156. Mata EG (1996) Phosphorus, Sulfur and Silicon 117:231
157. Kagan HB (1993) Asymmetric oxidation of sulfides. In: Ojima I (ed) Catalytic asymmetric synthesis. VCH, Weinheim, p 203
158. Colonna S, Gaggero N, Carrea G, Pasta P (1992) J Chem Soc Chem Commun 357
159. Ozaki SI, Ortiz de Montellano PR (1995) J Am Chem Soc 117:7056
160. Fu H, Kondo H, Ichikawa Y, Look GC, Wong CH (1992) J Org Chem 57:7265

161. Colonna S, Gaggero N, Casella L, Carrea G, Pasta P (1992) Tetrahedron: Asymmetry 3:95
162. Colonna S, Gaggero N, Carrea G, Pasta P (1994) Tetrahedron Lett 35:9103
163. Ozaki SI, Ortiz de Montellano PR (1994) J Am Chem Soc 116:4487
164. Colonna S, Gaggero N, Manfredi A (1990) Biochemistry 29:10,465
165. Colonna S, Gaggero N, Manfredi A, Casella L, Gullotti M (1988) J Chem Soc Chem Commun 1451
166. Colonna S, Gaggero N, Carrea G, Pasta P (1997) J Chem Soc Chem Commun 439
167. Schmidt MM, Schüler E, Braun M, Häring D, Schreiner P (1998) Tetrahedron Lett 39:2945

Received January 1998

Production of Chiral C3- and C4-Units by Microbial Enzymes

Sakayu Shimizu · Michihiko Kataoka

Division of Applied Life Sciences, Graduate School of Agriculture, Kyoto University, Kitashirakawa-Oiwakecho, Sakyo-ku, Kyoto 606–8502, Japan

Enzyme (biocatalysis) reactions display far greater specificities, such as substrate specificity, stereospecificity, regiospecificity and so on, than more conventional forms of organic reactions. Using these specificities of the enzymes, many useful compounds have been enzymatically produced. Compounds possessing C3- and C4-units with additional functional groups are promising materials for the synthesis of various useful compounds. In particular, optically active C3- and C4-synthetic units are quite important intermediates for the preparation of pharmaceuticals and fine chemicals. Microbial transformation with enzymes showing stereospecificities have been applied to the asymmetric synthesis of optically active substances. In this article the recent works on the practical production of chiral C3- and C4-synthetic units with microbial enzymes are described.

Keywords: Chiral synthon, Chiral building blocks, Stereospecific synthesis, Stereoselective degradation, Microbial transformation.

1 **Introduction** 110

2 **Production of Chiral 2,3-Dichloro-1-Propanol and Epichlorohydrin** 111

3 **Production of Chiral 3-Chloro-1,2-Propanediol and Glycidol** 113

3.1 Production of Chiral 3-Chloro-1,2-Propanediol and Glycidol by Microbial Stereoselective Degradation of Racemic 3-Chloro-1,2-Propanediol 113

3.2 Microbial Production of (*R*)-3-Chloro-1,2-Propanediol from Prochiral 1,3-Dichloro-2-Propanol 114

3.3 Production of (*R*)-Glycidol by Enantioselective Oxidation of Racemic Glycidol 116

4 **Production of Optically Active 4-Chloro-3-Hydroxybutanoate Esters** 116

4.1 Production of Chiral 4-Chloro-3-Hydroxybutanoate Ethyl Ester by Microbial Asymmetric Reduction of 4-Chloroacetoacetate Ethyl Ester 117

4.2 Production of Chiral 4-Chloro-3-Hydroxybutanoate Ethyl Ester by Microbial Stereoselective Dechlorination of Racemic 4-Chloro-3-Hydroxybutanoate Ethyl Ester 119

Advances in Biochemical Engineering/
Biotechnology, Vol. 63
Managing Editor: Th. Scheper

5 **Production of (*R*)-4-Chloro-3-Hydroxybutyronitrile** 120

6 **Production of Optically Active 1,2-Diols** 120

7 **Conclusion** 121

8 **References** 122

List of Symbols and Abbreviations

AR	aldehyde reductase
CAAE	4-chloroacetoacetate ethyl ester
CHBE	4-chloro-3-hydroxybutanoate ethyl ester
CHBN	4-chloro-3-hydroxybutyronitrile
3-CPD	3-chloro-1,2-propanediol
CR	carbonyl reductase
1,3-DCP	1,3-dichloro-2-propanol
2,3-DCP	2,3-dichloro-1-propanol
e.e.	enantiomeric excess
EP	epichlorohydrin
GDH	glucose dehydrogenase
GLD	glycidol
HDDase	halohydrin dehydro-dehalogenase
H-lyase	halohydrin hydrogen-halide lyase

1 Introduction

Since many useful organic compounds, such as pharmaceuticals and food additives, have asymmetric carbon atoms, there are enantiomers of them. In most cases, only one enantiomer is useful as a biologically active substance, the others not showing such activity and sometimes having a harmful effect. Therefore, a racemic mixture of such compounds which have been chemically synthesized cannot be used, especially for pharmaceuticals, and thus the troublesome optical resolution of the racemic mixture by means of a conventional organic synthetic process for such optically active substances is unavoidable.

Reactions catalyzed by enzymes or enzyme systems exhibit far greater specificities than more conventional organic reactions. Among these specificities which enzymatic reactions possess, stereospecificity is one of the most excellent. To overcome the disadvantage of a conventional synthetic process, i.e., the troublesome resolution of a racemic mixture, microbial transformation with enzymes possessing stereospecificities has been applied to the asymmetric synthesis of optically active substances [1–10]. C3- and C4-synthetic units (synthons, building blocks), such as epichlorohydrin (EP), 2,3-dichloro-1-propanol (2,3-DCP), glycidol (GLD), 3-chloro-1,2-propanediol (3-CPD), 4-chloro-

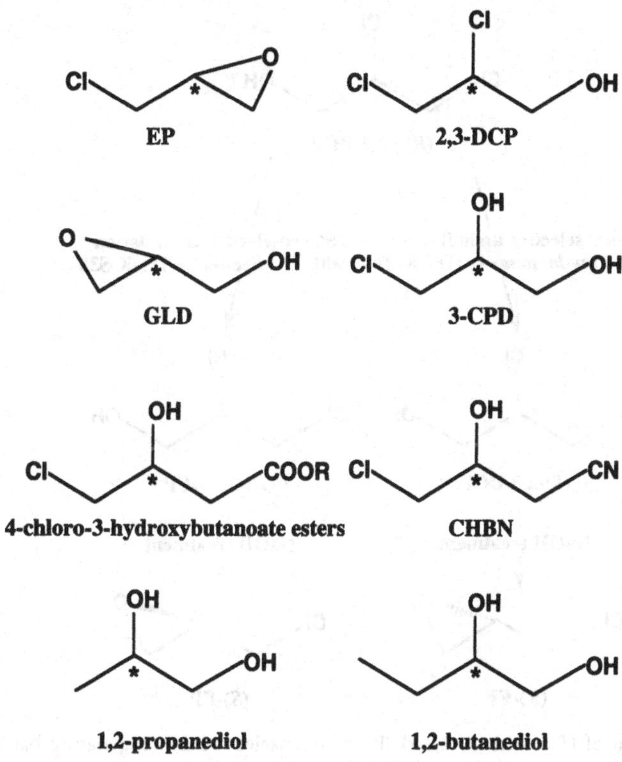

Fig. 1. Structures of various C3- and C4-chiral building blocks

3-hydroxybutanoate esters, 4-chloro-3-hydroxybutyronitrile (CHBN), 1,2-propanediol, and 1,2-butanediol (Fig. 1), exhibit high potential via conversion to various useful compounds. Therefore, if C3- and C4-units of high optical purity could be produced by means of microbial transformation, the synthesis of various optically active compounds might be easier.

This paper describes recent work on the practical production of chiral C3- and C4-units with microbial enzymes.

2 Production of Chiral 2,3-Dichloro-1-Propanol and Epichlorohydrin

Optically active EP is an important C3 chiral building block for the synthesis of chiral pharmaceuticals such as β-adrenergic blockers [11–13], vitamins [14, 15], pheromones [16], natural products [17], and new materials such as ferro-electric crystals [18]. Racemic EP can be made via 2,3-DCP and 1,3-dichloro-2-propanol (1,3-DCP) synthesized from propylene by organic synthesis [19]; however, a practical production method for optically active EP has not yet been established. Racemic 2,3-DCP, which is easily synthesized by the chemical

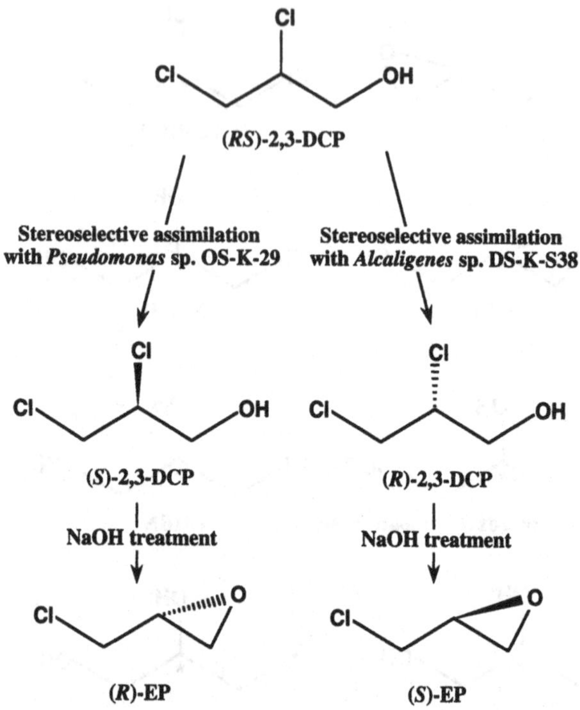

Fig. 2. Generation of (*R*)- and (*S*)-2,3-DCP by stereoselective dehalogenating bacteria, and their use in the preparation of (*S*)- and (*R*)-EP

method, is considered as one of the promising precursors for the preparation of optically active EP. Kasai et al. [20–22] isolated two bacterial strains, i.e., *Pseudomonas* sp. OS-K-29 and *Alcaligenes* sp. DS-K-S38, growing on a medium containing 2,3-DCP as the sole carbon source, and found that the former stereospecifically assimilated (*S*)-2,3-DCP and the latter assimilated (*R*)-2,3-DCP from the racemic mixture (Fig. 2). *Pseudomonas* sp. OS-K-29 degraded (*R*)-2,3-DCP via EP, 3-CPD, GLD, and glycerol. *Alcaligenes* sp. DS-K-S38 exhibited a similar assimilation route for (*S*)-2,3-DCP.

When cells of *Pseudomonas* sp. OS-K-29 immobilized on calcium alginate were incubated in 80 l of a synthetic medium containing 0.2 vol.% of racemic 2,3-DCP, optically pure (*S*)-2,3-DCP of 100% *e.e.* was obtained. The immobilized cells could be re-used for a series of continuous reactions, a bioreactor being used for 19 reactions over 50 days without any loss of activity (Fig. 3). Optically pure (*R*)-2,3-DCP (100% *e.e.*) was also isolated from the racemate by means of stereospecific assimilation by *Alcaligenes* sp. DS-K-S38. Highly pure (*R*)- and (*S*)-EP of 99.5% *e.e.* were prepared from (*S*)- and (*R*)-2,3-DCP by treatment with aqueous NaOH, respectively [20–22] (Fig. 2).

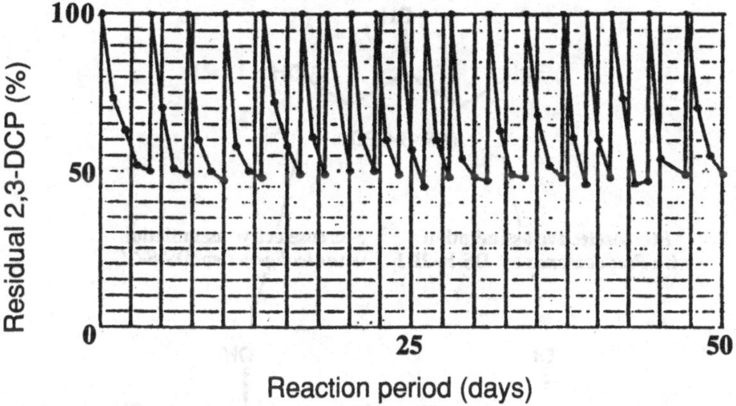

Fig. 3. Time course of continuous reactions with a bioreactor. Symbol: •, residual 2,3-DCP (%) in the bioreactor

3 Production of Chiral 3-Chloro-1,2-Propanediol and Glycidol

3.1 Production of Chiral 3-Chloro-1,2-Propanediol and Glycidol by Microbial Stereoselective Degradation of Racemic 3-Chloro-1,2-Propanediol

Optically active GLD is an important C3 chiral building block for chiral pharmaceuticals, such as *β*-adrenergic blockers [23, 24] and cardiovascular drugs [25, 26], phospholipids [27] and polymers [28]. A practical production method for obtaining optically active GLD with a high yield and optical purity has not been established yet. Optically active 3-CPD, which is a precursor of optically active GLD, is also useful as a chiral building block. Suzuki et al. found that *Alcaligenes* sp. DS-S-7G and *Pseudomonas* sp. DS-K-2D1 stereosepecifically assimilated (*R*)- and (*S*)-3-CPD, respectively [29–31] (Fig. 4).

Two enzymes (enzymes 1 and 2) involved in the dechlorination of (*R*)-3-CPD were isolated from *Alcaligenes* sp. DS-S-7G, which stereoselectively assimilated (*R*)-3-CPD from the racemate [32]. Enzyme 1 is a flavoprotein with a relative molecular mass of 70,000 Da and is composed of two kinds of polypeptides (58,000 and 16,000 Da). The enzyme exhibits the ability to convert (*R*)-3-CPD to hydroxyacetone with the liberation of chloride ions under aerobic conditions. On the other hand, enzyme 2, with a molecular mass of 86,000 Da, which is also composed of two kinds of polypeptides (33,000 and 53,000 Da), shows no dechlorination activity toward (*R*)-3-CPD. However, in the presence of NAD^+, when enzyme 1 was conjugated with enzyme 2 in the (*R*)-3-CPD-dechlorination reaction, the co-operative dechlorination activity was four to five times higher than that with enzyme 1 alone. (*R*)-3-CPD was finally degraded to acetic and formic acid through the joint actions of the two enzymes (Fig. 5). These facts indicate that (*R*)-3-CPD is oxidatively dechlorinated by the enzymes in the presence of NAD^+ in *Alcaligenes* sp. DS-S-7G. Enzyme 1 is designated as halohydrin dehydro-dehalogenase (HDDase).

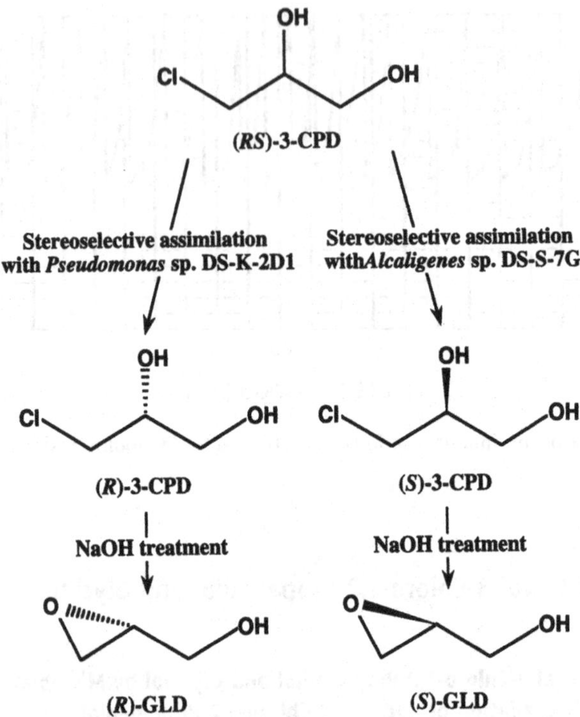

Fig. 4. Generation of (*R*)- and (*S*)-3-CPD by stereoselective dehalogenating bacteria, and their use in the preparation of (*R*)- and (*S*)-GLD

Alcaligenes sp. DS-S-7G and *Pseudomonas* sp. DS-K-2D1 grew in a medium containing racemic 3-CPD (0.2 vol.%) as the sole carbon source and stereoselectively degraded (*R*)- or (*S*)-3-CPD, liberating chloride ions, repectively. The residual 3-CPD was identified as the (*S*)-isomer (99.4% *e.e.*) or (*R*)-isomer (99.5% *e.e.*), with a final yield of 38 or 36%, respectively. Subsequently, highly optically active (*S*)-GLD (99.4% *e.e.*) or (*R*)-GLD (99.3% *e.e.*) was prepared from the (*R*)- or (*S*)-3-CPD by alkaline treatment, respectively [29–31] (Fig. 4).

3.2
Microbial Production of (*R*)-3-Chloro-1,2-Propanediol from Prochiral 1,3-Dichloro-2-Propanol

Prochiral 1,3-DCP is also a promising starting material for the synthesis of chiral 3-CPD (Eq. 1):

HO; HI, Cl (1,3-DCP) → −HCl → Cl, O, * (EP) → H_2O → OH; Cl, *, OH (3-CPD) (1)

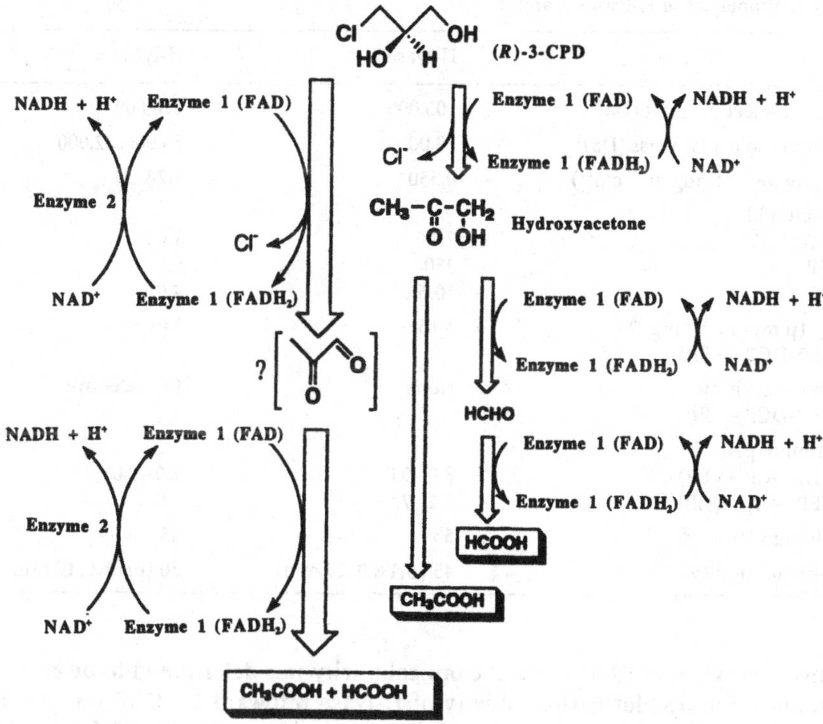

Fig. 5. Possible degradation routes for (*R*)-3-CPD in *Alcaligenes* sp. DS-S-7G

Corynebacterium sp. N-1074 was isolated as a microorganism converting 1,3-DCP to optically active (*R*)-3-CPD via EP [33, 34]. Two 1,3-DCP-dechlorinating enzymes (halohydrin hydrogen-halide lyases (H-lyases) A and B) and two EP-hydrolyzing enzymes (epoxide hydrolases A and B) are involved in the conversion of 1,3-DCP to 3-CPD [35]. The genes encoding the two H-lyases were cloned and expressed in *Escherichia coli*, and then H-lyases A and B purified from each recombinant were characterized (Table 1) [36–38]. H-lyases A and B catalyzed the interconversion of 1,3-DCP to EP. H-lyase A showed only low enantioselectivity, whereas H-lyase B exhibited considerable enantioselectivity, yielding (*R*)-enriched EP. No significant homology was observed, except in the C-terminal region, between the amino acid sequences of H-lyases A and B, and low level sequence identity between the H-lyases and several proteins belonging to the short-chain alcohol dehydrogenase family [39] was found in the C-terminal region. The two epoxide hydrolases, A and B, which catalyze the conversion of EP to 3-CPD, were isolated from *Corynebacterium* sp. N-1074 [40]. (*R*)-Enriched 3-CPD was formed from racemic EP with epoxide hydrolase B, but the resulting 3-CPD with epoxide hydrolase A was almost racemate.

The production of (*R*)-3-CPD from prochiral 1,3-DCP was performed with resting cells of *Corynebacterium* sp. N-1074 [33, 34]. Under optimized reaction conditions, (*R*)-3-CPD was formed from 77.5 mmol l^{-1} 1,3-DCP with a molar

Table 1. Properties of H-lyases A and B

	H-lyase A	H-lyase B
Native molecular mass (Da)	105,000	118,000
Subunit molecular mass (Da)	28,000	35,000, 32,000
A_{280} coefficient (mg ml^{-1} cm^{-1})	0.550	1.28
K_m (mmol l^{-1})		
1,3-DCP	2.44	1.03
EP	350	5.00
Cl^-	10.6	4.00
V_{max} (μ mol min^{-1} mg^{-1}) (1,3-DCP → EP)	3.13	148
Stereoselectivity (1,3-DCP → EP)	None	(*R*)-selective
Optimum pH		
(1,3-DCP → EP)	8.0–9.0	8.0–9.0
(EP → 1,3-DCP)	7.2–7.5	5.0
Optimum temp. (°C)	55	45
Thermal stability (°C)	45 (pH 8.0, 10 min)	30 (pH 8.0, 10 min)

conversion yield of 97.3%, and the optical purity was determined to be 83.8% *e.e.* Since the dechlorination activity of H-lyase B toward 1,3-DCP was much higher than that of H-lyase A in *Corynebacterium* sp. N-1074 cells, EP formed in the first step reaction might be almost all the (*R*)-isomer, resulting in the formation of (*R*)-3-CPD by the epoxide hydrolases.

3.3 Production of (*R*)-Glycidol by Enantioselective Oxidation of Racemic Glycidol

Several microbial alcohol oxidoreductases can catalyze the stereoselective oxidation of GLD. Quinohaemoprotein ethanol dehydrogenase of *Acetobacter pasteurianus* is able to oxidize stereospecifically (*S*)-GLD to (*R*)-glycidic acid in racemic GLD [41, 42]. When washed cells of *A. pasteurianus* were incubated with 4.8 mg ml^{-1} of racemic GLD, (*R*)-GLD was obtained, with an optical purity of 99.5% *e.e.* and 64% conversion.

4 Production of Optically Active 4-Chloro-3-Hydroxybutanoate Esters

Chiral 4-chloro-3-hydroxybutanoate esters are important chiral C4-building blocks [43–53]. For example, (*R*)- and (*S*)-isomers can be converted to L-carnitine and the hydroxymethyl glutaryl-CoA reductase inhibitor. Since these compounds are used as pharmaceuticals, a high optical purity is required. A practical enzymatic method for the production of chiral 4-chloro-3-hydroxybutanoate esters from prochiral carbonyl compounds, i.e., 4-chloroacetoacetate esters, or racemic 4-chloro-3-hydroxybutanoate esters is described.

Table 2. Properties of aldehyde reductase (AR) of *S. salmonicolor* and carbonyl reductase (CR) of *C. magnoliae*

	AR	CR
Native molecular mass (Da)	37,000	76,000
Subunit molecular mass (Da)	37,000	32,000
Number of subunit	1	2
K_m for CAAE (mmol l^{-1})	0.36	4.6
V_{max} for CAAE (μmol min^{-1} mg^{-1})	144	273
Stereoselectivity	(*R*)-CHBE	(*S*)-CHBE
Cofactor (K_m)	NADPH (37.5 μmol l^{-1})	NADPH (16.7 μmol l^{-1})
Optimum pH	7.0	5.5
Optimum temp. (°C)	60	55
Thermal stability	80% (60 °C, 10 min)	90% (45 °C, 30 min)
Inhibitor	dicoumarol, quercetin	quercetin
Reaction mechanism	ordered Bi-Bi	n.d.
Enzyme formation	constitutive	constitutive

n.d., not determined.

4.1 Production of Chiral 4-Chloro-3-Hydroxybutanoate Ethyl Ester by Microbial Asymmetric Reduction of 4-Chloroacetoacetate Ethyl Ester

The reduction of 4-chloroacetoacetate ethyl ester (CAAE) to 4-chloro-3-hydroxybutanoate ethyl ester (CHBE) usually proceeds stereospecifically [54, 55]. This activity is widely distributed in yeasts, molds and bacteria, most of which give the (*S*)-enantiomer. *Sporobolomyces salmonicolor* was found to produce the (*R*)-enantiomer predominantly (62% *e.e.*) with high molar conversion [54, 55], and several *Candida* yeasts formed (*S*)-CHBE of high optical purity (>90% *e.e.*) [54, 55].

The enzyme yielding (*R*)-CHBE was isolated in crystalline form from *S. salmonicolor* cells and was characterized in some detail [56–58] (Table 2). It is a novel NADPH-dependent enzyme, "aldehyde reductase (AR)", belonging to the aldo-keto reductase superfamily of enzymes [59]. It catalyzes the asymmetric reduction of CAAE, in addition to common substrates of aldo-keto reductase superfamily enzymes, such as *p*-nitrobenzaldehyde and pyridine-3-aldehyde (Table 3). The structural gene of the enzyme was cloned and the enzyme has been shown to exhibit similarity to the aldo-keto reductase superfamily enzymes in primary structure [60]. The enzyme catalyzing the asymmetric reduction of CAAE to (*S*)-CHBE was isolated from *C. magnoliae* cells [54] (Table 2). It is also a novel NADPH-dependent enzyme and it catalyzes the reduction of some carbonyl compounds such as diacetyl and glyceraldehyde, other than 4-haloacetoacetate esters, whereas typical substrates of AR do not act as substrates (Table 3). The NH_2-terminal amino acid sequence shows no significant homology with those of other oxidoreductases.

Table 3. Substrate specificities of aldehyde reductase (AR) of *S. salmonicolor* and carbonyl reductase (CR) of *C. magnoliae*

Substrate	Relative activity (%)		Substrate	Relative activity (%)	
	AR	CR		AR	CR
Cl, O, COOC$_2$H$_5$	100	100	O, H	14	0
Cl, O, COOCH$_3$	25	11	Cl, O, H	56	0
Cl, O, COOC$_8$H$_{17}$	240	36	Cl, O, H	52	0
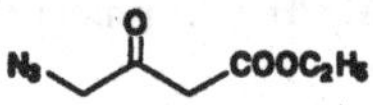 N$_3$, O, COOC$_2$H$_5$	65	n.d.	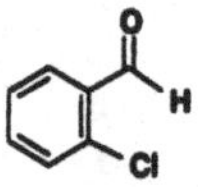 O, H, Cl	58	0
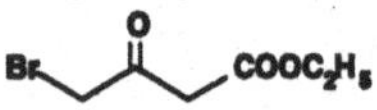 Br, O, COOC$_2$H$_5$	75	n.d.	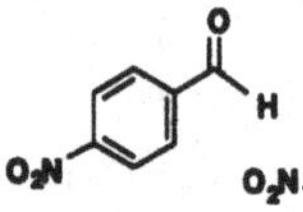 O$_2$N, O, H	468	0
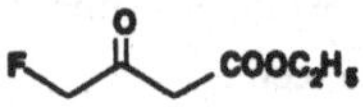 F, O, COOC$_2$H$_5$	153	n.d.	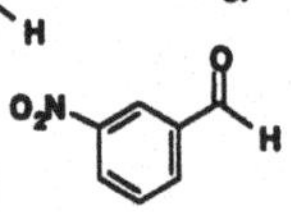 O$_2$N, O, H	63	0
O, COOC$_2$H$_5$, Cl	330	90	O, H, NO$_2$	14	0
O, COOCH$_3$, Cl	74	11	O, H, N	228	0
			O, H, N	54	0
H, O, H, O	74	0	HO, O, H, OH	64	37
H$_3$C, O, H, O	219	0	HO, O, H, OH	n.d.	65
H$_3$C, O, CH$_3$, O	75	19	HO, OH, OH, O, H, OH, OH	81	0
Cl, O, H	17	0	HO, OH, OH, O, H, OH, OH	24	0
H$_3$C, CH$_3$, O, H$_3$C, O	16	0	HO, OH, O, H, OH, OH	472	0
CH$_3$, H$_3$C, O, O, O	0	78	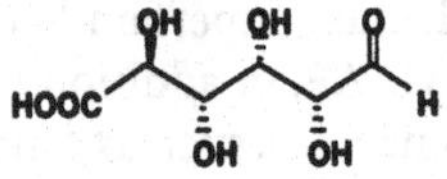 HOOC, OH, OH, O, H, OH, OH	173	0

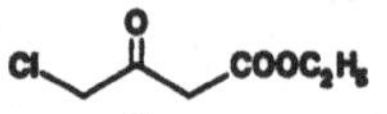

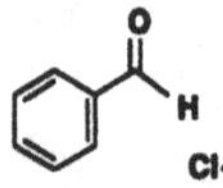

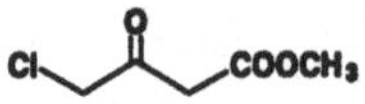

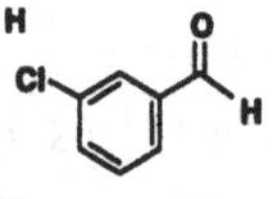

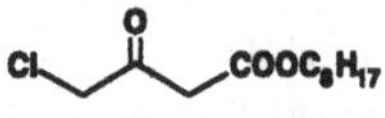

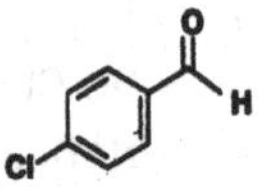

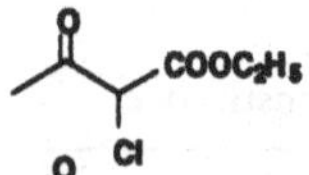

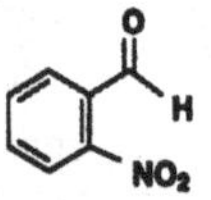

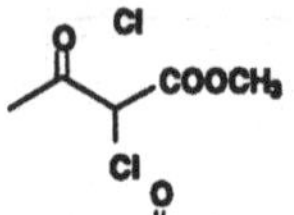

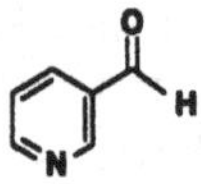

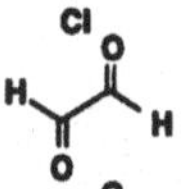

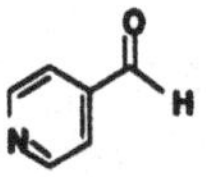

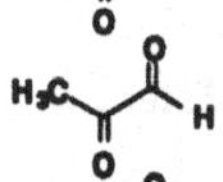

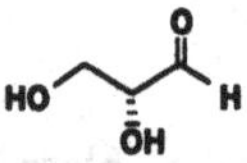

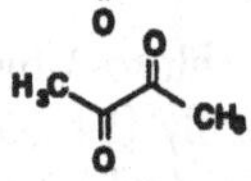

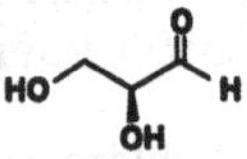

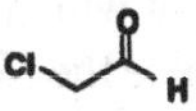

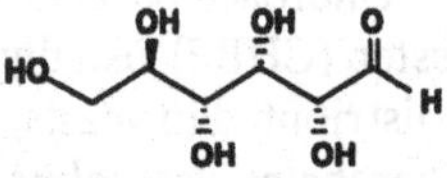

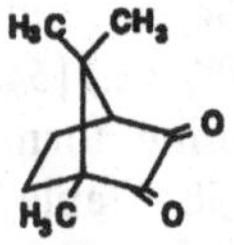

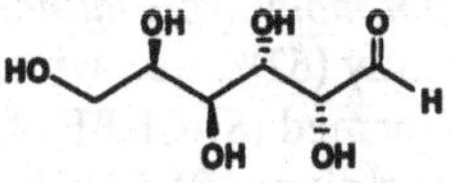

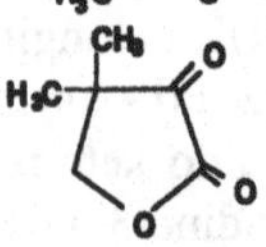

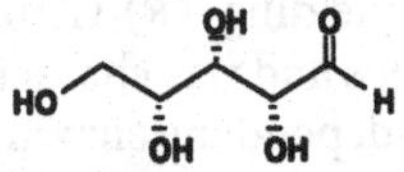

* See [9–12] for details. n.d., not determined.

Since the two CAAE-reducing enzymes described above require NADPH for their reactions, glucose dehydrogenase (GDH) was used as a cofactor regeneration system together with NADP$^+$ and glucose. A practical preparation is carried out in a water-organic solvent two-phase system, as shown in Fig. 6, because the substrate is unstable in an aqueous solution, and both the substrate and product strongly inhibit the enzyme reactions. *n*-Butyl acetate is the most suitable or-

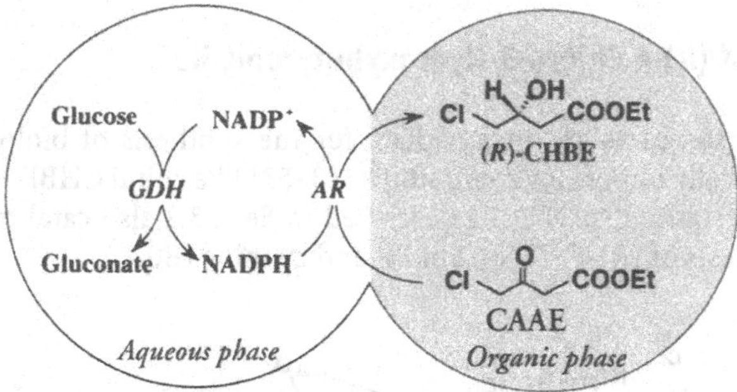

Fig. 6. Outline of the stereospecific reduction of CAAE by aldehyde reductase (AR) with glucose dehydrogenase (GDH) as the cofactor regenerator in an organic solvent-water two-phase system

ganic solvent, since it shows high partition coefficients for both the substrate and product, and both the enzymes are stable in the presence of this organic solvent. In a 1.6 l/1.6 l bench-scale two-phase reaction involving a crude extract from *S. salminocolor* after heat- and acetone-treatments (these treatments removed (*S*)-CHBE-forming enzyme(s) from the cell extract), 83.8 mg ml^{-1} of (*R*)-CHBE (86% *e.e.*) was produced from CAAE in a molar yield of 95.4% [61]. Furthermore, when *E. coli* cells overexpressing the AR gene from *S. salmonicolor* [60] and the GDH gene from *Bacillus megaterium* [62] were used as the catalyst, 300 mg ml^{-1} of CAAE was almost stoichiometrically converted to (*R*)-CHBE (92% *e.e.*) [63–65]. On the other hand, when washed cells of *C. magnoliae* were incubated with commercially available GDH in a similar two-phase system, 90 mg ml^{-1} of CAAE was stoichiometrically converted to (*S*)-CHBE (96% *e.e.*) in 50 h [54].

4.2
Production of Chiral 4-Chloro-3-Hydroxybutanoate Ethyl Ester by Microbial Stereoselective Dechlorination of Racemic 4-Chloro-3-Hydroxybutanoate Ethyl Ester

Racemic CHBE, as a starting material for chiral CHBE, was prepared from propene via EP using petroleum chemicals. *Pseudomonas* sp. OS-K-29 described in Sect. 2, which stereoselectively assimilates (*R*)-2,3-DCP, can also stereoselectively degrade (*R*)-CHBE in the racemate (1 vol.%) to give the (*S*)-enantiomer of high optical purity (98.5% *e.e.*) and a yield of 35% [66]. The (*R*)-CHBE was converted into ethyl 3,4-dihydroxybutyrate by *Pseudomonas* sp. OS-K-29. On the other hand, the preparation of (*R*)-CHBE using resting cells of *Pseudomonas* sp. DS-K-NR818, which was isolated as a mutant of *Pseudomonas* sp. OS-K-29, also gave an excellent enantiomeric purity (98.4% *e.e.*) with a yield of 42% [66]. The (*S*)-CHBE was converted into (*S*)-3-hydroxy-β-butyrolactone, with the release of ethanol and chloride ions, by *Pseudomonas* sp. DS-K-NR818.

5
Production of (*R*)-4-Chloro-3-Hydroxybutyronitrile

Chiral CHBN serves as an intermediate for the synthesis of biologically and pharmacologically important compounds [43–53] like chiral CHBE. The H-lyase B of *Corynebacterium* sp. N-1074 described in Sect. 3.2 also catalyzes the irreversible synthesis of (*R*)-CHBN from EP and cyanide (Eq. 2):

1,3-DCP →(H-lyase B, −HCl) (*R*)-EP →(HCN, H-lyase B) (*R*)-CHBN (2)

The synthesis of (*R*)-CHBN from prochiral 1,3-DCP was performed, since the enzyme can catalyze the stereospecific conversion of 1,3-DCP to EP (Eq. 2) [67, 68]. When 50 mM 1,3-DCP was incubated with 500 mM KCN and H-lyase B purified from *E. coli* cells overexpressing the H-lyase B gene from *Corynebacterium* sp. N-1074 at 20 °C for 2 h, 38 mmol l^{-1} (4.5 mg ml^{-1}) of (*R*)-CHBN was formed, with 95.2% *e.e.*

6
Production of Optically Active 1,2-Diols

Optically active 1,2-diols are very important as chiral building blocks for the syntheses of pharmaceuticals, agrochemicals, and natural products.

The HDDase of *Alcaligenes* sp. DS-S-7G described in Sect. 3.1 can also catalyze the enantioselective oxidation of (*S*)-1,2-butanediol and other long-chain alkane-1,2-diols [69, 70]. When washed cells of *Alcaligenes* sp. DS-S-7G grown on a medium containing 3-CPD as the sole carbon source were incubated with 1 vol.% 1,2-butanediol in the presence of NAD^+, the degradation of 1,2-butanediol ceased at about 50%, and the optical purity of the remaining 1,2-butanediol was 99.7% *e.e.* for the (*R*)-isomer. The resting cell degradation could be repeated eight times, which indicated that the resting cells were stable for at least 200 h (Fig. 7). As for 1,2-propanediol, stereoselective degradation by the HDDase of *Alcaligenes* sp. DS-S-7G also occurred, but the optical purity of the remaining (*R*)-1,2-diol was not so high (60.0% *e.e.*).

Kometani et al. [71] reported that baker's yeast catalyzed the asymmetric reduction of acetol to (*R*)-1,2-propanediol with ethanol as the energy source. The enzyme involved in the reaction was an NADH-dependent reductase, and NADH required for the reduction was supplied by ethanol oxidizing enzyme(s) in the yeast. When washed cells of baker's yeast were incubated with 10 mg ml^{-1} of acetol in an ethanol solution with aeration, (*R*)-1,2-propanediol was formed almost stoichiometrically with an optical purity of 98.2% *e.e.*

Candida parapsilosis was found to be able to convert (*R*)-1,2-butanediol to (*S*)-1,2-butanediol through stereospecific oxidation and asymmetric reduction reactions [72]. The oxidation of (*R*)-1,2-butanediol to 1-hydroxy-2-butanone and the reduction of 1-hydroxy-2-butanone to (*S*)-1,2-butanediol were cataly-

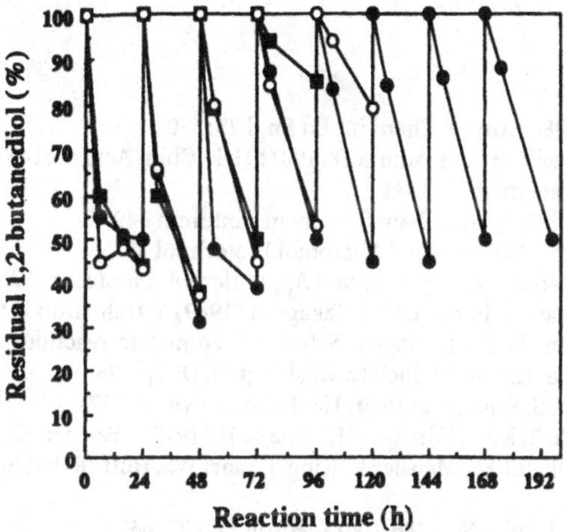

Fig. 7. Time course of repeated resolution. The repeated resolution of 1 vol.% 1,2-butanediol was carried out in the presence of 1 mmol l^{-1} NAD$^+$ (•) or phenazinemethosulfate (○), or in the absence of an electron acceptor (■) under aerobic conditions

zed by an NAD$^+$-dependent (*R*)-specific alcohol dehydrogenase and an NADPH-dependent (*S*)-specific 2-keto-1-alcohol reductase, respectively. Racemic 1,2-butanediol (10 mg ml^{-1}) was converted to (*S*)-1,2-butanediol with an optical purity of 79% *e.e.* by *C. parapsilosis*.

7 Conclusion

There are three ways of using microbial enzymes for the production of optically active compounds: (i) stereoselective degradation of one enantiomer (unwanted) in a racemic mixture of the target compounds; (ii) stereospecific conversion of prochiral compounds to the target compounds; and (iii) stereoselective inversion of the unwanted enantiomer to its mirror image in a racemic mixture of the target compounds. In the case of stereoselective degradation (i), there is the disadvantage that the yield is theoretically 50% at most; however, this method might be most applicable to the practical production of optically active compounds, which are synthesized by conventional optical resolution of racemic mixtures, because only the resolution step in the conventional synthetic process is replaced by an enzymatic reaction. Although both the stereospecific conversion of prochiral compounds (ii) and the stereoselective inversion of one enantiomer to another (iii) give theoretical yields of 100% as to the substrates, the former requires starting materials different from those used for conventional optical resolution, and the latter requires the discovery of quite unique and unusual enzymatic reactions. The high stereospecificities and yields of the reactions catalyzed by microbial enzymes overcome these disadvantages in microbial production processes, and the diversity and versatility of microorganisms reward our requirement by screening.

8 References

1. Sih CJ, Chen CS (1984) Angew Chem Int Ed Engl 23:570
2. Hungerbühler E, Seebach D, Wasmuth D (1981) Helv Chim Acta 64:1467
3. Mori K (1981) Tetrahedron 37:1341
4. Ohta H, Tetsukawa H (1978) J Chem Soc Chem Commun 849
5. Furuhashi K, Takagi M (1984) Appl Microbiol Biotechnol 20:6
6. Furuhashi K, Shintani M, Takagi M (1986) Appl Microbiol Biotechnol 23:218
7. Takahashi O, Umezawa J, Furuhashi K, Takagi M (1989) Tetrahedron Lett 30:1583
8. Chibata I (1982) In: Eliel EL, Otsuka S (eds) Asymmetric reactions and processes in chemistry. American Chemical Society, Washington, DC, p 195
9. Yamada H, Shimizu S, Yoneda K (1980) Hakko to Kogyo 38:937
10. Takahashi S, Ohashi T, Kii Y, Kumagai H, Yamada H (1979) J Ferment Technol 57:328
11. McClure DE, Engelhaldt EL, Mensler K, King S, Saari WS, Huff JR, Baldwin JJ (1979) J Org Chem 44:1826
12. Cimetiere B, Jacob L, Julia M (1986) Tetrahedron Lett 27:6329
13. Kawamura K, Ohta T, Otani G (1990) Chem Pharm Bull 38:2092
14. Takano S, Yanase M, Sekiguchi Y, Ogasawara K (1987) Tetrahedron Lett 28:1783
15. Kasai N, Sakaguchi K (1992) Tetrahedron Lett 33:1211
16. Takano S, Yanase M, Takahashi M, Ogasawara K (1987) Chem Lett 2017
17. Imai T, Nishida S (1990) J Org Chem 55:4849
18. Koden M, Kuratake T, Funada F, Awane K, Sakaguchi K, Shiomi Y, Kitamura K (1989) J Applied Physics 29:981
19. Weissermel K, Arpe HJ (1976) Industrielle Organische Chemie. Verlag Chemie GmbH, Weinheim
20. Kasai N, Tsujimura K, Unoura K, Suzuki T (1990) Agric Biol Chem 54:3185
21. Kasai N, Tsujimura K, Unoura K, Suzuki T (1992) J Ind Microbiol 9:97
22. Kasai N, Tsujimura K, Unoura K, Suzuki T (1992) J Ind Microbiol 10:37
23. Klunder JM, Ko SY, Sharpless KB (1986) J Org Chem 51:3710
24. Miyano S, Lu LD, Viti SM, Sharpless KB (1985) J Org Chem 50:4350
25. Danilewicz JC, Kemp JEG (1973) J Med Chem 16:168
26. Dukes M, Smith LH (1976) J Med Chem 14:326
27. Burgos CE, Ayer DE, Johnson RA (1987) J Org Chem 52:4973
28. Haouet A, Sepulchre M, Spassky N (1983) Eur Polym J 19:1089
29. Suzuki T, Kasai N (1991) Bioorg Med Chem Lett 1:343
30. Suzuki T, Kasai N, Yamamoto R, Minamiura N (1993) Appl Microbiol Biotechnol 40:273
31. Suzuki T, Kasai N, Yamamoto R, Minamiura N (1992) J Ferment Bioeng 73:443
32. Suzuki T, Kasai N, Yamamoto R, Minamiura N (1994) Appl Microbiol Biotechnol 42:270
33. Nakamura T, Yu F, Mizunashi W, Watanabe I (1991) Agric Biol Chem 55:1931
34. Nakamura T, Yu F, Mizunashi W, Watanabe I (1993) Appl Environ Microbiol 59:227
35. Nakamura T, Nagasawa T, Yu F, Watanabe I, Yamada H (1992) J Bacteriol 174:7613
36. Nakamura T, Nagasawa T, Yu F, Watanabe I, Yamada H (1994) Appl Environ Microbiol 60:1297
37. Nagasawa T, Nakamura T, Yu F, Watanabe I, Yamada H (1992) Appl Microbiol Biotechnol 36:478
38. Yu F, Nakamura T, Mizunashi W, Watanabe I (1994) Biosci Biotech Biochem 58:1451
39. Persson B, Krook M, Jornvall H (1991) Eur J Biochem 200:537
40. Nakamura T, Nagasawa T, Yu F, Watanabe I, Yamada H (1994) Appl Environ Microbiol 60:4630
41. Geerlof A, van Tol JBA, Jongerjan JA, Duine JA (1994) Biosci Biotech Biochem 58:1028
42. Geerlof A, Jongejan JA, van Dooren TJGM, Raemakers-Franken PC, van den Tweel WJJ, Duine JA (1994) Enzyme Microb Technol 16:1059
43. Zhou B, Gopalan AS, VanMiddlesworth F, Shieh WR, Sih CJ (1983) J Am Chem Soc 105:5925

44. Shieh WR, Gopalan AS, Sih CJ (1985) J Am Chem Soc 107:2993
45. Wong CH, Drueckhammer DG, Sweers HM (1985) J Am Chem Soc 107:4028
46. Grundwald J, Wirtz MP, Scollar MP, Klibanov AM (1986) J Am Chem Soc 108:6732
47. Kitamura M, Ohkuma T, Takaya H, Noyori N (1988) Tetrahedron Lett 29:1555
48. Jung ME, Shaw TJ (1980) J Am Chem Soc 102:6304
49. Bock K, Lundt I, Pedersen C (1983) Acta Chem Scand 37:341
50. Rossiter BE, Sharpless KB (1984) J Org Chem 49:3707
51. Pifferi G, Pinza M (1977) Farmaco Ed Sci 32:602
52. Pellegata R, Pinza M, Pifferi G (1978) Synthesis 614
53. Patel RN, McName CG, Benerjee A, Howell JM, Robinson RS, Szarka LJ (1992) Enzyme Microb Technol 14:731
54. Wada M, Kataoka M, Kawabata H, Yasohara Y, Kizaki N, Hasegawa J, Shimizu S (1998) Biosci Biotech Biochem 62:280
55. Shimizu S, Kataoka M, Morishita A, Katoh M, Morikawa T, Miyoshi T, Yamada H (1990) Biotechnol Lett 12:593
56. Yamada H, Shimizu S, Kataoka M, Sakai H, Miyoshi T (1990) FEMS Microbiol Lett 70:45
57. Kataoka M, Sakai H, Morikawa T, Katoh M, Miyoshi T, Shimizu S, Yamada H (1992) Biochim Biophys Acta 1122:57
58. Kataoka M, Shimizu S, Yamada H (1992) Arch Microbiol 157:279
59. Bohren KM, Bullock B, Wermuth B, Gabbay KH (1989) J Biol Chem 264:9547
60. Kita K, Matsuzaki K, Hashimoto T, Yanase H, Kato N, Chung MCM, Kataoka M, Shimizu S (1996) Appl Environ Microbiol 62:2303
61. Shimizu S, Kataoka M, Katoh M, Morikawa T, Miyoshi T, Yamada H (1990) Appl Environ Microbiol 56:2374
62. Makino Y, Negoro S, Urabe I, Okada H (1989) J Biol Chem 264:6381
63. Kataoka M, Rohani LPS, Yamamoto K, Wada M, Kawabata H, Kita K, Yanase H, Shimizu S (1997) Appl Microbiol Biotechnol 48:699
64. Kataoka M, Rohani LPS, Wada M, Kita K, Yanase H, Urabe I, Shimizu S (1998) Biosci Biotech Biochem 62:167
65. Kataoka M, Yamamoto K, Shimizu S, Kita K, Hashimoto T, Yanase H (1995) Nippon Nogei Kagaku Kaishi 69:270
66. Suzuki T, Idogaki H, Kasai N (1996) Tetrahedron:Asymmetry 7:3109
67. Nakamura T, Nagasawa T, Yu F, Watanabe I, Yamada H (1991) Biochem Biophys Res Commun 180:124
68. Nakamura T, Nagasawa T, Yu F, Watanabe I, Yamada H (1994) Tetrahedron 50:11,821
69. Suzuki T, Kasai N, Minamiura N (1994) Tetrahedron:Asymmetry 5:239
70. Suzuki T, Kasai N, Minamiura N (1994) J Ferment Bioeng 78:194
71. Kometani T, Yoshii H, Takeuchi Y, Matsuno R (1993) J Ferment Bioeng 76:414
72. Hasegawa J, Ogura M, Tsuda S, Maemoto S, Kutsuki H, Ohashi T (1990) Agric Biol Chem 54:1819

Received January 1998

Polyamino Acids as Man-Made Catalysts

Joanne V. Allen · Stanley M. Roberts · Natalie M. Williamson

Department of Chemistry, University of Liverpool, Liverpool, L69 7DZ, UK
E-mail: SJ11@liverpool.ac.uk

Polyamino acids are easy to prepare by nucleophile-initiated polymerisation of amino acid *N*-carboxyanhydrides. Polymers such as poly-(L)-leucine act as robust catalysts for the epoxidation of a wide range of electron-poor alkenes, such as γ-substituted α,β-unsaturated ketones. The optically active epoxides so formed may be transformed into heterocyclic compounds, polyhydroxylated materials and biologically active compounds such as diltiazem and taxol side chain.

Keywords: Polyamino acids, Optically active epoxyketones, Synthesis of diltiazem and taxol side-chain, Asymmetric oxidation.

1 Introduction and Background Information 126

2 Preparation of Polyamino Acids 128

2.1 Activation of the Amino Acid 128
2.2 Polymerisation 128
2.3 Downstream Processing 129

3 Mechanism of the Polyamino Acid-Catalysed Oxidations 130

4 Catalysis Evoked by Polyamino Acids 130

4.1 Epoxidation Reactions 130
4.1.1 Preparation of Substrates 131
4.1.2 Analysis of Products 132
4.1.3 Stereochemical Assignment 135
4.2 Other Michael-Type Reactions 136
4.3 Oxidation of Sulfide to a Sulfoxide 136

5 Transformations of Optically Enriched Epoxyketones 136

5.1 Reactions at the Carbonyl Group 137
5.2 Reactions at the Epoxide Unit 139
5.3 Reactions Adjacent to the Epoxide Moiety 142

6 Conclusions 143

7 References 143

Advances in Biochemical Engineering/
Biotechnology, Vol. 63
Managing Editor: Th. Scheper

1
Introduction and Background Information

Enzymes are natural biocatalysts that are becoming increasingly popular tools in synthetic organic chemistry [1]. The major areas of exploration have involved the use of hydrolases, particularly esterases and lipases [2]. These enzymes are readily available, robust and inexpensive. The second most popular area of investigation has been the reduction of carbonyl compounds to chiral secondary alcohols using either dehydrogenases (with co-factors) or a whole-cell system such as bakers' yeast [3].

Important oxidation reactions may be accomplished using natural biocatalysts; a selection of these are listed below:

- hydroxylation of aliphatic or alicyclic compounds, e.g. steroids [4]
- hydroxylation of aromatic and heteroaromatic compounds, e.g. phenols [5]
- Baeyer-Villiger oxidations [6]
- oxidation of benzene (and derivatives) to (substituted) cyclohexadienediols [7]

All of these oxidation reactions require the use of micro-organisms other than simple yeasts or, alternatively, enzymes that are not commercially available and that have to be isolated from the relevant micro-organism and partially purified [8].

A particularly difficult transformation to accomplish using natural biocatalysts is asymmetric epoxidation. Some organisms catalyse the oxidation of terminal alkenes to the corresponding oxiranes with a high degree of stereoselectivity but, generally, yields are modest to low [9]. Another problem is the high toxicity of the epoxides to living cells. More of a problem is the fact that 1,2-disubstituted alkenes and more heavily substituted olefins were not oxidised at all. Hence, the biocatalytic methodology is no competitor to the Sharpless [10] or Jacobsen [11] strategies for the oxidation of electron-rich alkenes.

In the early 1980's Juliá and Colonna published a series of papers which, to some extent, filled the gap left by the natural biocatalysts. The Spanish and Italian collaborators showed that α,β-unsaturated ketones of type **1** underwent asymmetric oxidation to give the epoxide **2** using a three-phase system, namely aqueous hydrogen peroxide containing sodium hydroxide, an organic solvent such as tetrachloromethane and insoluble poly-(L)-alanine, (Scheme 1) [12]. The reaction takes place via a Michael-type addition of peroxide anion (the Weitz-Scheffer reaction).

Juliá and Colonna showed that, as expected, poly-(L)-leucine and poly-(L)-isoleucine could be employed in place of poly-(L)-alanine, while poly-(L)-valine gave poorer results in terms of yields and stereoselectivity [13]. Some other enones were tried without success and nucleophiles other than peroxide anion were

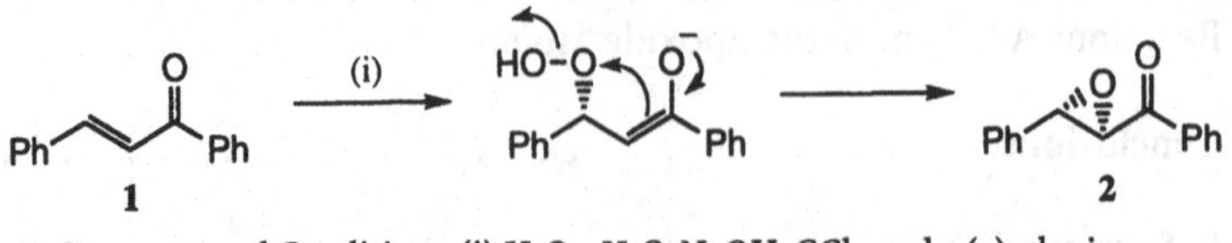

Scheme 1. *Reagents and Conditions:* (i) H_2O_2, H_2O, NaOH, CCl_4, poly-(L)-alanine

employed, again without significant asymmetric induction being achieved. Thus, though they had discovered a fascinating reaction, Juliá and Colonna believed that the synthetic utility of the reaction may be limited to the epoxidation of chalcone-type substrates. This impression was consolidated by other workers who used various chalcone derivatives in polyamino acid-catalysed asymmetric epoxidations. Flisak and Lantos used the methodology to prepare substituted phenyl glycidic esters [14] and then used the methodology to make a leukotriene antagonist **3** (Scheme 2) [15].

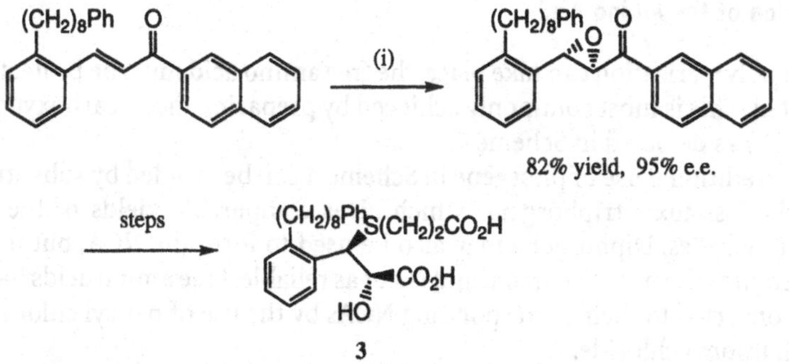

$(CH_2)_8Ph$ O (i) $(CH_2)_8Ph$ O O

82% yield, 95% e.e.

steps $(CH_2)_8Ph$ $S(CH_2)_2CO_2H$ CO_2H HO

3

Scheme 2. *Reagents and Conditions:* (i) NaOH, EDTA, *n*-hexane, poly-(L)-leucine, H_2O, H_2O_2

Two key points were made in the latter paper. First the polyamino acid-catalysed reactions could be conducted on a substantial scale (250 g). Secondly, the catalyst could be recovered and recycled with no loss in yield or enantioselectivity. In addition, the SK&F group emphasised the need to pre-swell the polymer with organic solvent and aqueous sodium hydroxide for about 20 h prior to its employment in the reaction.

A research team from Bloemfontein (South Africa) have also taken advantage of the Juliá and Colonna oxidation in elegant research aimed at the synthesis of optically active flavonoids. Bezuidenhoudt, Ferreira et al. have oxidised a range of chalcone derivatives using poly-(L)-alanine in the three phase system to afford optically active epoxides **4** which were readily cyclised to target compounds of the dihydroflavinol type **5**, (Scheme 3) [16].

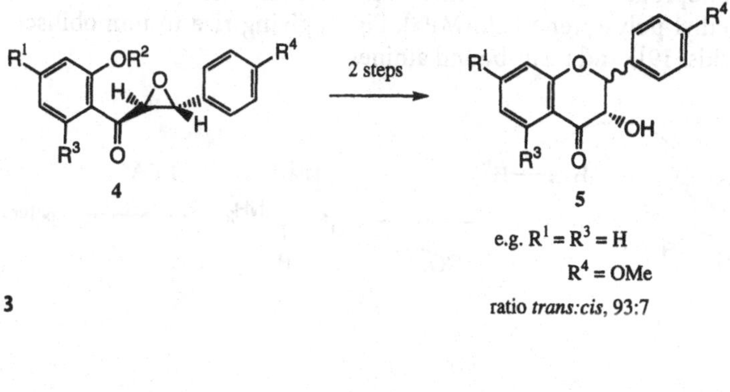

R^1 OR^2 H O H R^4 R^3 O **4** 2 steps R^1 O R^4 OH R^3 O **5**

e.g. $R^1 = R^3 = H$

$R^4 = OMe$

ratio *trans:cis*, 93:7

Scheme 3

In this review, recent advances in the preparation and use of polyamino acid catalysts are included; these latest results are a result of developments utilising as a starting point the seminal studies undertaken earlier by the research groups described above.

2 Preparation of Polyamino Acids

2.1 Activation of the Amino Acid

Before polymerisation can take place, the free amino acid must be protected and activated. This is most commonly achieved by preparing the N-carboxyanhydride (NCA) as depicted in Scheme 4.

The traditional use of phosgene in Scheme 4 can be avoided by substituting it with the less toxic triphosgene, which gives comparable yields of the *N*-carboxyanhydrides. Diphosgene may also be used to form the NCA, but the reaction requires the use of charcoal and is not as reliable. Free amino acids have also been converted to their corresponding NCAs by the use of benzyl chloroformate with thionyl chloride.

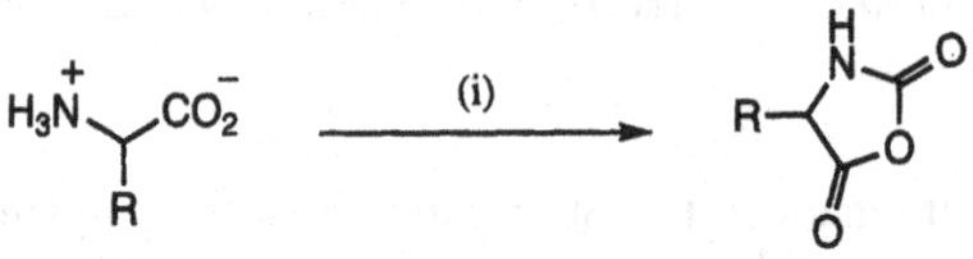

Scheme 4. *Reagents and Conditions:* (i) phosgene, diphosgene or triphosgene in THF

2.2 Polymerisation

Decarboxylative polymerisation of the *N*-carboxyanhydride can be brought about in two ways, the first being the use of a humidity cabinet [15] and the second involving the use of a suitable amine initiator, (Scheme 5).

A number of amines have been investigated for their suitability as polymerisation initiators, including aliphatic amines (such as butylamine [17] and 1,3-diaminopropane [18]), polymer supported amines (such as cross-linked aminomethyl polystyrene [CLAMPS], Fig. 1, giving rise to immobilised polyamino acids [19]) and resin-bound amines.

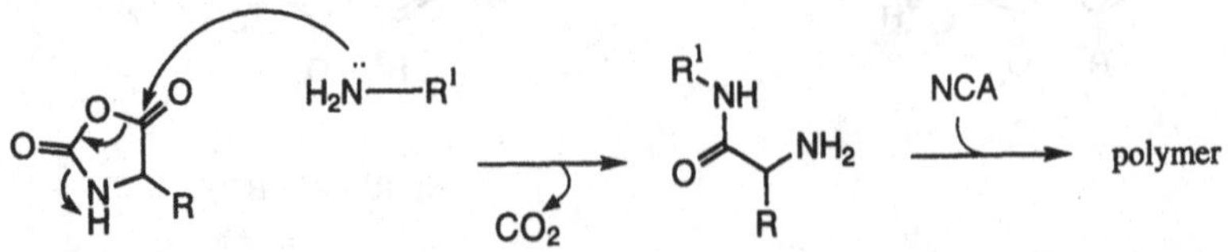

Scheme 5

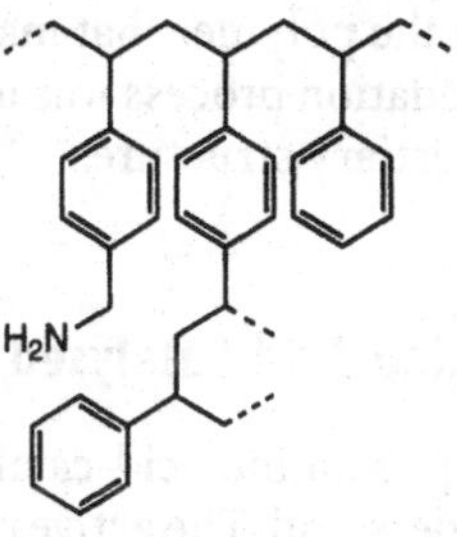

Fig. 1. A representative repeating unit of CLAMPS

Amine initiated polymerisation is achieved by stirring the NCA with an appropriate amount of initiator in a suitable solvent, such as tetrahydrofuran. Over a period of time, the polymer precipitates and can be isolated by filtration.

Although polymerisation using a humidity cabinet has been recommended, the amine-initiated method consistently gives better quality polymer and is therefore the method of choice. The humidity cabinet approach requires the use of high quality, crystalline NCA for success. The crystalline solid changes to an amorphous solid during the polymerisation process, making it likely that imperfections in the crystal may cause termination of the growth of the polymer chains, giving rise to material with a low molecular weight average and poor catalytic properties.

2.3 Downstream Processing

While immobilised polyamino acids can be recovered from a reaction and re-used in subsequent operations, it has been found that after repeated recycling the quality of the catalyst declines, resulting in increased reaction times and reduced stereoselectivity. The quality of the catalyst declines particularly quickly when it is used in the recently developed "biphasic" epoxidation conditions (see Sect. 4.1.2). This gradual "decay" of the polyamino acid catalyst led to the development of a regeneration procedure.

A suspension of the recycled polyamino acids in toluene is stirred vigorously with 4.0 M aqueous sodium hydroxide solution for 16 h. The procedure may be repeated to ensure optimum regeneration. The polyamino acid recovered from this procedure shows catalytic activity and stereoselectivity comparable to the freshly made polyamino acid.

A second regeneration procedure, similar to the above but using small amounts of aqueous hydrogen peroxide in addition to the sodium hydroxide solution, has also been developed. This latter procedure regenerates the recycled polyamino acid so effectively that usually only one treatment is required.

It is possible that the regeneration method succeeds by removing short chain, inactive polymer fragments, which will be soluble in the aqueous medium. Once these smaller polymer chains have been removed, the bulk of the polymer is then able to reform its preferred tertiary structure, thus restoring its catalytic activity. The aqueous nature of the procedure may also aid regeneration by replacing water molecules in the polymer that may have been leached out when the anhydrous biphasic epoxidation process was used. This would also aid in the reforming of the polymer's tertiary structure.

3 Mechanism of the Polyamino Acid-Catalysed Oxidations

The mechanism behind the polyamino acid-catalysed asymmetric epoxidation is particularly difficult to understand. The active catalyst exists as a paste or a gel following treatment with the organic solvent. Thus, studies on the helix/β-sheet structure of the amorphous solid, the form of the polyamino acid in the absence of solvent, are probably not meaningful in this context.

Colonna, Juliá et al. believed that polymer chain lengths having an average of ten or more leucine or alanine residues gave rise to active catalyst. We have carried out a full study using material prepared using an amino acid synthesiser and we can confirm that active catalyst is obtained when employing a decamer of (L)-leucine [20].

In a study involving mixed polymers of (L) and (D)-leucine it was clearly shown that the amino terminus of the polymer confers the stereochemical signature to the product. The experiments were conducted as follows. The initiator, diaminopropane was added to an amount of activated (D)-leucine in solvent sufficient to give an x-mer, say a 9-mer, as a living polymer. After this reaction was complete, sufficient (L)-leucine was added to give an 18-mer. Thus, if the (L)-leucine tail is labelled a y-mer then $x+y=18$. For polymer made up of $x=9, y=9$, the epoxide formed on epoxidation of the chalcone under standard conditions had the configuration corresponding to that formed from the 18-mer of (L)-leucine ($x=0, y=18$). Similarly, polymer with ($x=12, y=6$) configures epoxide with almost the same enantiomeric excess as that obtained with pure poly-(L)-leucine. Even polymer with the average make up ($x=15, y=3$) configured epoxide that is normally derived from poly-(L)-leucine, albeit with reduced enantiomeric excess (62% versus >90%) [21]. Further studies are underway using polyamino acids with mixed (L) and (D)-leucine sequences prepared using an amino acid synthesiser.

At this point it is impossible to guess the architecture of the active catalyst. The folding of 10-mers of leucine and alanine in organic solvents is clearly of critical importance, and studies are in progress to understand the preferred shapes. Obviously, in its active form the catalyst binds and activates peroxide anion and/or the electron-poor alkene near its chiral surface, perhaps in a chiral cavity, but the precise orientation of catalyst and reactants in the initial bond-forming Michael reaction remains unsolved.

4 Catalysis Evoked by Polyamino Acids

4.1 Epoxidation Reactions

To date the most useful and wide ranging reaction catalysed by polyamino acids is the asymmetric epoxidation of enones [22].

4.1.1
Preparation of Substrates

Many of the enone substrates used in polyamino acid-catalysed epoxidation reactions can be made via a simple aldol condensation, which leads directly to the desired enone after in situ dehydration. Enones that cannot be synthesised by the above route may often be synthesised using standard Wittig chemistry, (Scheme 6). The above methods of substrate synthesis provide compounds with a variety of groups R and R^1, enabling the incorporation of both aliphatic and aromatic moieties into the enone structure.

More complicated substrates such as some dienones and trienediones, required multi-stage syntheses. For example, the unsaturated keto-ester **6** was synthesised via sequential Wittig reactions, (Scheme 7).

The trienedione **7** was prepared using a sequence in which palladium-catalysed couplings were featured, (Scheme 8). In addition, the alkyne moiety of the chloroenyne **8** was partially reduced and the secondary alcohol group oxidised to give the dienone **9**.

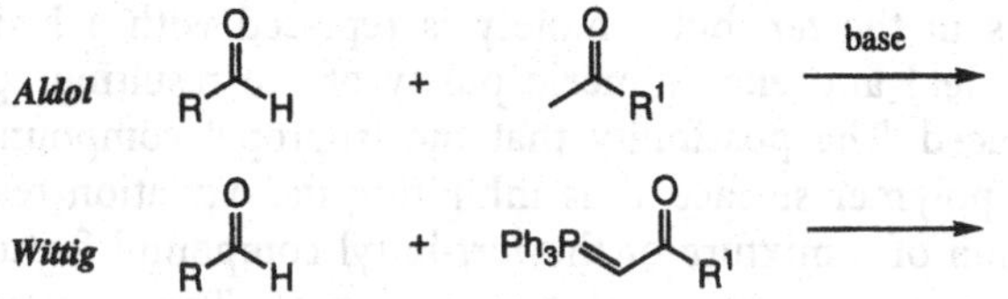

Scheme 6

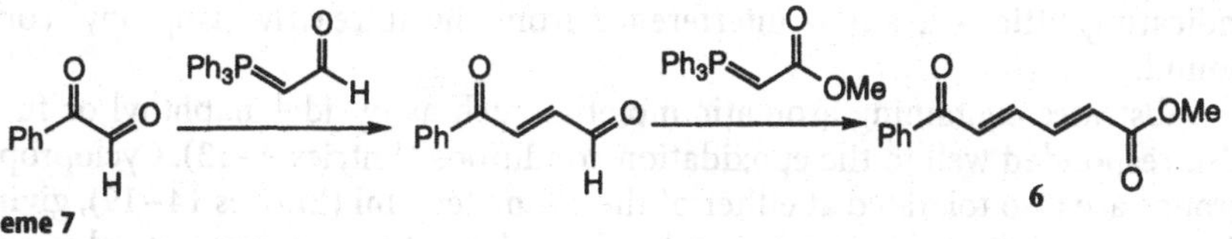

Scheme 7

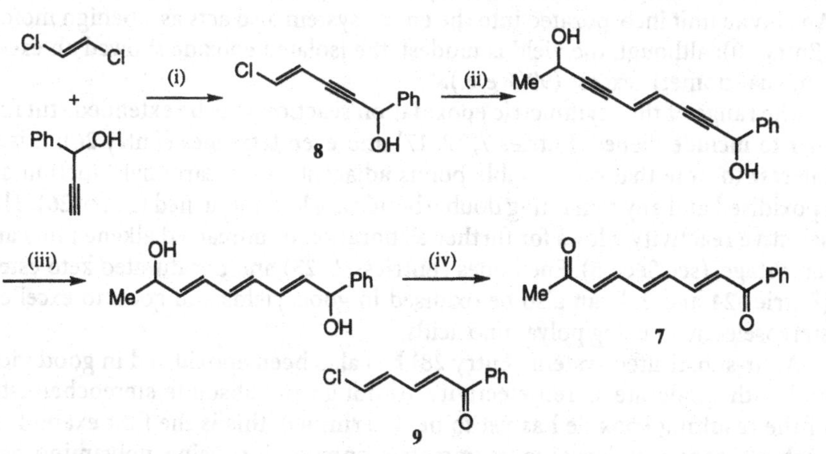

Scheme 8. *Reagents and Conditions:* (i) $Pd(PPh_3)_4$, CuI, piperidine, HF, (ii) MeCH(OH)CCH, $Pd(PhCN)_2Cl_2$, CuI, piperidine, (iii) Red-Al, (iv) MnO_2, CH_2Cl_2

4.1.2
Analysis of Products

Initial investigations into the use of polyamino acids as catalysts for asymmetric epoxidation concentrated on the conditions reported for the epoxidation of chalcone systems [22], involving basic peroxide in the presence of a polyamino acid. This protocol has become known as the "triphasic" method due to the three phases formed by the basic peroxide solution, the organic solvent and the polymer. Before epoxidising a substrate, the polymer is allowed to swell in the oxidant/solvent mixture for a period of at least 6 hours. Table 1 shows that the polyamino acid-catalysed epoxidation is not just limited to the transformation of chalcone itself (Entry 1). Chalcone systems with substituted aromatic rings also undergo oxidation smoothly (Entries 2 and 3). The substituent on the aromatic ring may either be electron-withdrawing (Entry 2) or electron-donating (Entry 3) without affecting the success of the reaction.

A wide range of non-chalcone based substrates also undergo oxidation. Either of the phenyl groups of chalcone may be replaced with a tertiary butyl group, with little – if any – loss of optical purity (Entries 4–6). If, however, one of the methyl groups in the *tert*-butyl moiety is replaced with a hydrogen atom (Entry 8), the yield and enantiomeric purity of the resulting epoxide are significantly reduced. The possibility that the isopropyl compound was tightly bound to the polymer surface, thus inhibiting the oxidation reaction, was tested by oxidation of a mixture of the *tert*-butyl compound featured in Entry 7 with the isopropyl compound shown in Entry 8. The former compound was oxidised at the expected rate and with the usual stereoselectivity, indicating little – if any – interference from the unreactive isopropyl compound.

Substrates containing aromatic moieties such as pyridyl, naphthyl or furyl also responded well to the epoxidation conditions (Entries 9–13). Cyclopropyl groups are also tolerated at either of the enone termini (Entries 14–19), giving the corresponding oxiranes in good yield and good to excellent optical purity. An alkyne unit incorporated into the enone system also acts as a benign moiety (Entry 20); although the yield is modest, the isolated epoxide showed an excellent enantiomeric excess (90% e.e.).

The range of the asymmetric epoxidation reaction may be extended still further to include dienes (Entries 7, 12, 17) and even tetraenes (Entry 26). It is of interest to note that only double bonds adjacent to the carbonyl function are epoxidised and any remaining double bonds are left untouched (Entry 26). This selective reactivity allows for further elaboration of unreacted alkene units at a later stage, (see Sect. 5). Enediones (Entries 21–23) and unsaturated keto esters (Entries 24 and 25) can also be oxidised in good yields and good to excellent stereoselectivity using polyamino acids.

An α-substituted system (Entry 28) has also been epoxidised in good yield and with moderate stereoselectivity. Although the absolute stereochemistry of the resulting epoxide has yet to be determined, this is the first example of such an enone undergoing asymmetric epoxidation using polyamino acid catalysis.

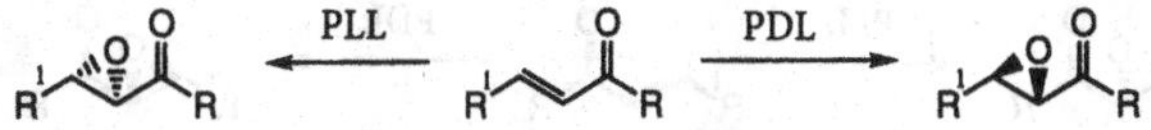

Table 1. Epoxidation of Enones using Triphasic Conditions

Entry	R	R′	Method[b]	Time (h)	Yield (%)	e.e. (%)	Ref.
1	Ph	Ph	A	N/A	78–85	78–86	[12]
2	Ph	p-$NO_2C_6H_4$	A	N/A	83	82	[12]
3	Ph	p-$OMeC_6H_4$	A	N/A	53	N/A	[12]
4	Ph	^tBu	B	18	85	90	[24]
5	^tBu	Ph	B	18	92	>98[a]	[15, 24]
6	^tBu	Ph	C	18	90	>98[a]	[24, 25]
7	^tBu	CH=CHPh	B	30	90	97	[18]
8	^iPr	Ph	B	168	60	62	[24, 25]
9	^tBu	4-pyridyl	B	30	70	72	[24]
10	Ph	4-pyridyl	B	18	60	75	–
11	2-pyridyl	2-naphthyl	B	16	75	>96	–
12	2-naphthyl	CH=CHPh	B	96	78	96	–
13	furyl-2-CH=CH	2-furyl	B	60	60	90	[18]
14	cyclopropyl	Ph	B	18	85	77	[24, 25]
15	cyclopropyl	2-naphthyl	C	29	61	90	[8, 9]
16	2-naphthyl	cyclopropyl	B	28	73	>98	[8, 9]
17	cyclopropyl	CH=CHPh	B	42	73	74	[8, 9]
18	cyclopropyl	CH=CHPh	D	19	52	98	[8, 9]
19	cyclopropyl	2-quinolyl	B	18	94	79	[8]
20	PhCC	Ph	E	96	57	90	[8, 9]
21	Ph	COPh	B	N/A	76	76	[8, 9]
22	p-ClC_6H_4	COPh	D	N/A	60	89	[8, 9]
23	^tBu	COPh	B	N/A	79	82	[8, 9]
24	^tBu	CO^tBu	C	N/A	100	>95	[8, 9]
25	Ph	$CO_2{}^tBu$	C	N/A	66	>95	[8, 9]
26	$Ph(CH{=}CH)_2$	CH=CHPh	B	N/A	50	80	[8, 9]
27	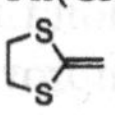	Ph	B	115	51	>94	[8]
28	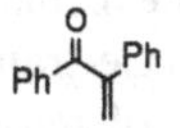	–	F	17	78	59	[8, 9]

[a] After one recrystallisation.

[b] *Methods A*: poly-(L)-alanine, H_2O_2, NaOH, H_2O, toluene. *B*: Diaminopropane-poly-(L)-leucine, H_2O_2, NaOH, H_2O, CH_2Cl_2. *C*: Diaminopropane-poly-(L)-leucine, H_2O_2, NaOH, H_2O, toluene. *D*: Diaminopropane-poly-(L)-leucine, $NaBO_3$, NaOH, CH_2Cl_2. *E*: Immobilised poly-(L)-leucine, H_2O_2, NaOH, H_2O, toluene. *F*: Immobilised poly-(L)-leucine, H_2O_2, NaOH, H_2O, toluene, Aliquat 336.

The oxidation procedure is amenable to a number of modifications. The solvent used may be varied; carbon tetrachloride, hexane, toluene and dichloromethane have all been used successfully, although the latter two are the solvents of choice. The most common oxidant is aqueous hydrogen peroxide, but other oxidants such as t-butyl hydroperoxide [4], sodium perborate (Table 1, Entries 18 and 22) and sodium percarbonate [4] have also been employed. It is interesting to note the effect of a change of oxidant. Using alkaline hydrogen peroxide

Table 2. Epoxidation of Enones using Biphasic Conditions

Entry	R	R′	Method[b]	Time (h)	Yield (%)	e.e. (%)	Ref.
1	Ph	Ph	A	0.5	85	94	[26]
2	Me	Ph	A	8	70	80	[26]
3	Ph	CO_2Me	A	2	63	55	[27]
4	Ph	$CO_2{}^tBu$	A	5	>95	47	[27]
5	Me	*p*-$NO_2C_6H_4$	A	60	60	33	[27]
6	Me	*p*-$OMeC_6H_4$	A	60	80	65	[27]
7	2-naphthyl	furyl-2-CH=CH	A	60	>95	85–90	[27]
8	Ph	$CH=CHCO_2Me$	A	4	88	90	[27]
9	Ph	$CH=CHCO_2{}^tBu$	A	2	95	93	[27]
10	Ph	CH=CHCl	A	1	55	>99[a]	[27]
11	Ph	$(CH=CH)_2COMe$	A	0.5	43	90	[27]
12	2-naphthyl	CH=CHPh	A	3.5	85	>99[a]	[27]
13	2-naphthyl	CH=CHPh	B	3.5	80	>99[a]	[27]
14	Ph	cyclohexyl	A	3	91	89	[26]
15	^tBu	Ph	A	12	76	94	[26]
16	^tBu	*p*-$OMeC_6H_4$	A	20	85	96	[26]
17	2-naphthyl	Ph	A	2	>85	>98	[27]

[a] After one recrystallisation.
[b] *Methods A*: Immobilised poly-(L)-leucine, urea hydrogen peroxide, DBU, THF.
B: Immobilised poly-(D)-leucine, urea hydrogen peroxide, DBU, THF.

the cyclopropyl-substituted system (Entry 17) required a reaction time of almost 2 days and the resulting epoxide showed only moderate enantiomeric excess. In contrast, with sodium perborate as the oxidant, the reaction time is halved to 19 hours and the enantiomeric excess of the epoxide is improved to 98% (Entry 18). Such a dramatic improvement, by merely changing the oxidant, is not always achieved, but is certainly worthy of consideration. A phase-transfer catalyst such as Aliquat 336 may be employed to increase the rate of reaction (Entry 28).

While the utility of the triphasic asymmetric oxidation system is obvious, the reaction times involved are often rather long, with some substrates taking up to 5 days to proceed to only 50% conversion (see Table 1, Entry 27). As a consequence of this observation, alternative reaction conditions were investigated, with the aim of reducing the reaction time required for epoxidation. This led to the development of "biphasic" conditions [26], using urea hydrogen peroxide complex (UHP) as the oxidant and diazabicycloundecene (DBU) as a non-nucleophilic organic base, with tetrahydrofuran as the solvent. Table 2 lists the results of epoxidations carried out using this new method.

It is immediately apparent that the main attraction of the biphasic method is the much shorter reaction times it affords. This is illustrated well by the chalcone system. Using triphasic conditions, the epoxidation of chalcone can take up to 8 hours, while the biphasic reaction (Table 2, Entry 1) is complete in 30 minutes with the resulting epoxide showing excellent optical purity. Epoxidation of a naphthyl- substituted dieneone (Entries 12 and 13) also shows a remarkable increase in reaction rate when biphasic conditions are employed, proceeding to completion in 3.5 hours. This is in comparison with the 96 hour reaction time observed when triphasic conditions are used (Table 1, Entry 12). Other substrates that show increased reaction rates when epoxidised using biphasic conditions include a methyl ketone (Entry 2, 70% yield, 80% e.e., in 8 hours) and the cyclohexyl-phenyl substituted system (Entry 14, 91% yield, 89% e.e., in 3 hours), both of which were totally unreactive in the triphasic system. The *p*-methoxy substituted aromatic ring of Entry 16 slows the reaction rate somewhat, but the resulting epoxide is obtained in excellent yield and optical purity.

The biphasic method, like the triphasic method, allows the epoxidation of a broad range of substrates accommodating both aliphatic and aromatic substituents at either end of the enone moiety. However, unlike the triphasic method, the biphasic conditions allow reaction of hydroxide-sensitive systems.

Just as for the triphasic system, the dienes and triene (Entries 7–13) undergo selective epoxidation at only one of the double bonds. The selective reactivity of the compound featured in Entry 11 is particularly noteworthy. Monoepoxidation of the double bond next to the phenyl-bearing carbonyl group was the only reaction observed, despite extended reaction times and use of excess oxidant.

A *p*-nitro substituent (Entry 5) may be expected to increase the reaction rate compared to the unsubstituted system; however, the former compound is epoxidised at approximately the same rate as the unsubstituted system (Entry 2) and at the same rate as the *p*-methoxy substituted system (Entry 6), in which the electron-donating group would be expected to decrease the reaction rate.

The dienedione system (Scheme 9) demonstrates the potential for regioselectivity in polyamino acid-catalysed epoxidations. The epoxide arising from oxidation of the di-substituted double bond does not react further[15].

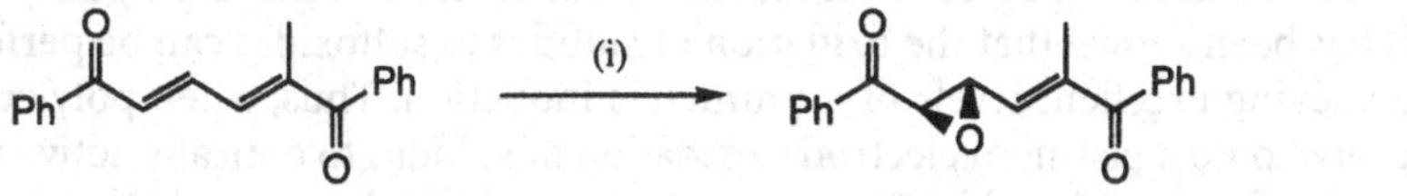

Scheme 9. *Reagents and Conditions:* (i) Immobilised poly-(L)-leucine, urea hydrogen peroxide, DBU, THF, 1.5 h

4.1.3
Stereochemical Assignment

When a poly-(L)-amino acid is used in the epoxidation of chalcone, the predominant optical isomer is laevorotatory [10]. The laevorotatory enantiomer of epoxychalcone has been shown [7] to have absolute configuration (2*R*, 3*S*). As anticipated, the use of a poly-(D)-amino acid as the epoxidation catalyst gives rise to the dextrorotatory enantiomer [10].

Given that all of the optically active epoxides in Tables 1 and 2 synthesised using poly-(L)-amino acids are laevorotatory, their absolute stereochemistry has

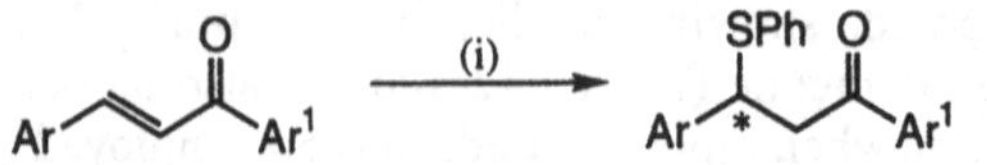

$$\text{Ar-CH=CH-CO-Ar}^1 \xrightarrow{\text{(i)}} \text{Ar-C}^*\text{H(SPh)-CH}_2\text{-CO-Ar}^1$$

Scheme 10. *Reagents and Conditions:* (i) thiophenol, polyamino acid, $CHCl_3$, H_2O, 0.1 eq NaOH

been assumed to be 2*R*, based upon the above findings. Circular dichroism measurements [24] have supported the absolute configuration of the 2-position as being *R*.

4.2 Other Michael-Type Reactions

Attempts to use polyamino acids as catalysts in other reactions have proved successful, notably in the coupling of thiophenol with chalcones. Employing poly-(L)-leucine, the β-phenylthioether was obtained in up to 45% e.e., (Scheme 10). Studies indicated that the best results were obtained by slow addition of the thiophenol to the reaction mixture.

In a similar fashion to those results obtained for the oxidation process, on switching from poly-(L)-leucine to poly-(D)-leucine the opposite configuration of the polyether was observed (absolute configuration of products unknown). Unexpected, however, was the observation that poly-(L)-phenylalanine furnished the opposite enantiomer to that observed employing poly-(L)-leucine. Thus, it has been shown that addition of nucleophiles other than peroxide anion can be catalysed by polyamino acids with significant stereocontrol [22].

4.3 Oxidation of Sulfide to Sulfoxide

In determining the diversity of reactions which may be catalysed by polyleucine, it has been shown that the oxidation of sulfides to sulfoxides can be performed, achieving excellent levels of asymmetric induction. Thus, when polyleucine is coated onto a platinum electrode, oxidation of sulfides to optically active sulfoxides has been achieved in 77% e.e. and 56% yield, (Scheme 11) [22]

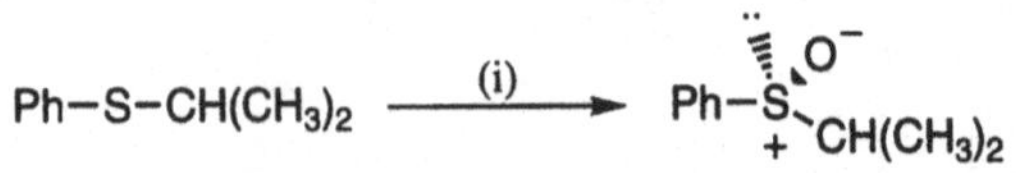

$$\text{Ph-S-CH(CH}_3)_2 \xrightarrow{\text{(i)}} \text{Ph-}\overset{+}{\text{S}}(\text{O}^-)\text{-CH(CH}_3)_2$$

Scheme 11. *Reagents and Conditions:* (i) Pt electrode, 0.1 M *n*-Bu_4NBF_4 in aq. acetonitrile, *ca* 14 h

5 Transformations of Optically Enriched Epoxyketones

The stereo- and chemoselective epoxidation α,β-unsaturated ketones is best achieved employing the biphasic conditions (PLL/UHP/DBU/THF), (see above).

The epoxidation occurs at electron deficient double bonds, thus it is only the double bond adjacent to a ketone functionality that is epoxidised. This phenomenon allows scope for further elaboration of our substrates, hence several transformations have been investigated in order to diversify the nature of the products. Indeed, substrates used in the epoxidation reaction intentionally featured several functionalities which could each be modified selectively dependent upon the reaction conditions employed. Such reactions include:

(i) reactions at the carbonyl group
(ii) reactions at the epoxide unit
(iii) reactions adjacent to the epoxyketone moiety

Each of these reaction types will be discussed in turn.

5.1 Reactions at the Carbonyl Group

Alkylations of epoxychalcones **2** has been successfully effected using organocerium reagents, displaying excellent yields (80–91%) and moderate diastereoselectivities (4:1–9:1) of the resultant tertiary alcohols, (Scheme 12).

These results were greatly enhanced by changing the organometallic agent to the corresponding Grignard reagent as the source of the alkyl group. The reaction times were comparable at 1.5 h, as were the yields (89%), however a dramatic increase in the diastereoselectivity of **10**:**11** was observed (>99:1). The use of additives in the reaction mixture and the effect of the reaction temperature were further examined to see if further improvements could be made on these results (Table 3).

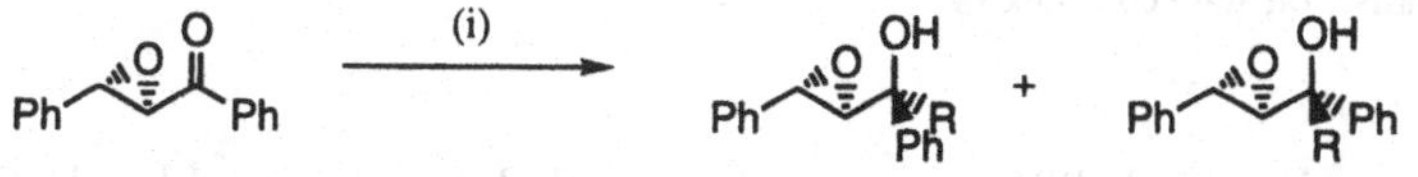

2 → (i) → **10** + **11**

R = Me 4:1 91%
R = Bu 9:1 80%

Scheme 12. *Reagents and Conditions:* (i) R_3Ce, –78 °C, 2 h

Table 3

Reagent	Time (h)	Yield (%)	10:11
BuMgBr (THF/HMPA)	1.5	37	>99:1
BuMgBr (THF/DMPU)	3.5	44	>99:1
BuMgBr (THF/DMPU) –78 °C	10	59	>99:1

As indicated, the diastereoselectivities are comparable in all cases to the initial Grignard conditions; however there is a dramatic decrease in the observed yields. Lowering the reaction temperature to –78 °C went some way to improving upon these lower yields and, although the diastereoselectivities remained unaffected, the reaction times were substantially longer.

Reaction of the epoxyketone **12** with methylmagnesium chloride resulted in one diastereomer **13** (Scheme 13) as identified by ^{1}H-NMR spectroscopy. The reasoning for the high diastereoselectivities observed has been explained by examination of the intermediate complex **14** (Fig. 2). Hence the Grignard reagent coordinates with both the epoxide and carbonyl oxygens, resulting in alkyl attack by a second molecule of reagent at the exposed face.

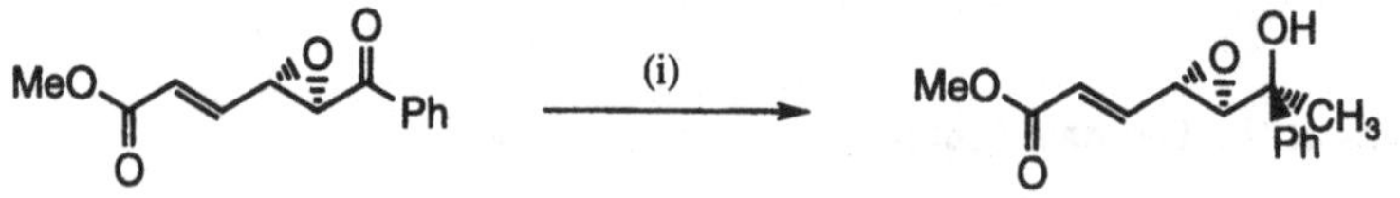

MeO O O Ph (i) MeO O O OH CH3 Ph

12 **13**

Scheme 13. *Reagents and Conditions:* (i) MeMgCl, 0 °C, 2 h, 72 %

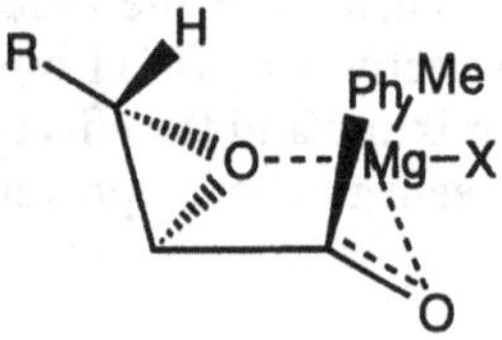

R H Ph Me O Mg–X O

Fig. 2. Transition state complex **14**

When using methyllithium instead of methylmagnesium chloride for the alkylation of **12**, the product is obtained in good yield (70 %); however the diastereoselective ratio drops to 5 : 1 in favour of alcohol **13**.

The Baeyer-Villiger reaction is well documented in the literature [28] and involves the oxidation of ketones using peroxy acids, such as *m*-chloroperoxybenzoic acid, to yield esters and lactones. The reactions are generally rapid and clean, giving high yields of desired product, though it is often necessary to add buffer to prevent side-product formation by transesterification. Application of this process to several substrates subsequent to the epoxidation reaction has proved extremely successful. A specific example is in the asymmetric synthesis of diltiazem, using KF and $NaHCO_3$ buffered *m*-CPBA oxidation conditions to prepare ester **16**, as shown in Scheme 14 [26].

Since the polyleucine epoxidation conditions are only favourable for highly electron-deficient unsaturated systems (i.e. ketones), use of the Baeyer-Villiger oxidation subsequent to the epoxidation reaction allows access to the optically active epoxyesters.

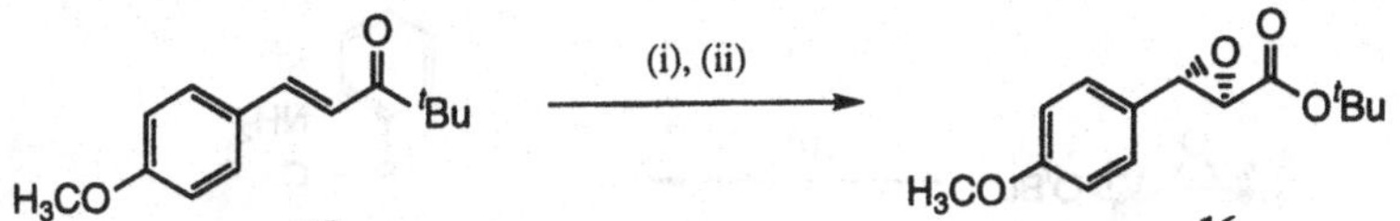

15 (i), (ii) 16

Scheme 14. *Reagents and Conditions:* (i) Immobilised poly-(L)-leucine, urea hydrogen peroxide, DBU, THF. (ii) KF, $NaCO_3$, *m*CPBA, CH_2Cl_2, 70% yield overall, > 96% e.e.

5.2 Reactions at the Epoxide Unit

Epoxides are extremely useful intermediates in organic synthesis since they react with a variety of nucleophiles suffering opening of the epoxide ring with retention or inversion of configuration at the carbon undergoing attack. Thus, the development of highly stereoselective methods for the synthesis of certain chiral epoxides, such as the methods under discussion, has enabled the asymmetric synthesis of a wide variety of 1,2-bifunctional compounds.

Epoxidation of amino chalcone **17**, followed by regioselective ring opening of the epoxide unit, demonstrates the formation of optically active lactam derivatives of type **18**, which are highly important structures for use within the pharmaceutical industry. The reaction proceeds in good yield and without loss of stereochemical integrity (Scheme 15).

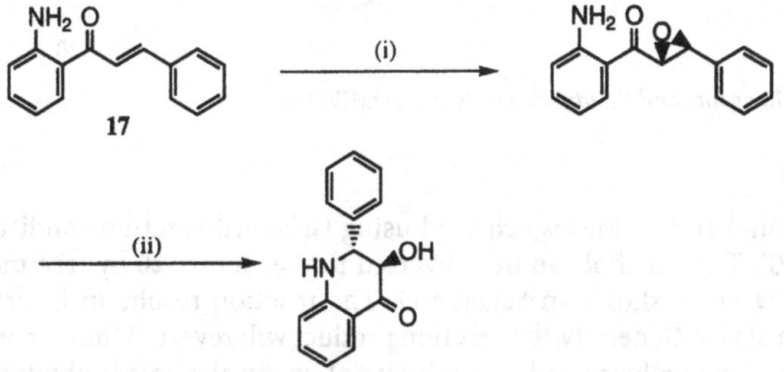

17 (i) (ii) 18

Scheme 15. *Reagents and Conditions:* (i) Immobilised poly-(L)-leucine, DBU, THF, 3 h. (ii) butanol, Δ, 15 min, 50% yield overall, > 99% e.e.

Returning to the enantioselective synthesis of the anti-angina drug diltiazem, an elegant example of oxirane ring opening with *retention* of configuration upon intermolecular nucleophilic attack at the stereogenic centre is observed, in the conversion of epoxyester **16** into the thioester **19** (Scheme 16) [26]. The stereochemical outcome of the latter reaction can be accounted for when considering the transition structure shown in Fig. 3. The mechanism of this epoxy-ring-opening reaction has been discussed by Schwartz et al. [30] and by Hashiyama et al. [31].

Payne rearrangements have been effectively demonstrated using derivatives of the epoxy chalcone products. Conversion of the epoxy ketone **2** to the 2,3-

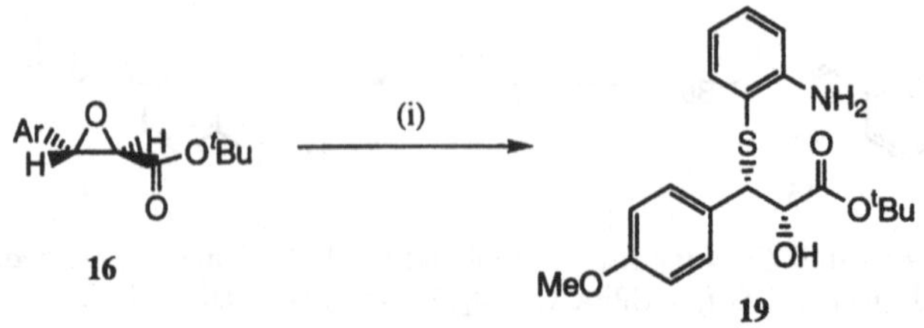

Scheme 16. *Reagents and Conditions:* (i) *o*-aminothiophenol, mesitylene, 90%

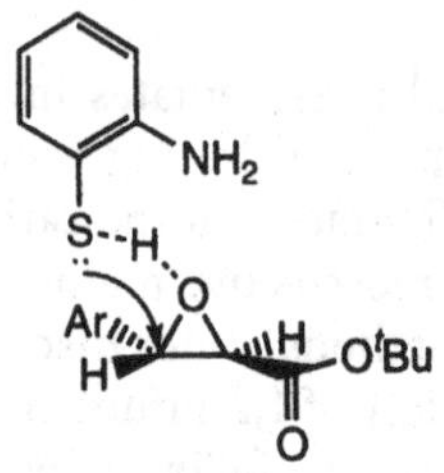

Fig. 3. Transition state complex

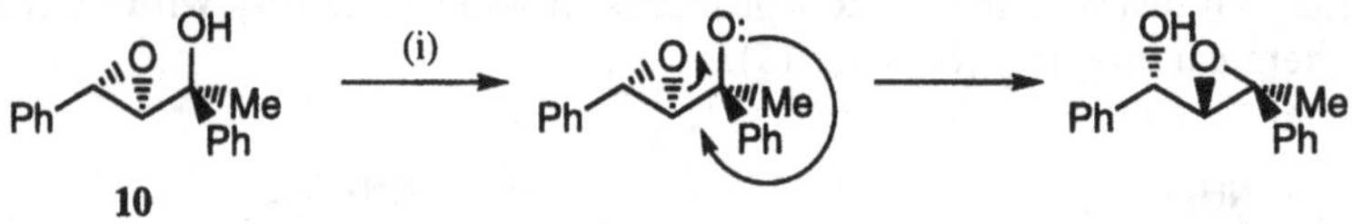

Scheme 17. *Reagents and Conditions:* (i) TBSCl, DMF, 71%

epoxy alcohol **10** (R=Me) is achieved using Grignard reaction conditions (see Scheme 10). This alcohol can be converted to the isomer **20** by treatment with TBSCl in DMF, as shown in Scheme 17. The reaction results in inverted configuration at C-2. Generally, the reaction product will revert to the starting material by the same pathway and so a mixture of epoxy alcohols is obtained. However, in this example, the more substituted, and so more thermodynamically favourable, epoxy alcohol **20** predominates.

A recent synthesis of the phenylisoserine side-chain of taxol is shown in Scheme 18. The enone **21** was obtained in high yield by condensation of benzaldehyde with pinacolone. Employing the non-aqueous two-phase epoxidation protocol, epoxide **22** was obtained in 76% yield and 94% e.e. Recrystallisation of the epoxide furnished the desired enantiomer in 97% e.e. Subsequent manipulations of the epoxy-ketone gave the taxol side-chain **23** with the required stereochemistry (Scheme 18).

The improved Juliá-Colonna epoxidation conditions have been successfully employed for poly-(D)-leucine.

Thus, epoxidation of the diene **6** using poly-(D)-leucine affords the epoxide ent-**12** in 90% yield and 95% e.e. (It is worth noting at this point that the per-

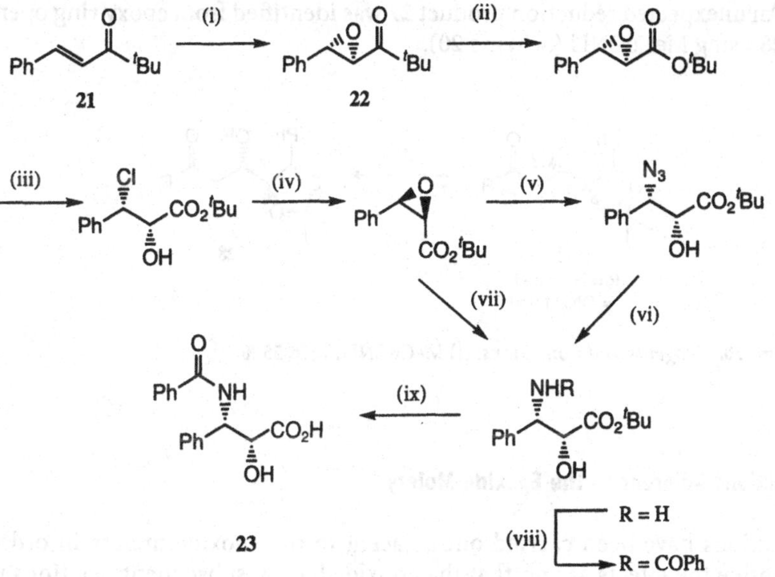

Scheme 18. *Reagents and Conditions:* (i) Immobilised poly-(L)-leucine, urea hydrogen peroxide, DBU, THF, 12 h, 76%, 94% e.e. (ii) mCPBA, CH_2Cl_2, 94%. (iii) HCl (g), CH_2Cl_2, 66%. (iv) Amberlite IRA-420 ($^-$OH), THF, 80%. (v) NaN_3, MeOH, H_2O, 94%. (vi) H_2, Pd/C, EtOAc. (vii) NH_3, MeOH. (viii) benzoyl chloride. (ix) trifluoroacetic acid, CH_2Cl_2, 74%

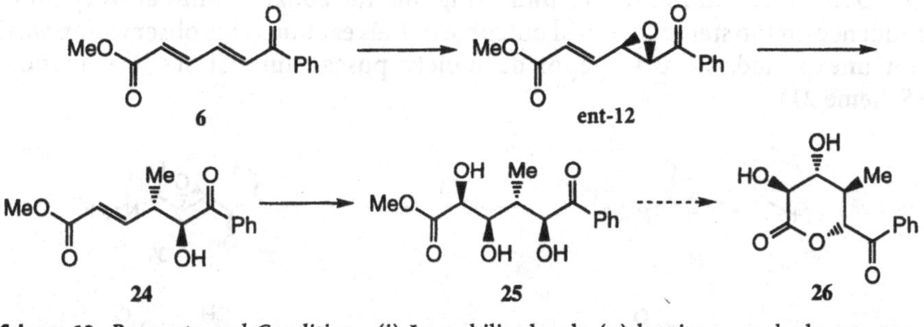

Scheme 19. *Reagents and Conditions*: (i) Immobilised poly-(D)-leucine, urea hydrogen peroxide, DBU, THF, 90%, 95% e.e. (ii) MeCuCNLi, THF, 30%. (iii)"super" AD mix β, 57%. (iv) base

centage yields and enantioselectivities observed using poly-(D)-leucine are comparable to those values observed when using poly-(L)-leucine (90% yield, 95% e.e.)). Subsequent opening of the epoxide moiety using a higher order organocuprate reagent (MeCuCNLi) yielded the secondary alcohol **24**, albeit in modest yield. Sharpless asymmetric dihydroxylation of the unsaturated ester gave the triol **25** in 67% yield and with a diastereoisomer ratio of 15:1. It is envisaged that base catalysed cyclisation of **25** will furnish the glycoside analogue **26** en route to a spongistatin synthon, (Scheme 19).

An unexpected reduction product **27** was identified from epoxy ring opening of **28** using MeCuCNLi (Scheme 20).

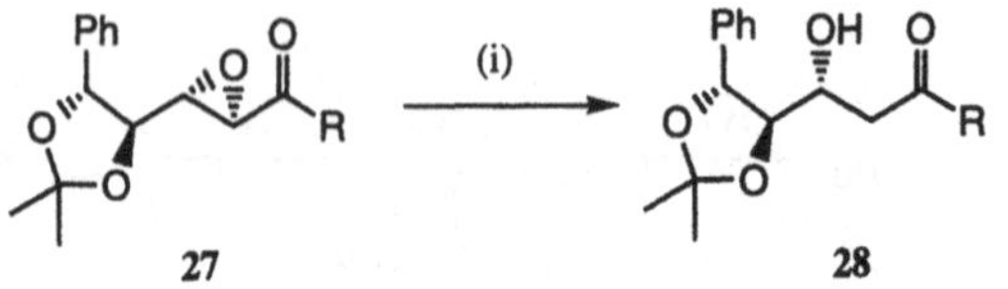

Ph O (i) Ph OH O

R R

27
R = Naphthyl
= ONaphthyl

28

Scheme 20. *Reagents and Conditions*: (i) MeCuCNLi, THF, 55%

5.3 Reactions Adjacent to the Epoxide Moiety

Reactions have been carried out adjacent to the epoxide moiety in order to examine the effects, if any, that the epoxide has on subsequent reactions with respect to the regio- and stereochemical outcome. Dihydroxylation using osmium tetraoxide and Sharpless asymmetric dihydroxylation reactions have been extensively studied using substrates **29** and **31**. Initial studies centred on the standard dihydroxylation conditions using *N*-methylmorpholine-*N*-oxide and catalytic osmium tetraoxide. The diastereomeric ratios were at best 3:2 for **29** and 2:1 for **31**, indicating that the epoxide unit had very little influence on the stereochemical outcome of the reaction. This observation was not unexpected, since the epoxide moiety poses minimal steric demands (Scheme 21).

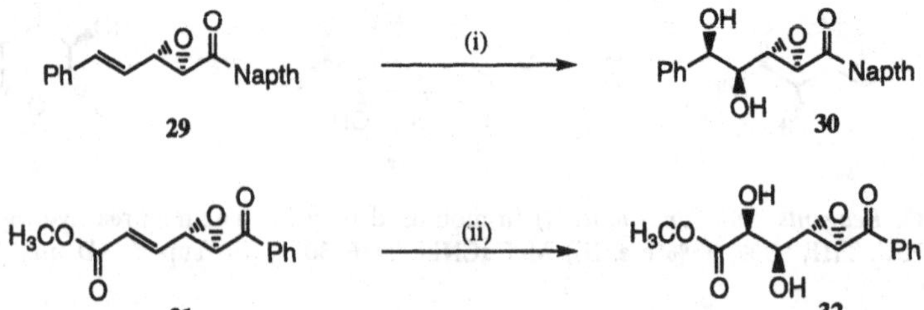

Ph Napth (i) OH O Ph Napth OH

29 **30**

H_3CO Ph (ii) H_3CO OH Ph O OH

31 **32**

Scheme 21. *Reagents and Conditions*: (i) "super" AD mix *β*, 95%. (ii) "super" AD mix *β*, 68%

In contrast, use of a "super" AD mix showed excellent levels of diastereoselectivity, such that diol **30** was obtained in a ratio of 38:1 and diolester **32** in a ratio of 15:1 (when using AD mix *β*). Use of AD mix *α* gave the expected (opposite) stereoselectivity.

6 Conclusions

The use of a polyamino acid such as polyleucine as a catalyst for the asymmetric epoxidation of α,β-unsaturated ketones is clearly established. The advantages and disadvantages of this methodology may be summarised as follows:

Advantages

- The catalyst is inexpensive and easy to prepare, on a large scale if required.
- The material is robust and is insensitive to the action of heat, light, oxygen, moisture etc.
- The catalyst is insoluble in the reaction media, hence may be recovered and reused.
- The oxidant (e.g. urea hydrogen peroxide) and the base (various) used in the reactions are readily available and cheap.
- Reactions are rapid (typical reaction time 0.5–5 h) and often highly enantioselective.
- The availability of enantiomeric catalysts (e.g. from (L) or (D)-leucine) ensures that epoxides of opposite configurations are readily accessed.

Disadvantages

- Relatively large amounts of catalyst are required (up to weight/weight with the substrate).
- Substrates that undergo asymmetric epoxidation are restricted to α,β-unsaturated ketones.

One or both of the disadvantages are likely to be overcome in due course. It is obvious that a clearer picture of the mechanism of the oxidation is mandatory before much progress can be made. Once it is understood how this very simple protein folds, in the presence of organic solvent, to form a chiral cavity or chiral surface that activates the peroxide and/or enone to accomplish the desired asymmetric oxidation then the reaction may be extended to other substrates, e.g. α, β-unsaturated esters, nitroalkenes, perhaps (under different conditions) electron-rich alkenes.

Furthermore, it is entirely reasonable that an understanding of the shapes of these simple polymers will allow discrete or radical modification and then extension of the range of oxidations to encompass Baeyer-Villiger reactions and sulfoxidations.

7 References

1. Wong CH, Whitesides GM (1994) Enzymes in organic synthesis, Pergamon, Oxford; Drauz K, Waldmann H (eds) (1995), Enzyme catalysis in organic synthesis, VCH, Weinheim
2. For an excellent review of the use of esterases and lipases see Faber K (1996), Biotransformations in Organic Chemistry, Springer-Verlag, Berlin

3. Servi S (1990) Synthesis: 1
4. Davies HG, Green RH, Kelly DR, Roberts SM (1989) Biotransformations in preparative organic chemistry. Academic Press, London, pp173–178 and references therein
5. For example see Dingler C, Ladner W, Krei GA, Cooper B, Hauer B (1996) Pestic Sci 46:33
6. Wan PWH, Roberts SM (1998) J Mol Cat B: Enzymatic 4:111
7. Hudlicky T, Abboud KA, Bolonik J, Maurya R, Stanton ML, Thorpe AJ (1996) J Chem Soc, Chem Commun :1717
8. For an introduction to the use of micro-organisms see Roberts SM, Turner NJ, Willetts AJ, Turner MK (1996) Introduction to biocatalysis using enzymes and micro-organisms. Chapter 2, Cambridge University Press
9. Mahnoudian M, Michael A (1993) J Biotechnol 27:173; Smet MJ de, Witholt B, Wynberg H (1989) Enzyme Microb Technol 5:352; Takahasi O, Umezawa J, Furuhshi K, Takagi M (1989) Tetrahedron Lett 30:1583
10. Kolb HC, Van Nieuwentize MS, Sharpless KB (1994) Chem Rev 94:2483
11. Larrow JF, Jacobsen EN (1994) J Am Chem Soc 116:12129; for a review of this and related methodology see Procter G (1996), Asymmetric synthesis. Oxford University Press, New York, pp 175–192
12. Banfi S, Colonna S, Molinari H, Juliá S, Guixer J (1984) Tetrahedron 40:5207 and references therein
13. Colonna S, Molinari H, Banfi S, Juliá S, Masana J, Alvarez A (1983) Tetrahedron 39:1635
14. Baures PW, Eggleston DS, Flisak JR, Gombatz K, Lantos I, Mendelson W, Remich JJ (1990) Tetrahedron Lett 31:6501
15. Flisak JR, Gombatz KJ, Holmes MM, Jarmas AA, Lantos I, Mendelson WL, Novack VJ, Remich JJ, Snyder L (1993) J Org Chem 58:6247
16. van Rensburg H, van Heerden PS, Bezuidenhoudt BCB, Ferreira D (1996) J Chem Soc, Chem Commun: 2747; Niel RJJ, van Heerden PS, van Rensburg H, Ferreira D (1998) Tetrahedron Lett 39:5623
17. Juliá S, Mosana J, Vega JC (1980) Angew Chem Int Ed Engl 19:929
18. Lasterra-Sánchez ME, Felfer U, Mayon P, Roberts SM, Thornton SR, Todd CJ (1996) J Chem Soc, Perkin Trans 1:343
19. Itsuno S, Sakkura M, Ito K (1990) J Org Chem 55:6047
20. Capp MW, Chen W-P, Flood RW, Liao YW, Roberts SM, Skidmore J, Smith JA, Williamson NM (1998) J Chem Soc Chen Commun 1159
21. Bentley P, Kroutil W, Littlechild JA, Roberts SM (1997) Chirality 9:198
22. Lasterra-Sánchez ME, Roberts SM (1997) Current Org Chem: 187–196, and references therein
23. Synthesis modified from Chemin D, Linstrumelle G (1994) Tetrahedron 50:5335
24. Kroutil W, Lasterra-Sánchez ME, Maddrell SJ, Mayon P, Morgan P, Roberts SM, Thornton SR, Todd CJ, Tüter M(1996) J Chem Soc, Perkin Trans 1:2837.
25. Kroutil W, Mayon P, Lasterra-Sánchez ME, Maddrell SJ, Roberts SM, Thornton SR, Todd CJ, Tüter M (1996) J Chem Soc, Chem Commun :845
26. Adger B, Bergeron S, Cappi MW, Flowerdew BE, Jackson M, McCague R, Nugent TC, Roberts SM (1997) J Chem Soc, Perkin Trans 1:3501 and unpublished results
27. Allen JV, Cappi MW, Kary PD, Roberts SM, Williamson NM, Wu LE (1997) J Chem Soc, Perkin Trans.1:3297 and unpublished results
28. Krow GR (1993) Org React vol 43, p 251, (NY)
29. Chen W-P, Egar AL, Hursthouse MB, Abdul Malik KM, Mathews JE, Roberts SM (1998) Tetrahedron Lett 39: in press
30. Schwartz A, Madan PB, Mohacsi E, O'Brien JP, Todaro LJ, Coffen DL (1992) J Org Chem 57:851
31. Hashiyama T, Inoue H, Konda M, Takeda M (1984) J Chem Soc, Perkin Trans 1:1725

Epoxide Hydrolases and Their Synthetic Applications

Romano V. A. Orru[1] · Alain Archelas[2] · Roland Furstoss[2] · Kurt Faber[1]

[1] Institute of Organic Chemistry, Graz University of Technology, Stremayrgasse 16, A-8010 Graz, Austria. *E-mail: faber@orgc.tu-graz.ac.at*

[2] Groupe Biocatalyse et Chimie Fine, ERS 157 CNRS, Faculté des Sciences de Luminy, Case 901, 163 Avenue de Luminy, F-13288 Marseille Cedex 9, France

Chiral epoxides and 1,2-diols, which are central building blocks for the asymmetric synthesis of bioactive compounds, can be obtained by using enzymes – i.e. epoxide hydrolases – which catalyse the enantioselective hydrolysis of epoxides. These biocatalysts have recently been found to be more widely distributed in fungi and bacteria than previously expected. Sufficient sources from bacteria, such as *Rhodococcus* and *Nocardia* spp., or fungi, as for instance *Aspergillus* and *Beauveria* spp., have now been identified. The reaction proceeds via an S_N2-specific opening of the epoxide, leading to the formation of the corresponding *trans*-configured 1,2-diol. For the resolution of racemic monosubstituted and 2,2- or 2,3-disubstituted substrates, various fungi and bacteria have been shown to possess excellent enantioselectivities. Additionally, different methods, which lead to the formation of the optically pure product diol in a chemical yield far beyond the 50% mark (which is intrinsic to classic kinetic resolutions), are discussed. In addition, the use of non-natural nucleophiles such as azides or amines provides access to enantiomerically enriched vicinal azido- and amino-alcohols. The synthetic potential of these enzymes for asymmetric synthesis is illustrated with recent examples, describing the preparation of some biologically active molecules.

Keywords: Epoxide, Vicinal diol, Epoxide hydrolase, Biocatalysis, Enantio-convergent.

1 **Introduction** 146

2 **Epoxide Hydrolases in Nature** 149

2.1 Biological Role 149
2.2 Distribution in Nature 150

3 **Epoxide Hydrolase Mechanism** 150

4 **Structural Features of Epoxide Hydrolases** 152

5 **Screening of Microorganisms for Epoxide Hydrolases** 152

6 **Biohydrolysis of Epoxides** 154

6.1 Classic Kinetic Resolution 154
6.1.1 By Fungal Cells 154
6.1.2 By Bacterial Cells 155
6.2 Asymmetrization of *meso*-Epoxides 157
6.3 Deracemization Methods 157

7 **Use of Non-Natural Nucleophiles** 160

Advances in Biochemical Engineering/
Biotechnology, Vol. 63
Managing Editor: Th. Scheper

8 Applications to Natural Product Synthesis 161

9 Summary and Outlook 165

10 References 165

1 Introduction

Chiral epoxides and vicinal diols (employed as their corresponding cyclic sulfate or sulfite esters as reactive intermediates) are extensively employed high-value intermediates in the synthesis of chiral compounds due to their ability to react with a broad variety of nucleophiles (Schemes 1 and 2). In recent years, extensive efforts have been devoted to the development of chemo-catalytic methods for their production [1, 2]. Thus, the Sharpless methods allowing for the asymmetric epoxidation of allylic alcohols [3] and the asymmetric dihydroxylation of alkenes [2] are now widely applied reliable procedures. In addition, asymmetric catalysts for the epoxidation of non-functionalized olefins [4–6] have been developed more recently. Although high stereoselectivity has been achieved for the epoxidation of *cis*-alkenes, the results obtained with *trans*- and terminal olefins were less satisfactory using the latter method. More recently, new chemical methods allowing for the desymmetrization of achiral epoxides have been reviewed [7]. Furthermore, a process has been described for the highly selective hydrolysis of terminal epoxides based on a kinetic resolution catalysed by cobalt-salen complexes [8].

On the other hand, a number of biocatalytic methods provide a useful arsenal of methods as valuable alternatives to the above-mentioned techniques [9–14]. One is where prochiral or racemic synthetic precursors of epoxides, such as halohydrins, can be asymmetrized or resolved using hydrolytic enzym-

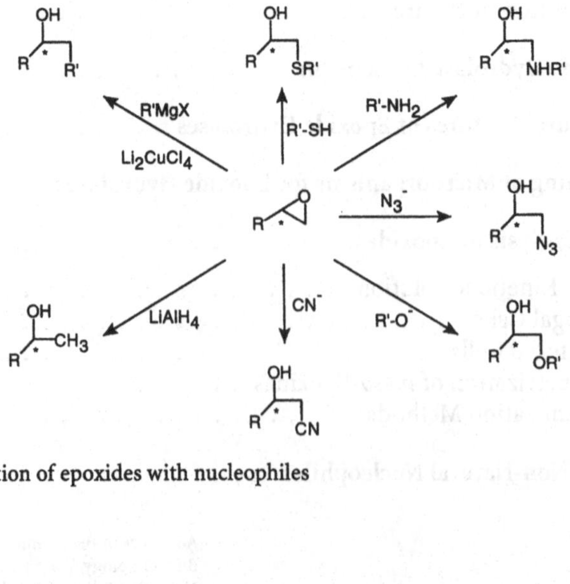

Scheme 1. Reaction of epoxides with nucleophiles

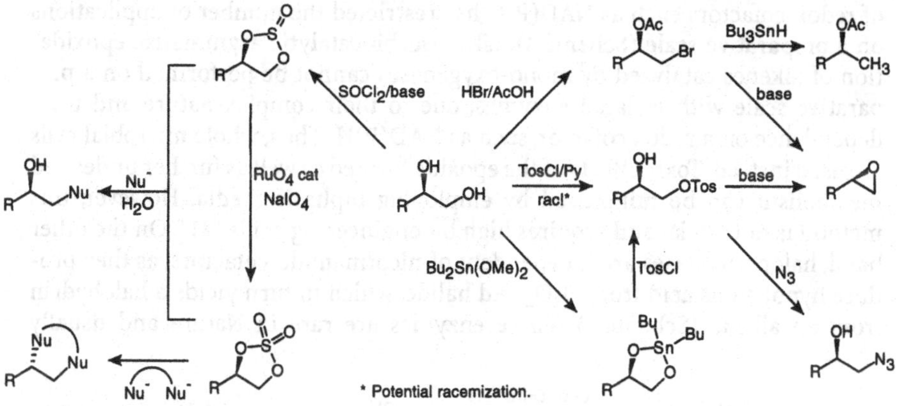

Scheme 2. Syntheses from chiral 1,2-diols

es [15, 16]. In particular esterases and lipases have been used for such a enantioselective ester hydrolysis or esterification (Scheme 3). This methodology is well developed and high selectivities have been achieved in particular for esters of secondary alcohols, but it is impeded by the requirement of regioisomerically pure halohydrins. Alternatively, α-haloacid dehalogenases catalyze the S_N2-displacement of a halogen atom at the α-position of carboxylic acids with a hydroxy function. This process leads to the formation of the corresponding α-hydroxy acid with inversion of configuration [17]. However, α-haloacid dehalogenation incurs two drawbacks, i.e. the instability of the substrates (particularly α-bromoacids) in aqueous systems and the limited substrate tolerance, as only short-chain haloacids are accepted [18].

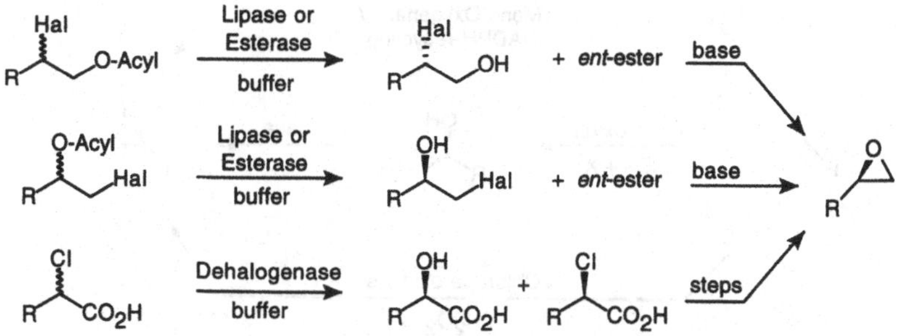

Scheme 3. Enzymatic syntheses of epoxides using hydrolytic enzymes

Asymmetric biocatalytic reduction of α-ketoacids [19] or α-keto-alcohols [20] using D- or L-lactate dehydrogenase or glycerol dehydrogenase, respectively, provides access to chiral α-hydroxyacids or 1,2-diols. The latter can be converted to the corresponding epoxide using conventional chemical methodology. Although excellent selectivities are generally achieved, the need for the recycling of redox-cofactors such as NAD(P)H has restricted the number of applications on a preparative scale (Scheme 4). Likewise, biocatalytic asymmetric epoxidation of alkenes catalyzed by mono-oxygenases cannot be performed on a preparative scale with isolated enzymes, due to their complex nature and their dependence on a redox cofactor, such as NAD(P)H. Thus, whole microbial cells are used instead. Toxic effects of the epoxide formed as well as further undesired metabolism can be minimised by employing biphasic media. However, this method is not trivial and requires high bioengineering skills [21]. On the other hand, haloperoxidases are independent of nicotinamide-cofactors, as they produce hypohalous acid from H_2O_2 and halide, which in turn yields a halohydrin from an alkene (Scheme 4). These enzymes are rare in Nature and usually

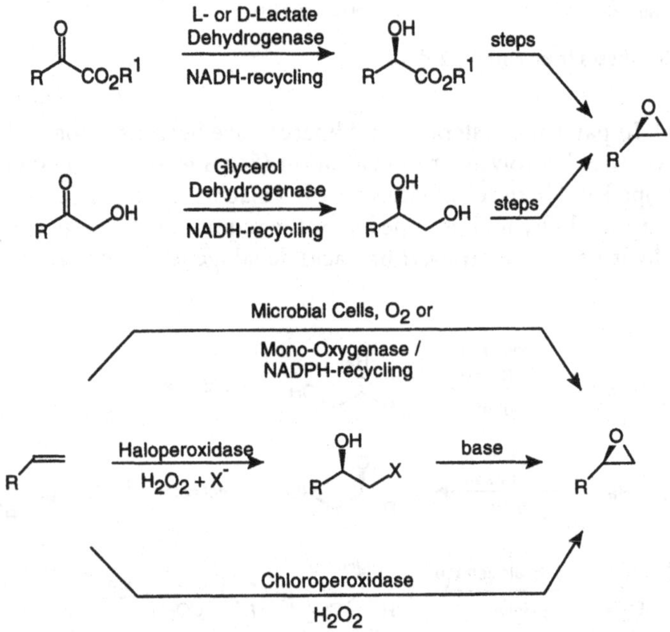

Scheme 4. Enzymatic syntheses of epoxides using dehydrogenases, mono-oxygenases and peroxidases

exhibit low selectivities due to the fact that the formation of halohydrins can take place not only in the active site of the enzyme but also without direct enzyme catalysis [22]. Similar low selectivities have been observed with halohydrin epoxidases, which act like a "biogenic chiral base" by converting a halohydrin into the corresponding epoxide [23]. In contrast, peroxidases, such as chloroperoxidase, are cofactor-independent and thus can also be used in isolated form for the enzymatic epoxidation of alkenes. Although excellent selectivities were obtained with internal *cis*-olefins, long-chain substrates and terminal alkenes were unreactive [24].

A valuable alternative to the above existing methods is the use of cofactor-independent epoxide hydrolases [EC 3.3.2.X]. Enzymes from mammalian sources – such as rat liver tissue – have been investigated in great detail during detoxification studies [25]. However, biotransformations on a preparative scale were hampered by the limited supply of enzyme and the described examples rarely surpass the mmol range [26]. In contrast, highly selective epoxide hydrolases were identified from microbial sources, which allows for an (almost) unlimited supply of these enzymes for preparative-scale applications. These interesting biocatalysts have recently received considerable attention and some of their merits have been reviewed [27–29].

2 Epoxide Hydrolases in Nature

2.1 Biological Role

In eukaryotes, such as mammals and fungi, epoxide hydrolases play a key role in the metabolism of xenobiotics, in particular of aromatic systems [30, 31]. On the other hand, in prokaryotes (e.g. bacteria) these enzymes are essential for the utilization of alkenes as carbon-source. In general, aromatics can be metabolized via two different pathways (Scheme 5): (i) dioxetane formation via dioxyge-

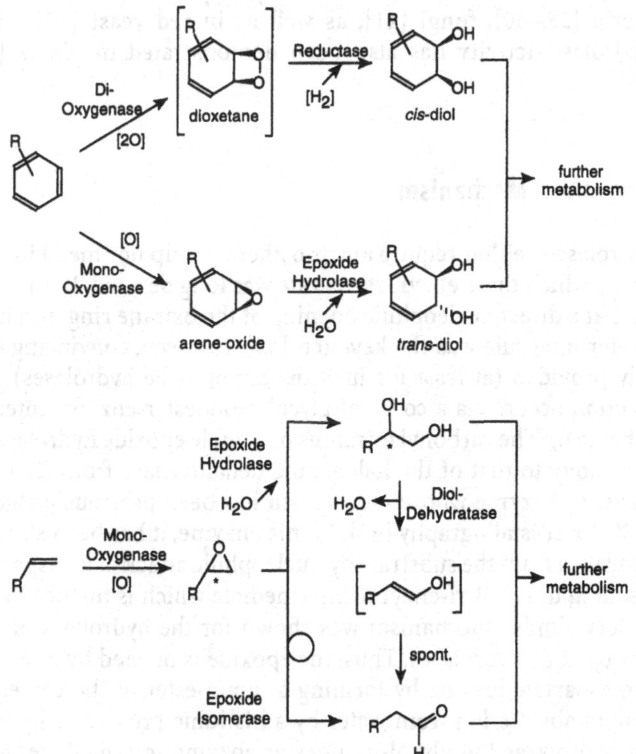

Scheme 5. Involvement of epoxide hydrolases in the biodegradation of aromatics and alkenes

nase-catalyzed (cyclo)addition of molecular oxygen onto the C=C bond yields a (putative) dioxetane species, which is then detoxified via reductive cleavage of the O-O bond furnishing a physiologically more innocuous *cis*-1,2-diol; (ii) formation of a highly reactive arene oxide via the introduction of a single O atom (from molecular oxygen) into the aromatic system is catalyzed by a mono-oxygenase. The latter epoxy-species is further metabolized via hydrolysis catalyzed by an epoxide hydrolase to yield a *trans*-1,2-diol. In lower organisms, alkenes can be metabolized in an analogous fashion, i.e. via an epoxide intermediate. This is in turn hydrolyzed to the corresponding 1,2-diol by an epoxide hydrolase. The latter product can be easily degraded either by oxidation or by elimination of water under catalysis of a diol dehydratase, yielding an aldehyde [32]. Alternatively, aldehydes are obtained via direct rearrangement of the epoxide catalyzed by an epoxide isomerase [33].

2.2 Distribution in Nature

For a long time it was generally accepted that epoxide hydrolases are predominantly found in mammals [25, 26], although epoxide hydrolase activities had been detected in bacteria [34, 35] or fungi [36, 37] quite some years ago. This view was too simplistic, since these enzymes have now been detected in many bacteria [38–40], fungi [41], as well as in red yeast [42]. Moreover, epoxide hydrolase activity has also been demonstrated in plants [43] and insects [44].

3 Epoxide Hydrolase Mechanism

Epoxide hydrolases neither require any prosthetic group nor metal ion, and the mechanism by which these enzymes operate was long debated. It was previously assumed that a direct nucleophilic opening of the oxirane ring by a histidine-activated water molecule was the key-step [45]. However, convincing evidence was recently provided (at least for mammalian epoxide hydrolases), showing that the reaction occurs via a covalent glycol-monoester-enzyme intermediate [46, 47] (Scheme 6). The carbonyl terminus of soluble epoxide hydrolase showed striking homology to that of the haloalkane dehalogenase from *Xanthobacter autotrophicus*, an enzyme whose mechanism has been previously studied and verified by X-ray crystallography [48]. In this enzyme, it has been shown that a halide is displaced from the substrate by nucleophilic attack of an aspartate residue, thus leading to an alkyl-enzyme intermediate which is further hydrolyzed [49, 50]. A very similar mechanism was shown for the hydrolysis of epoxides catalyzed by epoxide hydrolases. Thus, the epoxide is opened by a nucleophilic attack of an aspartate residue by forming a mono-ester of the corresponding 1,2-diol. Proton abstraction from water by a histidine provides a hydroxyl ion which then hydrolyzes the glycol-monoester-enzyme intermediate. As a consequence, the epoxide is generally opened in a *trans*-specific fashion with one

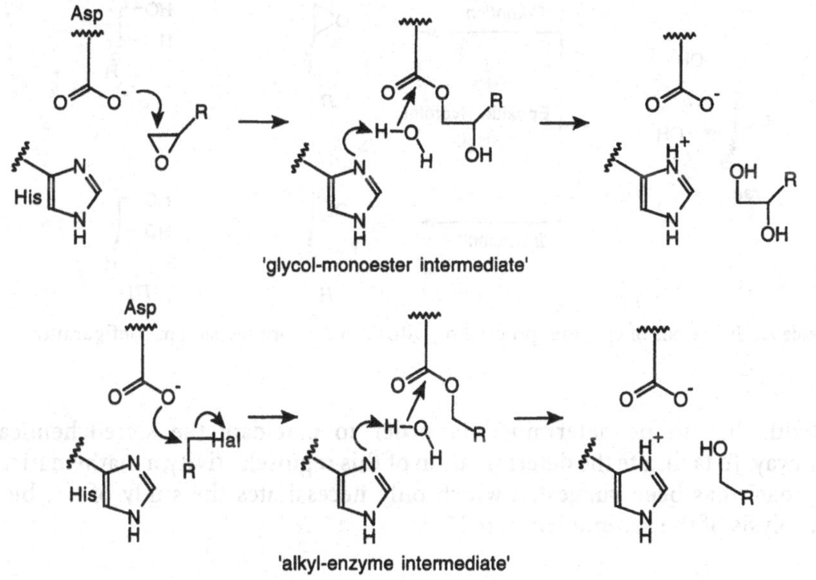

'glycol-monoester intermediate'

'alkyl-enzyme intermediate'

Scheme 6. Mechanism of microsomal epoxide hydrolase and of haloalkane dehalogenase

oxygen from water being incorporated into the product diol [51]. For instance, (±)-*trans*-epoxysuccinate was converted to *meso*-tartrate by an epoxide hydrolase isolated from *Pseudomonas putida* [51]. In a complementary fashion, *cis*-epoxysuccinate gave D- and L-tartrate (with a *Rhodococcus* sp.) albeit in low optical purity [52]. In addition, it was shown by $^{18}OH_2$-labelling experiments that only one O-atom originates from water when both mammalian or bacterial epoxide hydrolases [53] or whole fungal cells [54] were used as catalysts.

Although two cases for reactions proceeding via a formal *cis*-hydration process have been reported [55, 56], they seem to be rare exceptions and – given the present knowledge of the enzyme mechanism – attempts to explain this phenomenon remain rather speculative [56]. It is interesting to note that β-glycosidase acts via formation of a covalent glycosyl-enzyme intermediate by retaining the configuration at the anomeric centre [57]. A mechanistic similarity to epoxide hydrolase-catalyzed hydrolysis may exist.

The above-mentioned facts have important consequences on the stereochemical outcome of the kinetic resolution of asymmetrically substituted epoxides. In the majority of kinetic resolutions of esters (e.g. by ester hydrolysis and synthesis using lipases, esterases and proteases) the absolute configuration at the stereogenic centre(s) always remains the same throughout the reaction. In contrast, the enzymatic hydrolysis of epoxides may take place via attack on either carbon of the oxirane ring (Scheme 7) and it is the structure of the substrate and of the enzyme involved which determine the regioselectivity of the attack [53, 58–61]. As a consequence, the absolute configuration of *both the product and substrate* from a kinetic resolution of a racemic

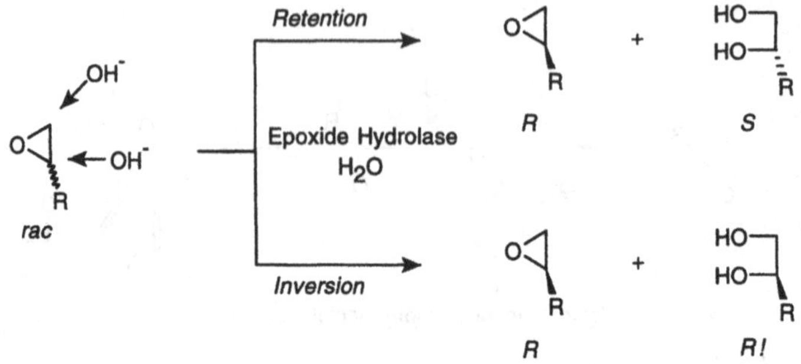

Scheme 7. Hydrolysis of epoxides proceeding with retention or inversion of configuration

epoxide has to be determined in order to elucidate the stereochemical pathway. To facilitate the determination of this regioselectivity, a mathematical approach has been suggested which only necessitates the study of the biohydrolysis of the racemic mixture [57].

4 Structural Features of Epoxide Hydrolases

In the past few years several membrane-bound and soluble epoxide hydrolases from various origins have been purified and at least partially sequenced. Some of them have also been cloned and overexpressed. This is the case for the soluble epoxide hydrolase from rat liver which has been overexpressed in *E. coli* [63, 64]. As pointed out previously, this enzyme (as well as the rat liver microsomal epoxide hydrolase) was shown to share amino acid sequence similarity to a region around the active site of a bacterial haloalkane dehalogenase [65], an enzyme with known three dimensional structure which belongs to the family of α/β-hydrolase fold enzymes [66]. Rat soluble epoxide hydrolase is a dimer where each monomer has two complete structural units, both with a distinct active site. The epoxide hydrolase activity is known to be located in the C-terminal unit, while the function of the N-terminal unit remains unknown. The C-and N-terminal units are linked by a segment of 18 residues. Comparison of the amino acid sequence to those of the related bromoperoxidase and haloalkane dehalogenase (whose X-ray structures are known) resulted in a model generated by atom-replacement structure-analogy [67]. Interestingly, these enzymes seem to bear a lid close to the active site similar to lipases. Further results – i.e. the first epoxide hydrolase X-ray structure – are obviously necessary in order to get a better understanding of the catalytic machinery of these enzymes.

5 Screening of Microorganisms for Epoxide Hydrolases

In spite of the potential interest of such enzymes for fine chemical synthesis, it was only recently that a detailed search for epoxide hydrolases from microbial sources has been undertaken by the groups of Furstoss [58, 68] and Faber [14, 52, 69], bearing in mind that the use of microbial enzymes allows an (almost) unlimited supply of biocatalyst. The screening was based on the following considerations: on the one hand, the catabolism of alkenes most likely implies the hydrolysis of an epoxy-intermediate and, on the other hand, efficient detoxification of the highly reactive epoxide is achieved via hydrolysis. Thus, fungi known to afford enantiopure vicinal diols from olefinic substrates were preferentially selected for screening. In a related approach, it was anticipated that highly selective epoxide hydrolases could be found, particularly in those bacterial strains that are known for the asymmetric epoxidation of alkenes. As a consequence, strains were selected after a careful literature search based on the above criteria [70].

Furthermore, a search for bacterial epoxide hydrolases was triggered by an unexpected result obtained during a study on the chemoselective hydrolysis of nitriles employing a crude enzyme preparation derived from *Rhodococcus* sp. [71, 72]. In this experiment an undesired enzymatic hydrolysis of an epoxide moiety was observed. Thus, a set of three representative substrates, a mono-substituted (**1a**), a 2,2-disubstituted (**2a**), and a 2,3-disubstituted (**3a**) oxirane (Chart 1) were subjected to microbial hydrolysis employing lyophilized whole bacterial cells in Tris-buffer at pH 7–8 [14, 62, 73]. The three standard epoxides were generally hydrolyzed with good activity (Table 1). However, small epoxides (**1b, 1c**), those having sterically hindered epoxide moieties (**1d, 4**) or *meso*-substrates (**5–8**) were not hydrolyzed by all *Rhodococcus* spp. [14]. Interestingly, *Corynebacterium glutamicum* and three *Pseudomonas* strains did not show any epoxide hydrolase activity at all.

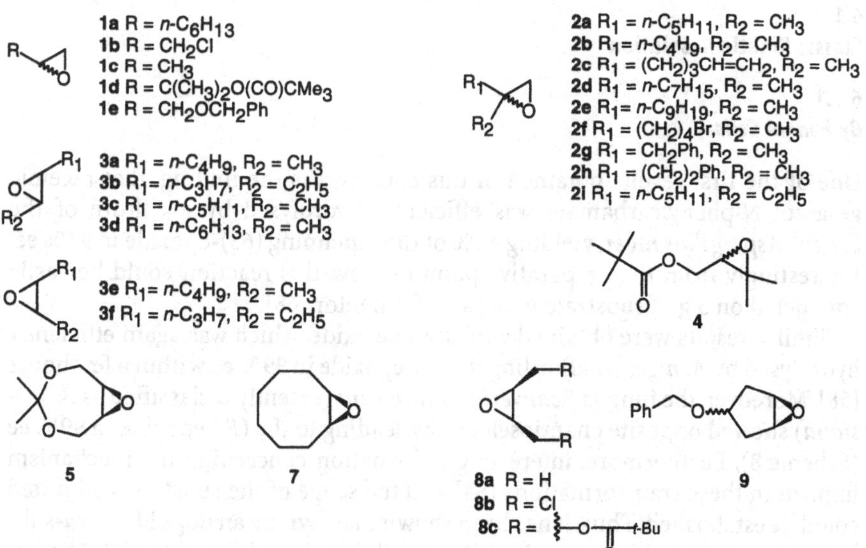

Chart 1. Substrates and non-substrates for bacterial epoxide hydrolases

Table 1. Screening of bacterial strains for activity[a]

Microorganism	(±)-**1a**	(±)-**2a**	(±)-**3a**	Reference
Rhodococcus sp. NCIMB 11215	+	+	n.d.	[14]
Rhodococcus sp. NCIMB 11216	+	+	+	[14, 62, 73]
Rhodococcus sp. NCIMB 11540	+	+	n.d.	[14]
Rhodococcus equi IFO 3730	+	+	+	[62, 73]
Rhodococcus ruber DSM 43338	+	+	+	[62, 73]
Nocardia H8	+	+	+	[62, 73]
Nocardia TB1	+	+	+	[62, 73]
Nocardia EH1	+	+	+	[62, 73]
Arthtrobacter sp. DSM 312	+	±	±	[62]
Mycobacterium paraffinicum NCIMB 10420	+	+	+	[62, 73]

[a] + = activity, – = no activity, ± = some activity, n.d. = not determined.

It is noteworthy that, in contrast to mammalian systems, the majority of bacterial strains exhibited sufficient activity even when the cells were grown under non-optimized conditions. Since enzyme induction is still a largely empirical task, cells were grown on standard media in the absence of inducers. Furthermore, all attempts to induce epoxide hydrolase activity in *Pseudomonas aeruginosa* NCIMB 9571 and *Pseudomonas oleovorans* ATCC 29347 by growing the cells on an alkane (decane) or alkene (1-octene) as the sole carbon source failed [27].

6 Biohydrolysis of Epoxides

6.1 Classic Kinetic Resolution

6.1.1 *By Fungal Cells*

One of the first results obtained in this context was the finding, that racemic geraniol *N*-phenylcarbamate was efficiently hydrolysed by a culture of the fungus *Aspergillus niger*, yielding 42% of the remaining (6*S*)-epoxide in 94% ee. Interestingly from the preparative point of view, this reaction could be easily conducted on 5 g of substrate using a 7 l fermentor [68].

Similar results were obtained with styrene oxide, which was again efficiently hydrolysed by *A. niger* by affording the (*S*)-epoxide in 99% ee within a few hours [58]. Moreover, the fungus *Beauveria sulfurescens* (recently reclassified as *B. bassiana*) showed opposite enantioselectivity, leading to the (*R*)-epoxide in 99% ee (Scheme 8). Furthermore, interesting information concerning the mechanism implied in these transformations [54] and the scope of the substrates admitted could be established. Thus, it has been shown that styrene analogs like *para*-substituted styrene oxide derivatives [74] or *β*-substituted derivatives [75] were accomodated by one or both of the fungi.

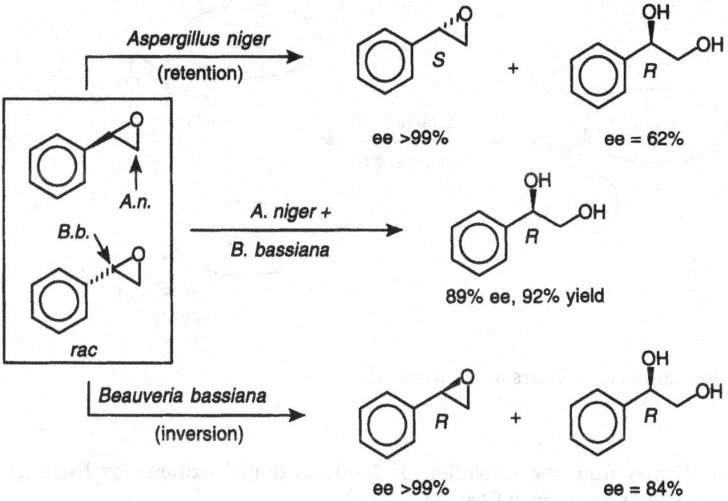

Aspergillus niger (retention) → S, ee >99% + R diol, ee = 62%
A.n. / B.b. / rac
A. niger + B. bassiana → R, 89% ee, 92% yield
Beauveria bassiana (inversion) → R, ee >99% + R diol, ee = 84%

Scheme 8. Resolution and deracemization of styrene oxide by fungal cells

In addition, racemic epoxyindene was rapidly hydrolyzed when submitted to a culture of *Beauveria sulfurescens*, leading to a 20% yield of recovered enantiomerically pure (ee >98%) (1*R*,2*S*)-epoxide, and to a 48% yield of the corresponding (1*R*,2*R*)-*trans*-diol showing 69% ee [76]. Since this diol has the absolute configuration required for the synthesis of the HIV protease inhibitor indinavir, this might be a good way to prepare this key-chiral building block intermediate by optimizing this biotransformation. In a more extensive study [77, 78], 80 fungal strains were evaluated for their ability to hydrolyze racemic epoxyindene enantioselectively. In a similar fashion, epoxydihydronaphtalene was hydrolyzed to the (1*R*,2*R*)-diol in excellent enantiomeric purity [76].

Many of these fungal epoxide hydrolases were found to be soluble enzymes, which could be conveniently stored in lyophilized form in a refrigerator without noticeable loss of activity. Thus, an easy-to-use water-soluble catalyst was developed which offered a solution to the problems usually encountered when an insoluble whole-cell mycelium was employed [79, 80].

6.1.2
By Bacterial Cells

The use of bacteria for preparative biotransformations is particularly attractive for the following reasons: (i) they do not tend to form dense mycelia, which may impede agitation of large-scale reactions when whole-cell (fungal) systems are employed and (ii) cloning of bacterial enzymes is generally less problematic.

However, disappointingly low selectivity was observed with monosubstituted aliphatic epoxides such as 1-epoxyoctane (**1a**, $E<5$) or benzyl glycidyl ether (**1e**, $E<2$) [51, 55]. On the other hand, sterically more demanding 2,2-disubstituted oxiranes (**2**) turned out to be much better substrates (Scheme 9, Table 2).

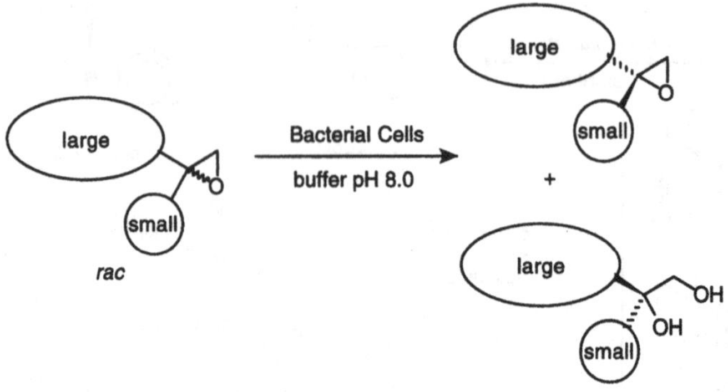

Scheme 9. Resolution of oxiranes by bacterial cells

Table 2. Selectivities from the resolution of 2,2-disubstituted oxiranes by bacterial cells (substrate structures are given in Chart 1)

Substrate	Biocatalyst	Selectivity (E)
2a	*Nocardia* sp. EH1	>200
2b	*Nocardia* sp. TB1	>200
2c	*Nocardia* sp. EH1	>200
2d	*Rhodococcus equi* IFO 3730	>200
2e	*Mycobacterium paraffinicum* NCIMB 10420	>200
2f	*Nocardia* sp. H8	>200
2g	*Nocardia* sp. EH1	123

Especially the substrates bearing a straight alkyl chain, even when functionalized with a double bond (**1c**) [81] or a terminal bromo-group (**1f**) [82], were transformed with virtually absolute enantioselectivity. As a consequence, the reactions ceased and did not proceed beyond a conversion of 50%. Interestingly, the enantiopreference was found to depend on the substrate structure, but not on the strain used. Thus, the (*R*)-enantiomer was preferred for substrates (±)-**1** (albeit at low selectivity), but the (*S*)-enantiomer was faster transformed for (±)-**2** [73]. When the epoxide bears a synthetically useful phenyl moiety (mimicking a masked carboxyl function) at the ω-position of the alkyl chain (**2g**), the selectivity was slightly reduced but still in a useful range ($E=123$) [82, 83]. Unexpectedly, when the carbon chain was extended by an additional CH_2-unit (**2h**) the selectivity declined ($E=13$). From earlier results with *Rhodococcus* sp. NCIMB 11216 it was concluded that the enantioselectivity largely depends on the relative difference in size of the two alkyl groups (Table 3). Thus increasing the difference in the relative size of the two alkyl substituents resulted in enhanced selectivities. The fact that the substrates bearing a phenyl group behave differently might be attributed to electronic effects. Further research on this aspect is necessary since the structure of the active site of the enzyme is still largely unknown. However, it should be noted that more recently a highly selective

Table 3. Selectivities from the resolution of 2,2-disubstituted oxiranes by whole cells of *Rhodococcus* sp. NCIMB 11216 (substrate structures are given in Chart 1)

Substrate	Small Substituent	Large Substituent	Selectivity *(E)*
1a	H	C_6H_{13}	2.8[a]
2a	CH_3	C_5H_{11}	105
2l	C_2H_5	C_5H_{11}	7
2d	CH_3	C_7H_{15}	125
2e	CH_3	C_9H_{19}	>200
2g	CH_3	CH_2-Ph	111
2h	CH_3	$(CH_2)_2$-Ph	9.5

[a] With opposite absolute configuration as shown in Scheme 9.

bacterial strain for all of these transformations has been isolated (Table 2) and all biohydrolyses can be performed on multigram scale.

6.2 Asymmetrization of *meso*-Epoxides

The asymmetrization of a *meso*-epoxide via regioselective attack at one of the (enantiomeric) stereocentres of the oxirane would be an elegant application of epoxide hydrolases since it leads to a single *trans*-diol in 100% theoretical yield. Such asymmetrization reactions have been demonstrated with epoxide hydrolases from mammalian origin which afforded the enantiomerically enriched corresponding diol [84, 85]. Unfortunately, only few of such reactions have been reported with microbial enzymes. For instance, cyclohexene oxide was hydrolyzed using *Corynesporium cassiicola* cells yielding *trans*-cyclohexane-1,2-diol with disappointing low ee (27%) [86]. It was only due to further metabolism involving an oxidation-reduction sequence by dehydrogenases present in the cells that the formed diol was transformed to optically pure (*S*,*S*)-cyclohexane-1,2-diol. In a related experiment, asymmetric hydrolysis of *cis*-epoxysuccinate using a crude enzyme preparation derived from *Rhodococcus* sp. led to D- and L-tartaric acid in almost racemic form [52]. Similar discouraging results were obtained using baker's yeast [87]. Obviously, microbial enzymes able to achieve this type of reaction in a satisfactory fashion are still to be discovered.

6.3 Deracemization Methods

In contrast to the asymmetrization of *meso*-epoxides, the kinetic resolution of racemic epoxides by whole fungal and bacterial cells has proven to be highly selective (see above). These biocatalysts supply both the unreacted epoxide enantiomer and the corresponding vicinal diol in high enantiomeric excess. This so-called classic kinetic resolution pattern of the biohydrolysis is often regarded as a major drawback since the theoretical chemical yield can never exceed 50% based on the racemic starting material. As a consequence, methods that offer a solution to this intrinsic problem are highly desirable. Several procedures have been reported in the last few years.

For instance, styrene oxide was resolved by whole cells of *Aspergillus niger* and *Beauveria bassiana* via two different pathways showing matching enantio- and regioselectivities with excellent results (Scheme 8). Combination of the two biocatalysts employing a deracemization process in a single reactor led to (*R*)-phenylethane-1,2-diol as the sole product in 98% ee and 85% isolated yield [58].

Another strategy for the achievement of an enantioconvergent process was set up using the combination of bio- and chemo-catalysis [82, 83, 88, 89]. For instance, 2,2-disubstituted epoxides are selectively resolved by lyophilized whole cells of *Nocardia* sp. The biohydrolysis proceeds via attack at the less substituted C-atom with excellent regioselectivity thus leading to *retention* of configuration at the stereogenic centre. On the other hand, acid-catalyzed hydrolysis of such epoxides usually proceeds at the more substituted oxirane carbon with *inversion.* Careful combination of both catalytic steps (Scheme 10) in a resolution-inversion sequence yields the corresponding (*S*)-1,2-diols in virtually enantiopure form and in high yields (>90%)[81, 82]. In a similar fashion, racemic *para*-nitrostyrene oxide was deracemized by using a crude enzyme extract from *Aspergillus niger* (Scheme 11). In this case a water:DMSO solvent

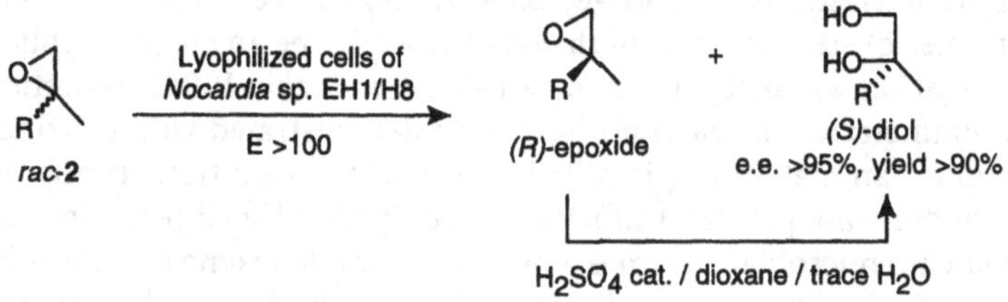

Scheme 10. Resolution-inversion sequence for the deracemization of 2,2-disubstituted oxiranes

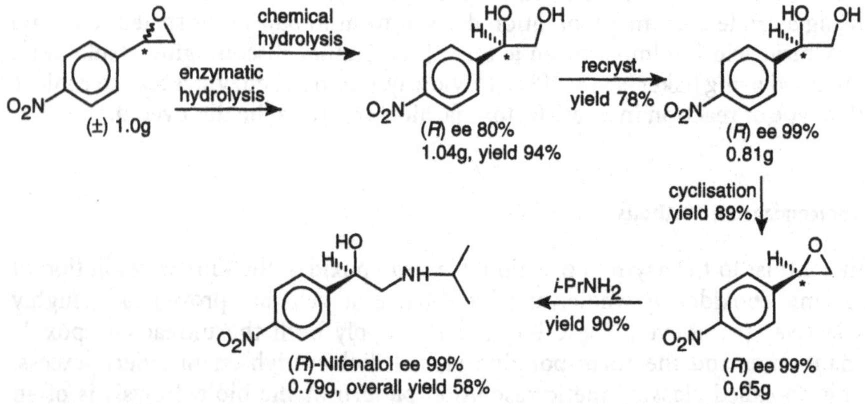

Scheme 11. Deracemization of *p*-nitrostyrene oxide and application to the synthesis of (*R*)-nifénalol

mixture (4:1) was used, showing that this enzyme is operative in the presence of water-miscible organic solvents. The resolution step was followed by the careful addition of acid, leading to (*R*)-*para*-nitrostyrene diol in good yield (94%) and ee (80%). Due to the reduced enantioselectivity and the fact that racemization occurred to a certain extent during the acidic hydrolysis, it was necessary to tune both catalytic steps carefully. A mathematical method was refined that made it possible to select the optimum conversion at which the acid hydrolysis step should be started [89, 90]. Careful mechanistic analysis of the acidic hydrolysis reaction, using different solvents and mineral acids, made it possible to select general conditions for the resolution-inversion procedure [82]. As a consequence, large scale deracemization of epoxides became feasible [91].

Cases for a non-classic deracemization of racemic epoxides using one single biocatalyst demand high requirements on matching regio- and enantioselectivities, and are therefore rare. For instance, the enantioconvergent hydrolysis of (±)-3,4-epoxytetrahydropyran [92] and several *cis*-β-alkyl substituted styrene oxides [93] by hepatic microsomal epoxide hydrolase have been reported on an analytical scale. Similarly, soybean epoxide hydrolase converted (±)-*cis*-9,10-epoxy-12(*Z*)-octadecenoic and (±)-*cis*-12,13-epoxy-9(*Z*)-octadecenoic acid into the corresponding (*R*,*R*)-dihydroxy acids as the sole products [94]. However, enantioconvergent hydrolysis on a synthetically useful scale was only reported recently. Thus, the fungus *Beauveria bassiana* transformed (±)-*cis*-β-methyl styrene oxide in an enantioconvergent manner to afford (1*R*,2*R*)-1-phenylpropane-1,2-diol in 85% yield and 98% ee [75]. In a related fashion, 2,3-disubstituted epoxides (Table 4) were hydrolyzed by using the bacterium *Nocardia* EH1 [62, 95]. Thus, the biohydrolysis of *cis*-2,3-epoxyheptane (**3a**) furnished (*R*,*R*)-*threo*-2,3-heptane diol in 79% isolated yield and 91% ee on a gram-scale. In the latter study, the four stereochemical pathways and the enzyme mechanism were elucidated by $^{18}OH_2$-labelling experiments. The hydrolysis reaction was shown to proceed by attack of a (formal) hydroxyl ion at the (*S*)-configurated oxirane carbon atom with concomitant *inversion* of configuration at *both enantiomers with opposite regioselectivity* (Scheme 12). In addition, a mathematical model for the kinetics which allows the optimization of such enantio-convergent processes in preparative applications was developed.

Table 4. Selectivities in the deracemization of 2,3-disubstituted oxiranes by bacterial cells (substrate structures are given in Chart 1)

Substrate	Biocatalyst	Configuration of Product Diol	Ee [%]
3a	*Nocardia* EH1	2*R*,3*S*	90
3b	*Arthrobacter* sp. DSM312	2*R*,3*S*	63
3c	*Rhodococcus* sp. NCIMB 11216	2*R*,3*S*	77
3d	*Rhodococcus* sp. NCIMB 11216	2*R*,3*S*	78
3e	*Nocardia* EH1	2*R*,3*R*	97
3f	*Nocardia* TB1	2*R*,3*R*	77

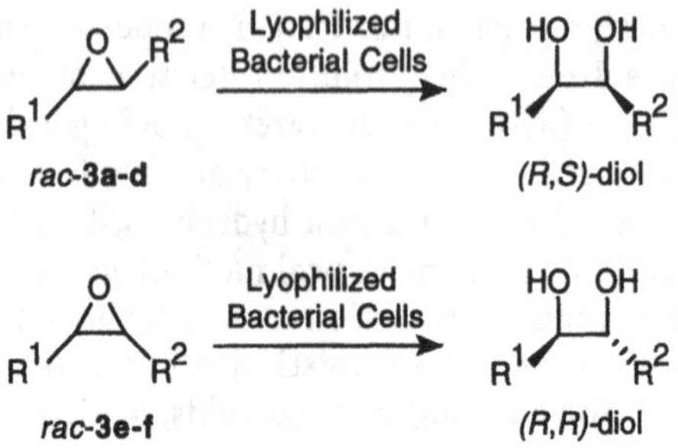

Scheme 12. Resolution and deracemization of 2,3-disubstituted oxiranes by bacterial cells

7 Use of Non-Natural Nucleophiles

In reactions catalyzed by hydrolytic enzymes of the serine-hydrolase type, which form covalent acyl-enzyme intermediates during the course of the reaction, it has been shown that the "natural" nucleophile (water) can be replaced with "foreign" nucleophiles [96] such as an alcohol, amine, hydroxylamine, hydrazine and even hydrogen peroxide. As a consequence, a wealth of synthetically useful reactions, which are usually performed in organic solvents at low water content, can be performed in a stereoselective manner. Although one requirement is fulfilled with epoxide hydrolases – i.e. a covalent enzyme-substrate intermediate is formed – the sensitivity of epoxide hydrolases to most of the water-miscible or -immiscible organic solvents [40, 91] poses a general problem towards the use of non-natural nucleophiles in enzymatic epoxide hydrolysis. However, two types of transformations, i.e. the aminolysis and azidolysis of an epoxide, have been reported as being carried out in an aqueous system (Scheme 13).

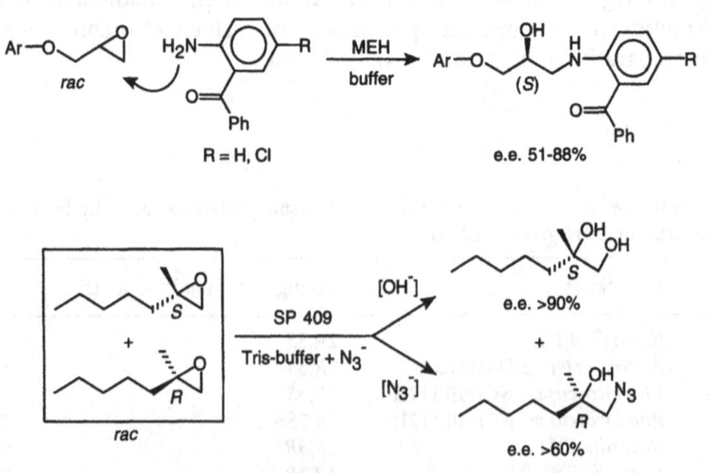

Scheme 13. Enzyme-catalyzed aminolysis and azidolysis of epoxides

When racemic aryl glycidyl ethers were subjected to aminolysis in aqueous buffer catalyzed by hepatic microsomal epoxide hydrolase from rat, the corresponding (*S*)-configurated amino-alcohols were obtained in 51–88% ee [97]. On the other hand, when azide was employed as nucleophile for the asymmetric opening of 2-methyl-1,2-epoxyheptane (±)-(**2a**) in the presence of an immobilized crude enzyme preparation derived from *Rhodococcus* sp., the reaction revealed a complex picture [98]. The (*S*)-epoxide from the racemate was hydrolyzed (as in the absence of azide), and the less readily accepted (*R*)-enantiomer was transformed into the corresponding azido-alcohol (ee > 60%). Although at present only speculation can be made about the actual mechanism of both the aminolysis and azidolysis reaction, in both cases it was proven that the reaction was catalyzed by an enzyme, and that no reaction was observed in the absence of biocatalyst or by using a heat-denaturated preparation. However, a recent related report on the aminolysis of epoxides employing crude porcine pancreatic lipase [99] may likewise be explained by catalysis of a chiral protein surface rather than true lipase-catalysis since the latter enzyme – being a serine hydrolase – is irreversibly deactivated by epoxides. In view of these facts, it remains questionable whether the use of non-natural nucleophiles will be of general applicability.

8
Applications to Natural Product Synthesis

Although the use of an epoxide hydrolase was already claimed for the industrial synthesis of L- and *meso*-tartaric acid in 1969 [51], it was only recently that applications to asymmetric synthesis appeared in the literature. This fact can be attributed to the limited availability of these biocatalysts from sources such as mammals or plants. Since the production of large amounts of crude enzyme is now feasible, preparative-scale applications are within reach of the synthetic chemist. For instance, fermentation of *Nocardia* EH1 on a 70 l-scale afforded > 700 g of lyophilized cells [100]).

One of the first applications of the microbial hydrolysis of epoxides for the synthesis of a bioactive compound is based on the resolution of a 2,3-disubstituted oxirane having a *cis*-configuration (Scheme 14). Thus, by using an enzyme preparation derived from *Pseudomonas* sp., the (9*R*,10*S*)-enantiomer was hydrolyzed in a *trans*-specific fashion (i.e. via inversion of configuration at C-10) yielding the (9*R*,10*R*)-*threo*-diol. The remaining (9*S*,10*R*)-epoxide was converted into (+)-disparlure, the sex pheromone of the gypsy moth in > 95% ee [101].

Another illustration of the use of such a biocatalytic approach was the synthesis of either enantiomer of α-bisabolol, one of these stereoisomers (out of four) which is of industrial value for the cosmetic industry. This approach was based on the diastereoselective hydrolysis of a mixture of oxirane-diastereoisomers obtained from (*R*)- or (*S*)-limonene [68]. Thus, starting from (*S*)-limonene, the biohydrolysis of the mixture of (4*S*,8*RS*)-epoxides led to unreacted (4*S*,8*S*)-epoxide and (4*S*,8*R*)-diol. The former showed a diastereomeric purity (> 95%) and was chemically transformed into (4*S*,8*S*)-α-bisabolol. The formed diol

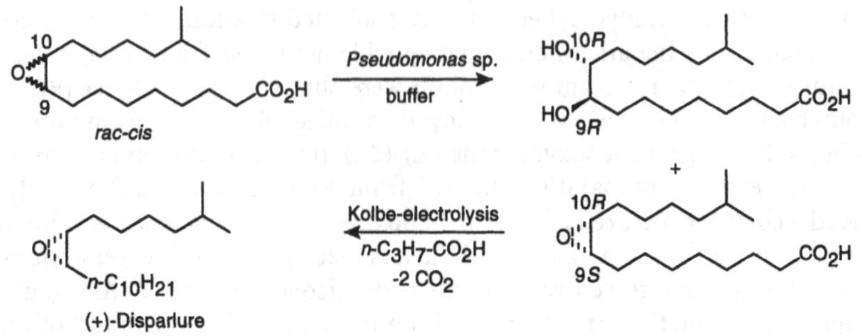

Scheme 14. Resolution of a *cis*-2,3-disubstituted epoxide and synthesis of (+)-disparlure

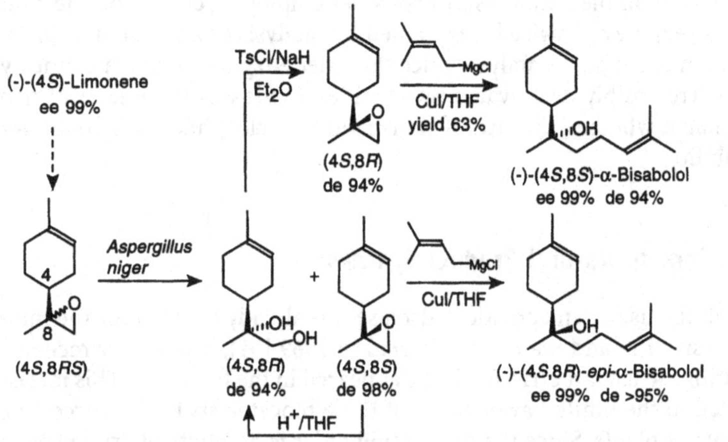

Scheme 15. Chemoenzymatic synthesis of α-bisabolol using fungal epoxide hydrolase

(de>94%) could be cyclised back to give the corresponding (4*S*,8*R*)-epoxide, thus affording access to another stereoisomer of α-bisabolol (Scheme 15). In addition, the two remaining stereoisomers of bisabolol could be prepared in a similar manner starting from (*R*)-limonene.

More recently, the deracemization of (±)-**2c** (Scheme 16) with *Nocardia* EH1 and sulfuric acid in dioxane containing a trace amount of water (see above) afforded (*S*)-2-methyl-hept-6-ene-1,2-diol in 97% yield and 99% ee [83]. This intermediate was successfully applied in a short synthesis of (*S*)-(–)-frontalin, a central aggregation pheromone of pine beetles of the *Dendroctonus* family [81].

Optically pure *trans*- and *cis*-linalool oxides, constituents of several plants and fruits, are among the main aroma components of oolong and black tea. These compounds were prepared from 2,3-epoxylinalyl acetate (**9**) (Scheme 17) [102]. The key step consist of a separation of the diastereomeric mixture of **9** by employing an epoxide hydrolase preparation derived from *Rhodococcus* sp. NCIMB 11216, yielding the product diol and remaining epoxide in excellent diastereomeric excess (de>98%). Further follow-up chemistry gave both linalool

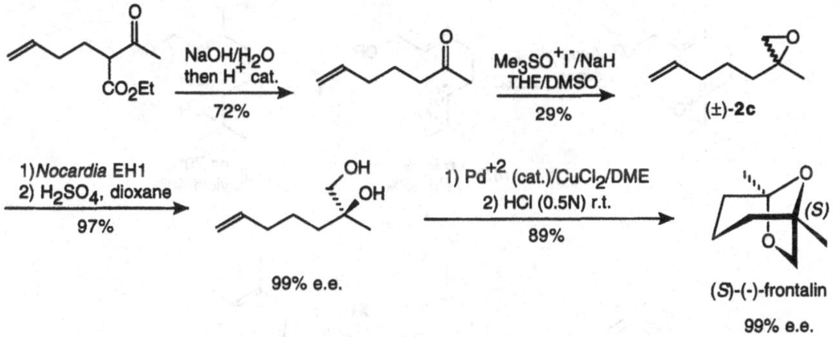

Scheme 16. Chemoenzymatic synthesis of (*S*)-(−)-frontalin using bacterial epoxide hydrolase

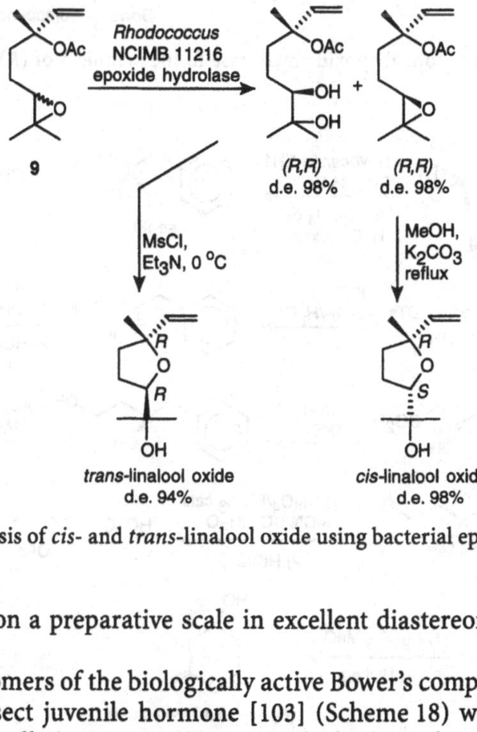

Scheme 17. Synthesis of *cis*- and *trans*-linalool oxide using bacterial epoxide hydrolase

oxide isomers on a preparative scale in excellent diastereomeric and enantiomeric purities.

Both enantiomers of the biologically active Bower's compound, a potent analogue of an insect juvenile hormone [103] (Scheme 18) were prepared using *Aspergillus* sp. cells in 96% ee. Interestingly, biological tests showed that the (6*R*)-antipode was about ten times more active than the (6*S*)-counterpart against the yellow meal worm *Tenebrio molitor*.

Aspergillus niger was the biocatalyst of choice for the biohydrolysis of *para*-nitrostyrene oxide (see above). A selective kinetic resolution using a crude enzyme extract of this biocatalyst followed by careful acidification of the cooled crude reaction mixture afforded the corresponding (*R*)-diol in high chemical yield (94%) and good ee (80%). This key intermediate could then be transformed via a four-step sequence (Scheme 11) into enantiopure (*R*)-nifenalol, a molecule with β-blocker activity, which was obtained in 58% overall yield [88].

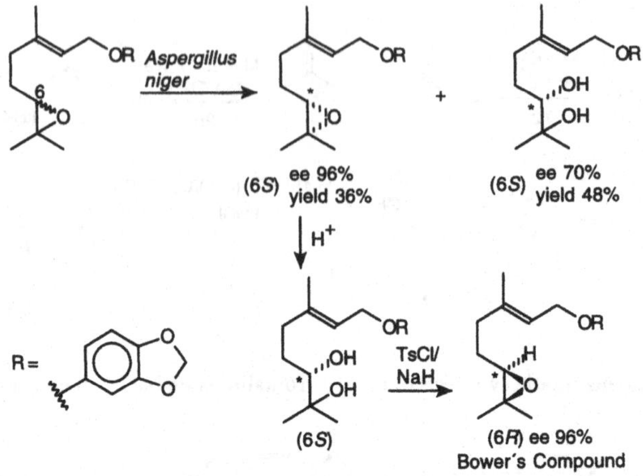

Scheme 18. Application of fungal epoxide hydrolase to the synthesis of (*R*)- and (*S*)-Bower's compounds

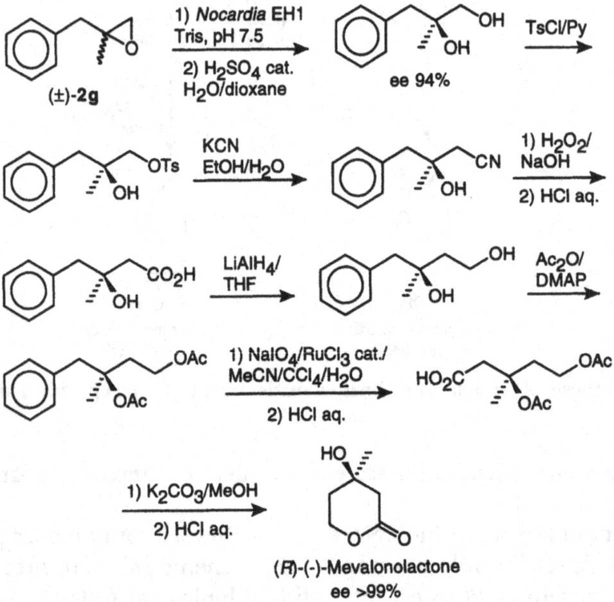

Scheme 19. Synthesis of (*R*)-(–)-mevalonolactone

Finally, natural (*R*)-(–)-mevalonolactone, a key intermediate from a broad spectrum of cellular biological processes and their regulation, was synthesised via eight steps in 55% overall yield and > 99% ee (Scheme 19). In the key step, the aforementioned enantioconvergent chemoenzymatic deracemization route was applied. Thus, 2-methyl-2-benzyl-oxirane (±)-2 g was deracemized on a large scale (10 g) using lyophilized cells of *Nocardia* EH1 and sulfuric acid. The product (*S*)-diol was isolated in 94% chemical yield and 94% optical purity [91].

9 Summary and Outlook

Epoxide hydrolases from microbial sources have recently received much attention as highly versatile biocatalysts for the preparation of enantiopure epoxides and 1,2-diols and the near future will certainly bring a number of useful applications of these systems to the asymmetric synthesis of chiral bioactive compounds. These biocatalysts are easy to use due to their independence of cofactors and the availability of a sufficient number of microbial strains from culture collections, permitting an ample supply of enzyme. An additional benefit of these enzymes is the fact that, in contrast to epoxide hydrolases from mammalian systems, enzyme induction does not seem to be necessary. Furthermore, whole lyophilized cells, as well as crude enzymatic extracts, which can be stored in a refrigerator for several months without significant loss of activity, can be used instead of isolated enzymes. As for all enzymes, the enantioselectivity of microbial epoxide hydrolases can be low to excellent depending on the substrate structure. As this field is currently still in a developmental stage, more information is needed to enable predictions of suitable microbial strains possessing epoxide hydrolase activity for a given substrate. This will certainly be helped once the first three-dimensional X-ray structure of such an enzyme is solved. Given the data presented above, a much more extensive use – and possibly also industrial applications – of microbial epoxide hydrolases for the preparation of enantiopure epoxides and/or vicinal diols on a preparative scale can be anticipated in the near future.

10 References

1. Schurig V, Betschinger F (1992) Chem Rev 92:873
2. Kolb HC, VanNieuenhze MS, Sharpless KB (1994) Chem Rev 94:2483
3. Johnson RA, Sharpless KB (1993) In: Ojima I (ed) Catalytic asymmetric synthesis. Verlag Chemie, New York, p 103
4. Jacobsen EN, Zhang W, Muci AR, Ecker JR, Deng L (1991) J Am Chem Soc 113:7063
5. Katsuki T (1995) Coord Chem Rev 140:189
6. Konishi K, Oda K, Nishida K, Aida, T, Inoue S (1992) J Am Chem Soc 114:1313
7. Hodgson DM, Gibbs AR, Lee GP (1996) Tetrahedron 52:14361
8. Tokunaga M, Larrow JF, Kakiuchi F, Jacobsen EN (1997) Science 277:936
9. de Bont JAM (1993) Tetrahedron: Asymmetry 4:1331
10. Leak DJ, Aikens PJ, Seyed-Mahmoudian M (1992) Trends Biotechnol 10:256
11. Onumonu AN, Colocoussi A, Matthews C, Woodland MP, Leak DJ (1994) Biocatalysis 10:211
12. Besse P, Veschambre H (1994) Tetrahedron 50:8885
13. Pedragosa-Moreau S, Archelas A, Furstoss R (1995) Bull Soc Chim Fr 132:769
14. Mischitz M, Kroutil W, Wandel U, Faber K (1995) Tetrahedron: Asymmetry 6:1261
15. Kim M-J, Choi YK (1992) J Org Chem 57:1605
16. Ader U, Schneider MP (1992) Tetrahedron: Asymmetry 3:205

17. Onda M, Motosugi K, Nakajima H (1990) Agric Biol Chem 54:3031
18. Allison N, Skinner AJ, Cooper RA (1983) J Gen Microbiol 129:1283
19. Casy G, Lee TV, Lovell H (1992) Tetrahedron Lett 33:817
20. Nakamura K, Yoneda T, Miyai T, Ushio K, Oka S, Ohno A (1988) Tetrahedron Lett 29:2453
21. Furuhashi K (1986) Chem Econ Eng Rev 18(7/8):21
22. Franssen MCR (1994) Biocatalysis 10:87
23. Geigert J, Neidleman SL, Liu T-NE, De Witt SK, Panschar BM, Dalietos DJ, Siegel ER (1983) Appl Environ Microbiol 45:1148
24. Allain EJ, Hager LP, Deng L, Jacobsen EN (1993) J Am Chem Soc 115:4415
25. Oesch F (1972) Xenobiotica 3:305
26. Berti G (1986) In: Schneider MP (ed) Enzymes as catalysts in organic synthesis. Reidel, Dordrecht, NATO ASI Series C 178:349
27. Faber K, Mischitz M, Kroutil W (1996) Acta Chem Scand 50:249
28. Archelas A, Furstoss R (1997) Annu Rev Microb 51:491
29. Archer IVJ (1997) Tetrahedron 53:15617
30. Epoxide hydrolases have also occasionally been termed "epoxide hydratases" or "epoxide hydrases"
31. Lu AYH, Miwa GT (1980) Annu Rev Pharmacol Toxicol 20:513
32. de Bont JAM, van Dijken JP, van Ginkel KG (1982) Biochim Biophys Acta 714:465
33. Weijers CAGM, Jongejan H, Franssen MCR, de Groot A, de Bont JAM (1995) Appl Microbiol Biotechnol 42:775
34. Niehaus WG, Kisic A, Torkelson A, Bednarczyk DJ, Schroepfer GJ (1970) J Biol Chem 245:3802
35. Michaels BC, Ruettinger RT, Fulco A (1980) Biochem Biophys Res Commun 92:1189
36. Hartmann, GR, Frear, DS (1963) Biochem Biophys Res Commun 10:366
37. Wackett LP, Gibson DT (1982) Biochem J 205:117
38. Nakamura T, Nagasawa T, Yu F, Watanabe I, Yamada H (1994) Appl Environ Microbiol 60:4630
39. Jacobs MHJ, van den Wijngaard AJ, Pentenga M, Janssen DB (1991) Eur J Biochem 202:1217
40. Mischitz M, Faber K, Willets A (1995) Biotechnol Lett 17:893
41. Moussou P, Archelas A, Furstoss R (1998) Tetrahedron 54:1563
42. Weijers CAGM (1997) Tetrahedron: Asymmetry 8:639
43. Blée E, Schuber F (1993) Plant J 4:113
44. Wojtasek H, Prestwich GD (1996) Biochem Biophys Res Comm 220:323
45. Dubois GC, Apella E, Levin W, Lu AYH, Jerina DM (1978) J Biol Chem 253:2932
46. Lacourciere GM, Armstrong RN (1993) J Am Chem Soc 115:10466
47. Hammock BD, Pinot F, Beetham JK, Grant DF, Arand ME, Oesch F (1994) Biochem Biophys Res Commun 198:850
48. Janssen DB, Fries F, van der Ploeg J, Kazemier B, Terpstro P, Witholt B (1989) J Bacteriol 171:6791
49. Janssen DB, Pries F, van der Ploeg JR (1994) Annu Rev Microbiol 48:163
50. Verschueren KHG, Seljee F, Rozeboom HJ, Kalk KH, Dijkstra BW (1993) Nature 363:693
51. Allen RH, Jakoby WB (1969) J Biol Chem 244:2078
52. Hechtberger P, Wirnsberger G, Mischitz M, Klempier N, Faber K (1993) Tetrahedron: Asymmetry 4:1161
53. Bellucci G, Chiappe C, Cordoni A, Marioni F (1994) Tetrahedron Lett 35:4219
54. Pedragosa-Moreau S, Archelas A, Furstoss R (1994) Bioorg Med Chem 2:609
55. Kolattukudy PE, Brown L (1975) Arch Biochem Biophys 166:599
56. Suzuki Y, Imai K, Marumo S (1974) J Am Chem Soc 96:3703
57. Withers SG, Warren RAJ, Street IP, Rupitz K, Kempton JB, Aebersold R (1990) J Am Chem Soc 112:5887
58. Pedragosa-Moreau S, Archelas A, Furstoss R (1993) J Org Chem 58:5533
59. Escoffer B, Prome J-C (1989) Bioorg Chem 17:53
60. Mischitz M, Mirtl C, Saf R, Faber K (1996) Tetrahedron: Asymmetry 7:2041

61. Moussou P, Archelas A, Baratti J, Furstoss R (1998) J Org Chem 63:3532
62. Kroutil W, Mischitz M, Faber K (1997) J Chem Soc Perkin Trans 1:3629
63. Knehr M, Thomas H, Arand M, Gebel T, Zeller HD, Oesch F (1993) J Biol Chem 268:17623
64. Bell PA, Kasper CB (1993) J Biol Chem 268:14011
65. Janssen DB, Fries F, van der Ploeg J, Kazemier B, Terpstro P, Witholt B (1989) J Bacteriol 171:6791
66. Franken SM, Rozeboom HJ, Kalk KH, Dijkstra BW (1991) Embo J 10:1297
67. Jones A, Zou J Unpublished, personal communication
68. Chen X-J, Archelas A, Furstoss R (1993) J Org Chem 58:5528
69. Wandel U, Mischitz M, Kroutil W, Faber K (1995) J Chem Soc Perkin Trans 1:735
70. Only the strains showing promising activity are reported
71. Novo nitrilase SP 409 or SP 361
72. DeRaadt A, Klempier N, Faber K, Griengl H (1992) In: Servi S (ed) Microbial reagents in organic synthesis. Kluwer, Dordrecht, NATO ASI Series C 381:209
73. Osprian I, Kroutil W, Mischitz M, Faber K (1997) Tetrahedron: Asymmetry 8:65
74. Pedragosa-Moreau S, Morisseau C, Zylber J, Archelas A, Baratti JC, Furstoss R (1996) J Org Chem 61:7402
75. Pedragosa-Moreau S, Archelas A, Furstoss R (1996) Tetrahedron 52:4593
76. Pedragosa-Moreau S, Archelas A, Furstoss R (1996) Tetrahedron Lett 37:3319
77. Zhang J, Reddy J, Roberge C, Senanayake C, Greasham R, Chartrain M (1995) J Ferment Bioeng 80:244
78. Chartrain M, Senanayake CH, Rosazza JPN, Zhang J (1996) Int Patent No WO96/12818; Chem Abstr 125:P56385w
79. Nellaiah H, Morisseau C, Archelas A, Furstoss R, Baratti JC (1996) Biotechnol Bioeng 49:70
80. Morisseau C, Nellaiah H, Archelas A, Furstoss R, Baratti JC (1997) Enz Microbiol Technol 20:446
81. Kroutil W, Osprian I, Mischitz M, Faber K (1997) Synthesis:156
82. Orru RVA, Mayer SF, Kroutil W, Faber K (1998) Tetrahedron 54:859
83. Orru RVA, Kroutil W, Faber K (1997) Tetrahedron Lett 38:1753
84. Jerina DM, Ziffer H, Daly JW (1970) J Am Chem Soc 92:1056
85. Bellucci G, Capitani I, Chiappe C (1989) J Chem Soc, Chem Commun 1170
86. Carnell AJ, Iacazio G, Roberts SM, Willetts AJ (1994) Tetrahedron Lett 35:331
87. Takeshita M, Akagi N, Akutsu N, Kuwashima S, Sato T, Ohkubo Y (1990) Tohoku Yakka Daigaku Kenyo Nempo 37:175
88. Archer IVJ, Leak DJ, Widdowson DA (1996) Tetrahedron Lett 37:6619
89. Pedragosa-Moreau S, Morisseau C, Baratti J, Zylber J, Archelas A, Furstoss R (1997) Tetrahedron 53:9707
90. Vänttinen E, Kanerva LT (1995) Tetrahedron: Asymmetry 6:1779
91 Orru RVA, Osprian I, Kroutil W, Faber K (1998) Synthesis (in press)
92. Bellucci G, Berti G, Catelani G, Mastrorilli E (1981) J Org Chem 46:5148
93. Bellucci G, Chiappe C, Cordoni A (1996) Tetrahedron: Asymmetry 7:197
94. Blée E, Schuber F (1995) Eur J Biochem 230:229
95. Kroutil W, Mischitz M, Plachota P, Faber K (1996) Tetrahedron Lett 37:8379
96. Faber K (1997) Biotransformations in organic chemistry, 3rd edn. Springer, Berlin Heidelberg New York, p 270
97. Kamal A, Rao AB, Rao MV (1992) Tetrahedron Lett 33:4077
98. Mischitz M, Faber K (1994) Tetrahedron Lett 35:81
99. Kamal A, Damayanthi Y, Rao MV (1992) Tetrahedron: Asymmetry 3:1361
100. Kroutil W, Genzel Y, Pietzsch M, Syldatk C, Faber K (1997) J Biotechnol 61:143
101. Otto PPJHL, Stein F, van der Willigen CA (1988) Agric Ecosys Environ 21:121
102. Mischitz M, Faber K (1996) Synlett:978
103. Archelas A, Delbecque J-P, Furstoss R (1993) Tetrahedron: Asymmetry 4:2445

Received December 1998

Microbial Models for Drug Metabolism

R. Azerad
Laboratoire de Chimie et Biochimie Pharmacologiques et Toxicologiques,
Unité Associée au CNRS N° 400, Université René Descartes- Paris V, 45 rue des Saints-Pères,
75270 – Paris Cedex 06, France. *E-mail: azerad@bisance.citi2.fr*

This review describes microbial transformation studies of drugs, comparing them with the corresponding metabolism in animal systems, and providing technical methods for developing microbial models. Emphasis is laid on the potential for selected microorganisms to mimic all patterns of mammalian biotransformations and to provide preparative methods for structural identification and toxicological and pharmacological studies of drug metabolites.

Keywords: Microorganisms, Fungi, Hydroxylation, Screening, Biotransformation, Metabolites, Phase I, Phase II, Glycoconjugates, Toxicology, Pharmacology.

1 **Introduction** 169

2 **Mammalian Biotransformation of Drugs** 170

3 **Microbial Biotransformation of Drugs** 174

3.1 Comparison Between Mammalian and Microbial Metabolism . . . 174
3.2 Methodology Used in Microbial Model Development 198
3.3 Selected Examples and Applications of the Microbial Metabolism of Drugs 201
3.3.1 Warfarin and Related Anticoagulants 201
3.3.2 RU27987 (Trimegestone) 203
3.3.3 Tricyclic Antidepressants 204
3.3.4 Artemisinin and Analogous Antimalarial Compounds 206
3.3.5 Taxol and Related Antitumour Compounds 208
3.3.6 Miscelleanous Examples 210

4 **Conclusion and Perspectives** 212

5 **References** 213

1 Introduction

Before any drug can be approved for use in humans, extensive studies are required to establish its efficacy and safety. The elucidation of drug metabolism constitutes an important and necessary step in this evaluation. The biotransformation of the drug is generally considered as a "detoxification" reaction, leading to more polar substances which are more easily eliminated from the organism. However, in some cases, this metabolism can lead to an "activation", producing either pharmacologically more active substances or toxic reactive metabolites, thus justifying the need for a detailed pharmacological and toxicological study of the metabolic fate of the drug. Traditionally, metabolic studies are realized on codified animal models, perfused organs and normal or malignant cell cultures. Microbial models may constitute an alternative or, at least, a complement to the use of animal systems, provided they can mimic the mammalian metabolism and afford any pertinent information about the metabolic fate of the drug. Beside the experimental facilities characterizing microbial culture and use, which will be later developed, such a methodology would have the advantage of reducing the demand for animals, particularly in the early phases of drug development. Initially, microbial transformations of drugs, particularly steroids and antibiotics, were essentially performed in an effort to obtain more active or less toxic substances [1-9], and this method is still currently in use with this intention. The information thus acquired has resulted in the formalization of the so-called "microbial models of mammalian drug metabolism", introduced in the mid-1970s by Smith and Rosazza [10-12], and more systematically exploited since then, a concept which has been the subject of several reviews and updates [9, 13-23]. This review presents a comprehensive overview of the field, and, through classical or more recent applications, attempts to delineate the scope and perpectives of this methodology. Our ambition is to contribute to a more widespread acceptance of such techniques which are not yet in current use, probably because, up to now, only few researchers engaged in drug pharmacological and toxicological studies are familiar with the practice of microbial transformations.

2 Mammalian Biotransformation of Drugs

The biotransformations involved in the study of drug metabolism (Fig. 1) are mainly represented by "detoxification" reactions, due to the fact that drugs (mainly hydrophobic substances easily permeating through the cell membrane)

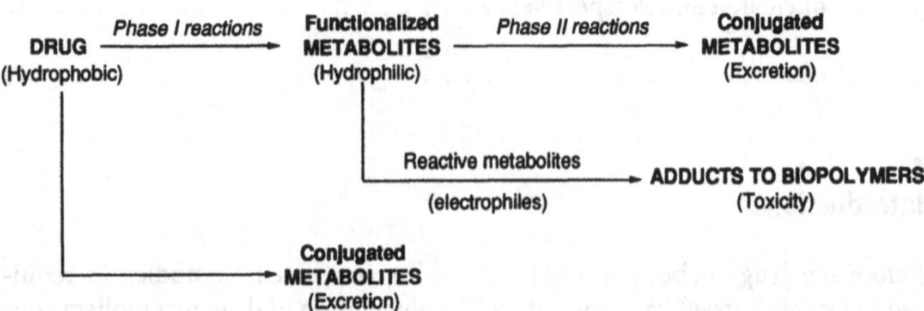

Fig. 1. Metabolic transformation of a drug

are metabolized in liver cells, kidney and other organs to more polar (hydrophilic) derivatives, in order to exclude them from the inside of the cells and to prevent them from penetrating again, resulting finally in their elimination from the organism. Such a process involves two types of transformations, classified as Phase I (functionalization) and Phase II (conjugation) reactions (Table 1).

Phase I reactions consist of oxidation, reduction and hydrolysis reactions. Hydrolysis reactions are catalyzed by esterases and can generally be easily mimicked by one of the numerous commercially available purified enzyme preparations. Most reduction reactions involve the reduction by an NAD(P) H-dependent dehydrogenase of a ketonic compound to one of the corresponding stereomeric secondary alcohols. The oxidative Phase I reactions in animals involve various types of reactions, essentially catalyzed by monoamine oxidases and flavine or cytochrome P450 monooxygenases. The latter are responsible for the largest number of Phase I metabolites, catalyzing the introduction of one atom of oxygen into a number of different substrates (Table 1). Some of the Phase I reactions can give rise to "activated" metabolites eventually responsible for higher activity (for example the hydrolysis of prodrugs), while others (for example epoxidation) can result in reactive electrophilic species and increased toxicity. Phase II reactions are synthetic reactions involving the conjugation of the drug or primary metabolites with common endogenous substances, such as glucuronic acid, glycine, acetate, methyl groups or inorganic acids, generally contributing to a further increase of hydrophilicity and facilitating the renal excretion of the drug and its metabolites. The final result of such biotransformations of complex drug molecules, operating either in sequential or parallel pathways, is often the formation of tens of metabolites.

Traditionally, in vivo drug metabolism studies involve the administration of the drug to a series of laboratory animals (rat, dog, cat, guinea pig, rabbit, etc.), used as model systems. The body fluids (plasma and urine) of these animals are then examined for the presence and identification of metabolites. In vitro studies are generally used to complement and specify the data obtained, using perfused organs, tissue or cell cultures and microsomal preparations [13]. Such methods suffer from a number of limitations, such as the cost of animals used for experimentation, the ethical controversies about the use of animals in biomedical research and the toxicity of drugs, limiting the amount administered and thus the quantities of metabolites isolated (frequently a few micrograms for the minor ones). Despite the progress in highly sensitive analytical techniques (radiochemical techniques, GC/MS and HPLC/MS, high-field NMR spectroscopy and the recently developed HPLC/NMR methods), structural identification on minute amount of metabolites, and particularly identification of (stereo)isomeric structures, remains difficult. Such an approach frequently needs, for comparison, the chemical synthesis of several possible metabolites, a time-consuming and often laborious process.

Table 1. Mammalian biotransformations of drugs

Reaction types	Substrates	Enzymes involved (cosubstrates)	Products
Phase I reactions			
– Hydrolysis	Alcohol and phenol esters	Esterases	Alcohols or phenols
	Amides	Amidases	Amines
	Epoxides	Epoxide hydrases	Diols
– Reduction	Ketones	Alcohol-dehydrogenases	Alcohols
	Olefins	Dehydrogenases	Saturated compounds
	Nitro- and Azo-compounds	Nitro and azo reductases	Nitroso, hydroxylamino and amino compounds
– Oxydation Hydroxylation	Aromatic, allylic, benzylic or unactivated carbon atoms	Cytochromes P450 Flavine monooxygenases	Phenols Alcohols
Epoxidation	Aromatic or aliphatic double bonds		Epoxides
N-,O-,S-dealkylation	N-, O- or S-alkyl derivatives		Amino, hydroxy or thiol compounds
N-, S-oxidation	Secondary and tertiary amines, S-alkyl derivatives		N-oxides Sulfoxides, sulfones
N-hydroxylation	Secondary and tertiary amines		Hydroxyamines
– Oxidative deamination	Primary amines	Monoamine oxidases	Aldehydes
Baeyer-Villiger reaction	Aldehyde, ketones	Flavine monooxygenase	Esters, lactones

Table 1 (continued)

Reaction types	Substrates	Enzymes involved (cosubstrates)	Products
Phase II reactions			
– Glucuronidation	Alcohols or phenols Carboxylic acids Amines Thiols	Glucuronyltransferases (UDP-glucuronic acid)	α- or β-Glucuronides
– Glucosylation	Alcohol or phenol Carboxylic acid Amines Thiols	Glucosyltransferases (UDP-glucose)	α- or β-Glucosides
– Thiol conjugation	Epoxides	Glutathione S-transferases (glutathione or N-acetylcysteine)	Glutathione or N-acetyl cysteine thioethers
– Glycine conjugation	Carboxylic acids	Acyl-CoA:amino acid N-transferase	Glycinamide conjugates
– Carbamoylation	Alcohols		O-carbamoyl derivatives
– Acetylation	Primary amines Hydrazines	Acetyltransferases (Acetylcoenzyme A)	N-acetylated derivatives
– O-methylation	Phenols	Methyltransferases (S-adenosylmethionine)	Methyl arylethers
– Sulfatation	Alcohols or phenols Amines	Sulfotransferases (PAPS)	Sulfate esters Sulfonamides

3
Microbial Biotransformation of Drugs

3.1
Comparison Between Mammalian and Microbial Metabolism

The hydrolytic and reductive capabilities of microorganisms (bacteria, yeast and fungi) have been known for a long time and are currently used in preparative reactions [24–32]. Concerning oxidative reactions, beside the classical microbial 11α (or 11β)-hydroxylation of steroids [33, 34], used for the first time as a preparative route for the entry into the corticosteroid chemistry, and which clearly parallels the same reaction in mammals, a number of early observations on fungal microorganisms [35, 36] indicated that monooxygenase enzymes were present and proved to be mechanistically similar to mammalian hepatic monooxygenases. The systematic examination of microbial hydroxylations on a variety of model aromatic compounds [10], followed by a comparison of *O*- and *N*-dealkylation reactions [11, 12], induced Smith and Rosazza to propose that microbial transformation systems could closely mimic most of the Phase I transformations of drugs observed in mammals [13–18]. There are clearly a number of practical advantages in the use of microbial systems as models for drug metabolism (Fig. 2): i) simple culture media are easily prepared at low cost, and screening for the ability of a large number of strains to metabolize the drug is a simple repetitive process, requiring only a periodical sampling of incubation media; ii) concentrations used (generally ranging from 0.2 to 0.5 g/l) are much higher than those employed in other cell or tissue models; as metabolic capabilities of microorganisms can be high, the amount of metabolites formed is currently in the 20–200 mg/l scale, allowing easier detection, isolation and

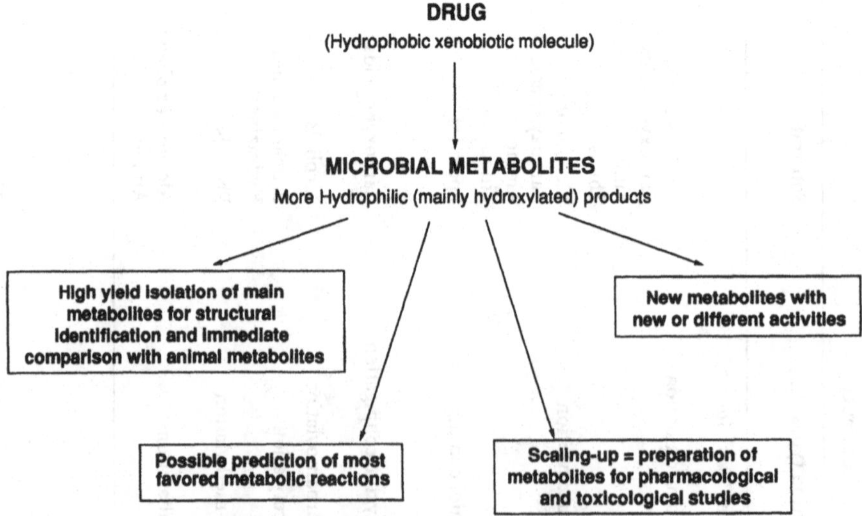

Fig. 2. Potential achievements in the use of microbial models of drug metabolism

structural identification. The only constraint is the maintenance of stock cultures of microorganisms, which is clearly simpler and cheaper than the maintenance of cell or tissue cultures or laboratory animals.

The biotransformations involved in the study of drug metabolism in microorganisms are mainly represented by the earlier "detoxification" reactions of hydrophobic substrates, i.e. Phase I reactions, generally sufficient to produce more polar derivatives, in order to exclude them from the microbial cell: a clear relationship between substrate hydrophobicity and the ability of microorganisms to metabolize them has been demonstrated [23] (Fig. 3). However negatively charged molecules are known to penetrate with difficulty into the cells, and are generally poorly metabolized. Positively charged amino compounds (primary, secondary or tertiary) which show a higher hydrophobicity behave differently and generally freely permeate; as a consequence they are excellent substrates for microbial conversions.

Whereas the hydrophobic substrate is rapidly adsorbed into the cell lipid phase (probably the cell membrane) and often completely disappears from the aqueous medium, one advantage of such reactions is that most of the derivatives produced are excreted out of the biomass as soon as they are formed, and thus

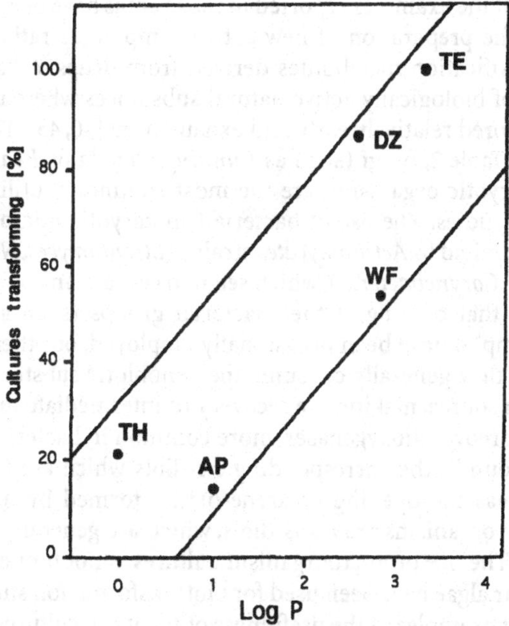

Fig. 3. Relationship between substrate lipophilicity (log P) and ability of screened microorganisms to metabolize model substrates (AP = Aminopyrine; DZ = Diazepam; TE = Testosterone; TH = Theophylline; WF = Warfarin). Log P is the logarithm of the partition coefficient of a compound between n-octanol and a pH 7.0 aqueous buffer. Ordinates represent the % of cultures tested able to transform the substrates, independently of the extent of metabolism and number of products formed (from [185])

do not suffer any further metabolism and can be isolated in high yield from the incubation medium by applying the usual extraction methods.

The use of microorganisms for the simulation of the mammalian metabolism of a large number of molecules of pharmacological importance is well documented in the 1975–1990 literature [7, 11, 12, 14, 15, 17–19, 22, 37]. In Table 2 are listed a series of recent typical biotransformations of drugs or related molecules, either in animal (in vivo or in vitro studies on cell lines, microsome preparations, etc.) or in microbial systems, emphasizing the high similarity observed in both types of metabolism. It should be noted that most of the microbial data reported in the literature have generally been selected as those giving a maximum number of identified mammalian metabolites, in sufficient yields, or as the particular ones allowing the best preparation of a single major metabolite. In all cases, as the result of the preliminary screening implemented, many other strains would have deserved complementary studies, either for their general metabolizing capabilities, or because some of them were shown to form new unidentified metabolites.

In Table 2 antibiotic and steroid biotransformations are only poorly represented compared to their actual frequency: more than 50 examples of antibiotic transformations were already reported in 1975 [9, 38, 39]; a listing of steroid microbial transformations, published in 1981, covers 853 substrates [40], while more than 1000 examples have been listed for the 1979–1992 period [41–44]. However, most of the examples reported in those areas have been essentially oriented towards the preparation of new active compounds rather than towards a study of detoxification metabolites derived from drugs. Alkaloids represent another group of biologically active natural substances where microbial metabolism was explored relatively early and exhaustively [40, 45–47].

As shown in Table 2, fungi (such as *Cunninghamella* or *Beauveria* species), which are eukaryotic organisms, are the most commonly utilized microorganisms in these studies. The use of bacteria (prokaryotic microorganisms) has been generally limited to *Actinomycetes* strains (*Streptomyces, Nocardia, Actinoplanes, Myco-* or *Corynebacteria*), which seem to contain an enzymic equipment very similar to that of fungi. Other bacterial groups (such as *Pseudomonas* strains, for example) have been occasionally employed, but their use is limited by the fact that they generally consume the xenobiotic substances as a carbon and/or nitrogen source, making the recovery of intermediate metabolites more problematic. Moreover dioxygenases, more common in bacteria, produce, from aromatic compounds, the corresponding *cis*-diols which are then oxidized to catechols, whereas the opening of arene-oxides formed by monooxygenases from eukaryotic organisms are *trans*-diols, which are generally further converted to phenols. The use of microorganism cultures is not limited to bacteria or fungi: unicellular algae have been used for biotransformation studies [132–134], and some recent examples of the usefulness of plant cell cultures have been provided [135].

An examination of data in Table 2 from the point of view of reaction types shows a striking similarity in the reactions catalyzed by animal and microbial systems, especially in the monooxygenase-catalyzed oxidation reactions. They illustrate the large variety of reactions involved in Phase I (and occasionally

Table 2. Selected examples of mammalian *vs.* microbial metabolism of pharmacologically active drugs

Drug	Pharmacological activity	Chemical structure	Mammalian metabolism	Microbial metabolism (microorganism)	Ref.
Acronycine	Antitumor	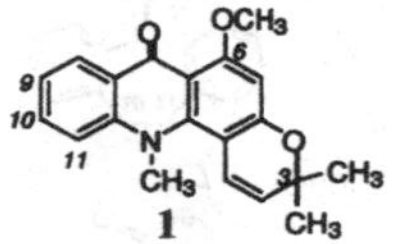 1	3-Methyl, C-9 and C-11 hydroxylations; 6-O-demethylation	C-9 Hydroxylation (*Cunninghamella echinulata, Aspergillus alliaceus*)	[14, 48, 49]
				3-Methyl and C-11 hydroxylations (*Streptomyces spectabilis*)	[49]
Alosetron	5HT3 receptor antagonist	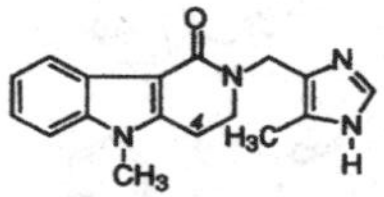 2	N-Demethylation; C-4 hydroxylation; N-dealkylation	N-Demethylation; C-4 hydroxylation; N-dealkylation with cleavage of the imidazole part; hydroxylation at C-2 of the imidazole ring, rearrangement to hydantoin and reductive cleavage (*Streptomyces griseus*)	[50]
Besipirdine	Cholinergic and noradrenergic agonist	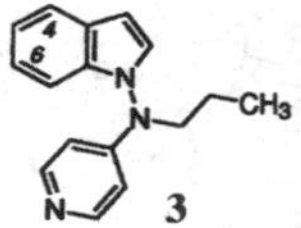 3	N-dealkylation C-4, C-5 (and C-6 ?) aromatic hydroxylations; C-2 oxidation	N-Dealkylation; C-4 (or C-6) and C-5 aromatic hydroxylations (*Cunninghamella elegans*)	[51]
Bisprolol	*β*-Blocking agent	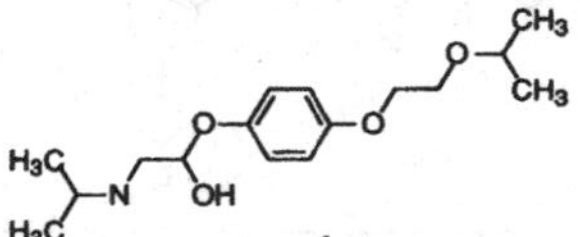 4	O-dealkylations and oxidation to carboxylic acid; Methyl group oxidation	O-Deisopropylation (*Cunninghamella echinulata, Gliocladium deliquescens*) and oxidation to carboxylic acid (*G. deliquescens*)	[52]

Table 2 (continued)

Drug	Pharmacological activity	Chemical structure	Mammalian metabolism	Microbial metabolism (microorganism)	Ref.
Bleomycin B-2	Antitumor	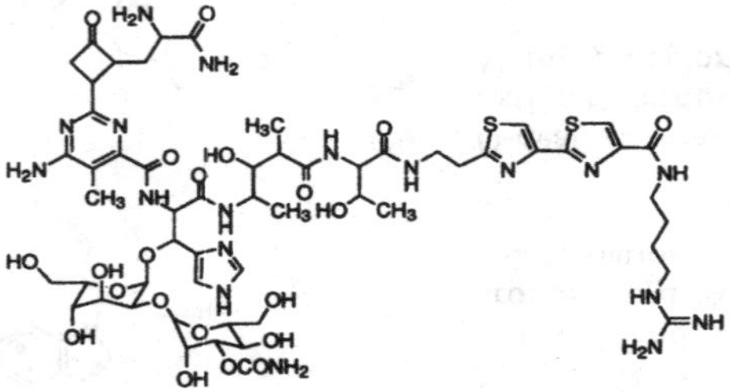 5		Hydrolytic cleavage of the terminal agmatine residue (*Fusarium anguoides*)	[14, 53]
Bornaprine	Anticholinergic	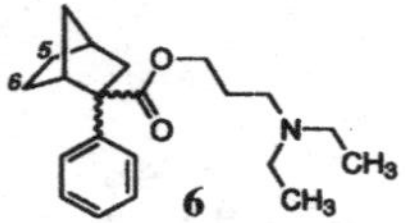 6	C-5 and C-6 hydroxylations and sulphoconjugation; N-dealkylation	C-5 and C-6 hydroxylations (*Cunninghamella echinulata*)	[54]
Carba-mazepine	Antiepileptic	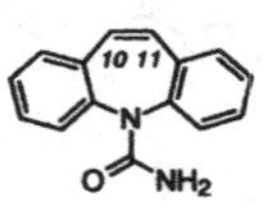 7	10,11-epoxidation; 10-O-Glucuronidation	10,11-epoxidation (*Streptomyces violascens*)	[55]
10,11-Dihydro-carbamazepine	Antiepileptic	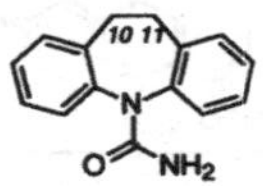 8	10-hydroxylation then oxidation to the 10-keto derivative	C-10 hydroxylation (*S. violascens* and *S. griseus*)	[55]

Table 2 (continued)

Drug	Pharmacological activity	Chemical structure	Mammalian metabolism	Microbial metabolism (microorganism)	Ref.
Codeine	Analgesic	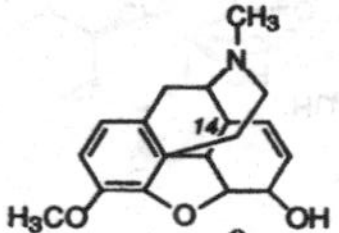 9	N- and O-demethylation; Glucuronidation; Oxidation to ketone	N-Demethylation; C-14 hydroxylation (*Streptomyces griseus*)	[56-58]
Colchicine	Gout pain relief Antitumor	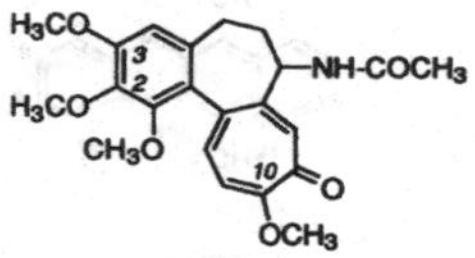 10	N-Deacetylation (?) O-Demethylations	10-O-Demethylation (*Streptomyces griseus*)	[1]
				N-Deacetylation followed by transamination and oxidation (*Arthrobacter colchovorum*)	[59]
				2- And 3-O-demethylation (*Streptomyces spectabilis, S. griseus*)	[60]
Crisnatol	Antitumor	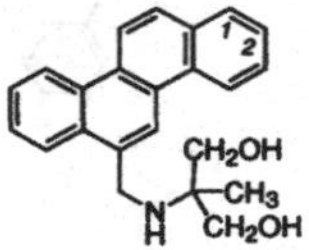 11	C-1 Aromatic hydroxylation; 1,2-epoxidation and hydrolysis to the *trans* 1,2-dihydro-diol; primary alcool oxidation to carboxylic acid	C-1 hydroxylation (*Cunninghamella elegans*)	[61]
Cyproheptadine	Antihistamine	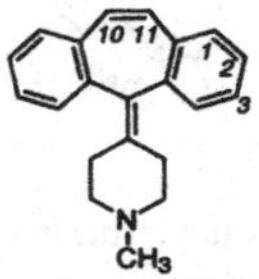 12	N-Demethylation; 10,11-epoxidation; aromatic hydroxylation; N-oxidation; N-glucuronidation	N-Demethylation; 10,11-epoxidation; aromatic hydroxylations at C-1, C-2 or C-3; N-oxidation (*C. elegans*)	[62]

Table 2 (continued)

Drug	Pharmacological activity	Chemical structure	Mammalian metabolism	Microbial metabolism (microorganism)	Ref.
Diazepam	Tranquilizer	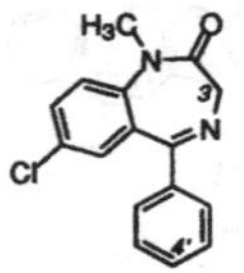 **13**	N-Demethylation and C-3 hydroxylation; C-4′ aromatic hydroxylation	N-Demethylation and C-3 hydroxylation; aromatic hydroxylation at C-4′ (*Aspergillus terreus, Beauveria bassiana*)	[4, 23, 63]
Ebastine	Antihistamine	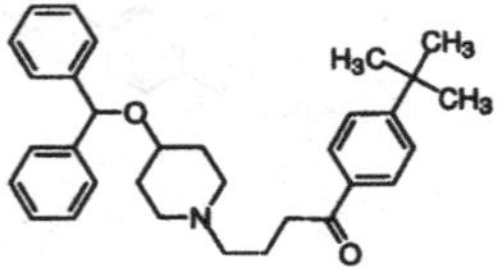 **14**	Terminal methyl hydroxylation and oxidation to carboxylic acid; aromatic hydroxylation	Terminal methyl hydroxylation and oxidation to carboxylic acid (*Cunninghamella blakesleana*)	[64]
Ellipticine	Antitumor	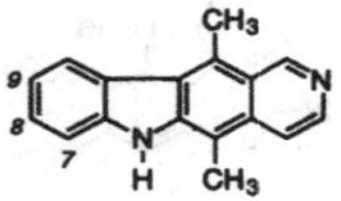 **15**	C-7- and C-9-hydroxylations	C-8- and C-9-hydroxylations (*Aspergillus alliaceus*)	[65, 66]
Erythromycin A	Antibiotic	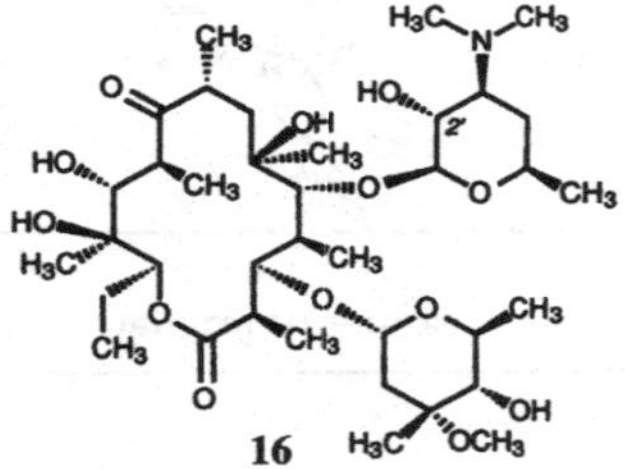 **16**	N-Demethylation	2′-O-Phosphorylation 2′-O-Glucosylation (*Streptomyces vendargensis*)	[67] [68]

Table 2 (continued)

Drug	Pharmacological activity	Chemical structure	Mammalian metabolism	Microbial metabolism (microorganism)	Ref.
Furosemide	Diuretic	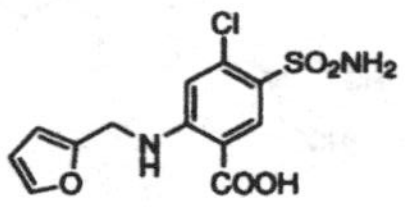 17	N-Dealkylation; arene oxide formation; 1-O-acyl glucuronide formation	N-Dealkylation; 1-O-acyl glucoside formation (*Cunninghamella elegans*)	[69, 70]
GR 117289	Angiotensin II antagonist	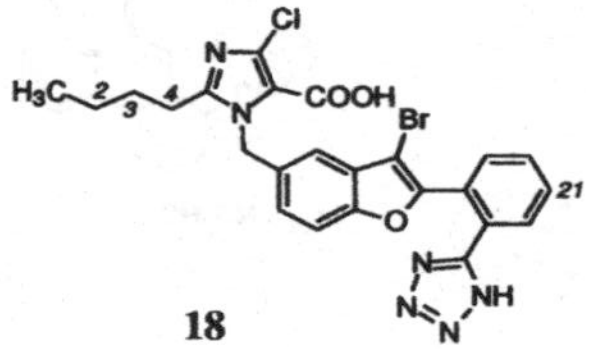 18	C-1 and C-2 hydroxylations; N^2-tetrazole glucuronide conjugate	C-2, C-3 and C-4 hydroxylations; C-2 oxidation to ketone; C-21 hydroxylation (*Streptomyces rimosus*)	[71]
Ibuprofen	Anti-inflammatory	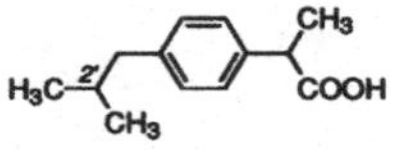 19	Metabolic stereoinversion (*R* to *S*); C-2′ hydroxylation; terminal methyl group oxidation	Metabolic stereoinversion ; (*R* to *S*) C-2′ hydroxylation (*Verticillium lecanii*)	[72, 73]
L-696,474	HIV Protease inhibitor	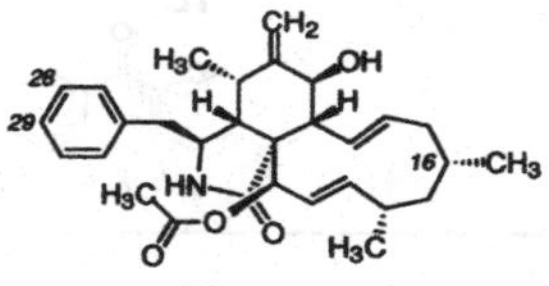 20		C-16 Methyl group hydroxylation; additional C-hydroxylations at C-28 and/or C-29, and C-16 (*Actinoplanes sp.*)	[74]

Table 2 (continued)

Drug	Pharmacological activity	Chemical structure	Mammalian metabolism	Microbial metabolism (microorganism)	Ref.
Lapachol	Antitumor Antibiotic Antimalarial	**21**	Allylic methyl group hydroxylation and oxidation to carboxylic acid; glucuronide formation	2,3-Epoxide formation and hydrolytic B-ring opening (*Penicillium notatum*)	[14, 75–78]
				Oxidative cyclization to chromene (*Curvularia lunata*)	[76]
				Allylic methyl group hydroxylation and oxidation to carboxylic acid (*Beauveria bassiana*)	[79]
				Allylic methyl group hydroxylation and lactate ester formation; glucosidation (*Cunninghamella echinulata*)	[77]
Lergotrile	Antiparkinson Antiprolactin	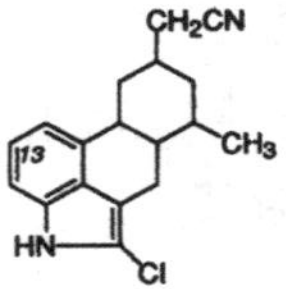 **22**	N-demethylation, hydration of the nitrile group; C-13 hydroxylation	N-demethylation (*Streptomyces platensis*)	[80]
Lucanthone	Schistosomicidal	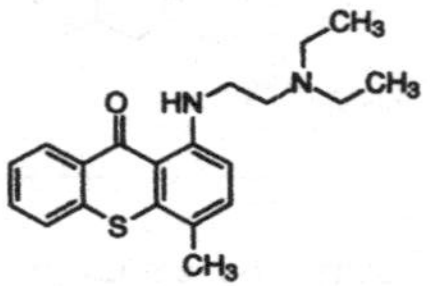 **23**	S-Oxidation to sulfoxide and sulfone; aromatic methyl group hydroxylation	Aromatic methyl group hydroxylation and oxidation to aldehyde and carboxylic acid (*Aspergillus sclerotiorum, Aspergillus sp.*)	[2, 3, 81]

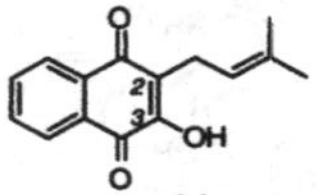

Table 2 (continued)

Drug	Pharmacological activity	Chemical structure	Mammalian metabolism	Microbial metabolism (microorganism)	Ref.
(±) Mexiletine	antiarrhythmic	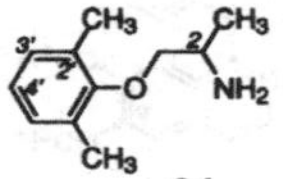 24	Enantioselective metabolism: oxidative deamination and reduction to 2-ol (*R*); benzylic 2′-methyl hydroxylation; C-3′ or 4′ aromatic hydroxylations (*S*)	Oxidative deamination and reduction to 2-ol (*S* ≥ *R*); C-4′ aromatic hydroxylation (*R* > *S*) (*Cunninghamella echinulata)*	[82]
MK 954	Angiotensin A II receptor antagonist	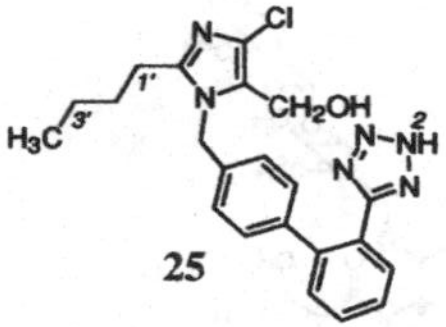 25	C-1′ or C-3′ hydroxylations; N^2-glucuronidation	C-1′ or C-3′ hydroxylations (*Actinoplanes sp.*, *Streptomyces sp.* ATCC 55293) N^2-glucuronidation (*Streptomyces sp.* ATCC 55043)	[83, 84]
Mycophenolic acid	Antitumor Immuno-suppressor	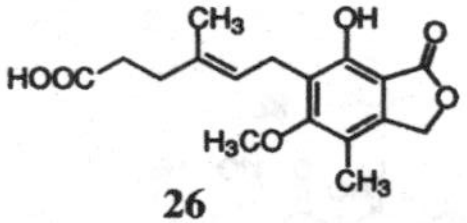 26	7-O-glucuronidation	C-3 Hydroxylation to lactol; aromatic CH_3 hydroxylation (*Diplodia sp.*) Conjugation with glycine or alanine (*Mucor ramanianus*)	[85]
Novobiocin	Antibiotic	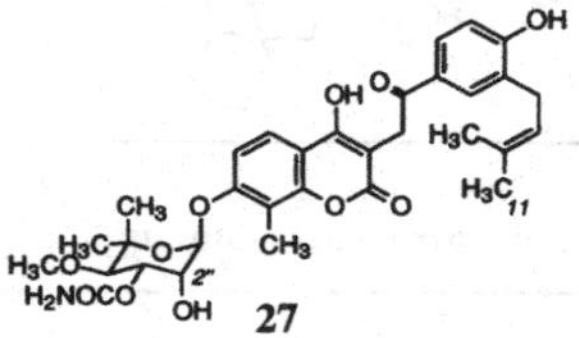 27		C-11 Hydroxylation (*Sebekia benihana*) 2″-O-carbamylation (*Streptomyces niveus*)	[86] [87]

Table 2 (continued)

Drug	Pharmacological activity	Chemical structure	Mammalian metabolism	Microbial metabolism (microorganism)	Ref.
NSC 159628 (Viridicatum-toxin)	Antitumor	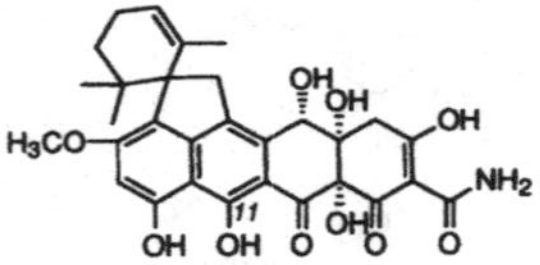 **28**		11-O-β-Glycosylation (*Rhizopus sp.*)	[88]
Olivomycin-A	Antitumor	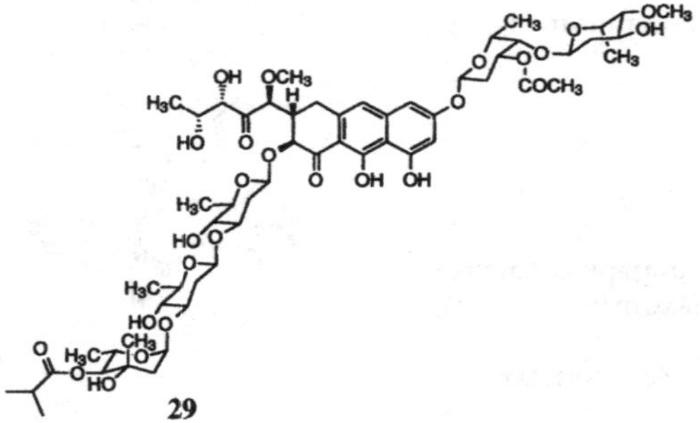 **29**		Deisobutyrylation and deacetylation (*Whetzelinia sclerotiorum*)	[14, 89]
Papaverine	Vasodilatator Smooth muscle relaxant Antitumor	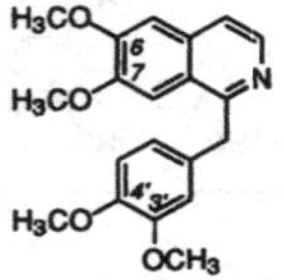 **30**	4'-, 6-, 7- and 4',6-O-demethylation	6-O-Demethylation (*Aspergillus alliaceus*) 4'-O-Demethylation (*Cunninghamella echinulata*) 4'- and 6-O-Demethylation (*Silene alba* plant cell culture)	[90] [91]

Table 2 (continued)

Drug	Pharmacological activity	Chemical structure	Mammalian metabolism	Microbial metabolism (microorganism)	Ref.
Parbendazole	Antihelminthic	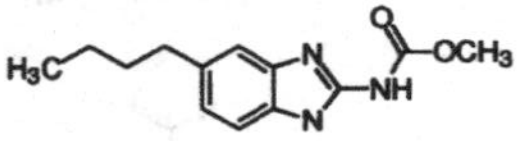 31	ω-Methyl group oxidation	ω-Methyl group oxidation (*Cunninghamella echinulata*)	[92, 93]
Pentoxifyline	Haemorheological agent	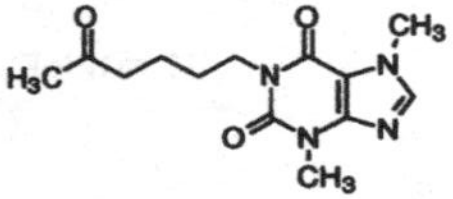 32	Oxidative cleavage of the ketonic side chain; reduction of the keto group	Oxidative cleavage of the ketonic side chain (*Curvularia falcata, Streptomyces griseus*)	[94]
				Reduction of the keto group to the *S*-alcohol (*Rhodotorula rubra*)	[95]
Pergolide	Antiprolactinemic Antiparkinson	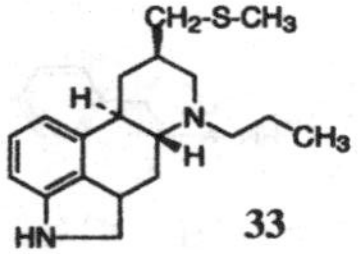 33	Oxidation to sulfoxide and sulfone	Oxidation to sulfoxide (*Heminthosporium sp.*) Oxidation to sulfone (*Aspergillus alliaceus*)	[96]
Phenacetin	Analgesic	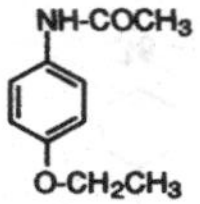 34	O-Dealkylation, N-hydroxylation and arene oxide formation Glutathione conjugates	O-Dealkylation (*Cunninghamella elegans*)	[97]
Phenazo-pyridine	Analgesic	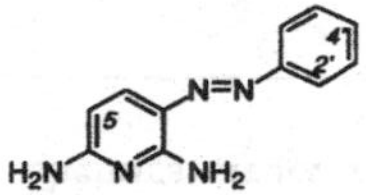 35	2′-, 4′- and 5-hydroxylations; cleavage to 4-aminophenol, followed by N-acetylation	4′-Hydroxylation and 4′-O-sulfate conjugation (*Cunninghamella echinulata*)	[98]

Table 2 (continued)

Drug	Pharmacological activity	Chemical structure	Mammalian metabolism	Microbial metabolism (microorganism)	Ref.
Phencyclidine	Hallucinogen	**36**	C-4 and C-4′ hydroxylations; N-didealkylation (piperidine ring cleavage)	C-4 and C-4′ hydroxylations; piperidine ring cleavage by N-monodealkylation and oxidation (*Beauveria bassiana*, *Cunninghamella echinulata*)	[99–101]
Phenelzine	Antidepressant	**37:** (R = H)	Cleavage to 2-phenylethylamine, phenylacetaldehyde and phenylacetic acid; N-acetylation	N^1-Acetylation (*Cunninghamella elegans*) N^1-Acetylation; cleavage to N-acetyl-2-phenylethylamine (*Mycobacterium smegmatis*)	[102]
Pheniprazine	Antidepressant	**38:** (R = CH_3)		N^1-Acetylation (*Cunninghamella elegans*) N^1-Acetylation; cleavage to N-acetyl-amphetamine (*Mycobacterium smegmatis*)	[102]
Pheniramine	Antihistamine	**39:** R = H	N-Mono- and didemethylation; N-oxidation	N-Monodemethylation; N-oxidation (*Cunninghamella elegans*)	[103]
Brompheniramine		**40:** R = Br	–id–	–id–	[103]
Chlorpheniramine		**41:** R = Cl	–id–	–id–	[103]

Table 2 (continued)

Drug	Pharmacological activity	Chemical structure	Mammalian metabolism	Microbial metabolism (microorganism)	Ref.
Primaquine	Antimalarial	**42:** (R = H) 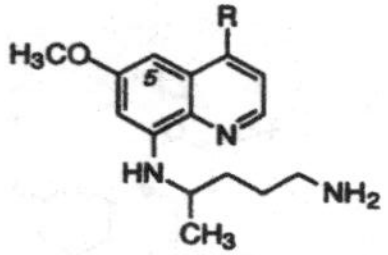	N-Dealkylation (?) Oxidative deamination to carboxylic acid	N-Acetylation (*Streptomyces roseochromogenus*) Oxidative deamination to carboxylic acid (*Aspergillus flavus*) 5,5′-Dimerization (*Streptomyces rimosus, Candida tropicalis*)	[19, 104–106] [107, 108]
				Oxidative deamination to carboxylic acid and amide formation (*Streptomyces rimosus*) Oxidative deamination to carboxylic acid (*Aspergillus ochraceus*)	[106]
4-Methyl-primaquine		**43:** (R = CH3)	Oxidative deamination to carboxylic acid		[109]
Praziquantel	Antihelminthic	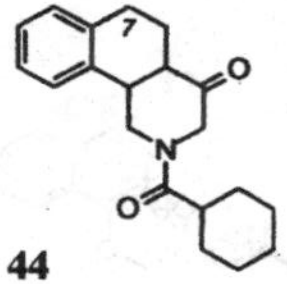 **44**	Numerous unidentified B, C and D-ring mono- and dihydroxylation products	C-7 Hydroxylation and D-ring (unidentified position) hydroxylation (*Cunninghamella echinulata, Beauveria bassiana*)	[19]

Table 2 (continued)

Drug	Pharmacological activity	Chemical structure	Mammalian metabolism	Microbial metabolism (microorganism)	Ref.
Propanolol	β-Blocker	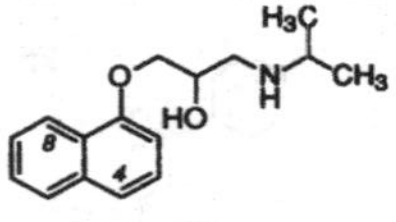 **45**	N-Deisopropylation; oxidative deamination to alcohol or carboxylic acid; ring C-4 hydroxylation and glucuronidation	N-Deisopropylation; oxidative deamination to alcohol or carboxylic acid; ring C-4 and C-8 hydroxylations and undetermined conjugation (*Cunninghamella echinulata*)	[110, 111]
Rifamycin B	Antitumor	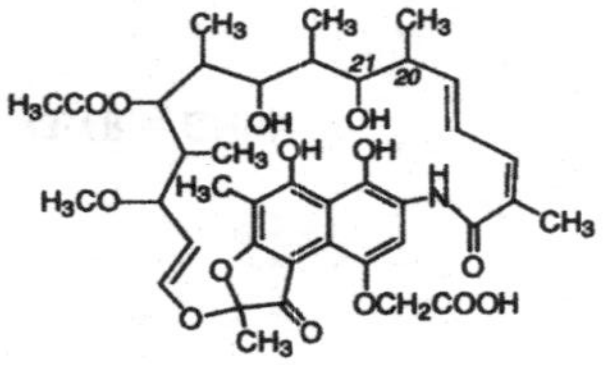 **46**		Hydroxylation at C-20 and oxidation at C-21 (*Streptomyces mediterranei*)	[112]
				Hydrolysis and oxidation to quinone (*Curvularia lunata*)	[113]
Spironolactone	Antihypertensive Diuretic	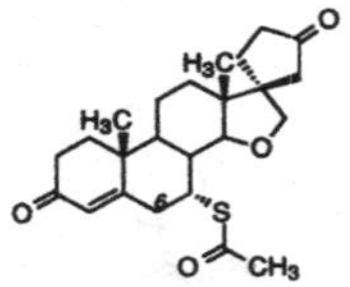 **47**	Conversion to the thiomethyl analogue, then S-oxidation and C-6 oxidation Thioelimination to a 6,7-dehydroderivative	(From the thiomethyl analogue) S-oxidation and C-6 oxidation (*Chaetomium cochloides*)	[5, 8]
Sulindac	Antiinflammatory	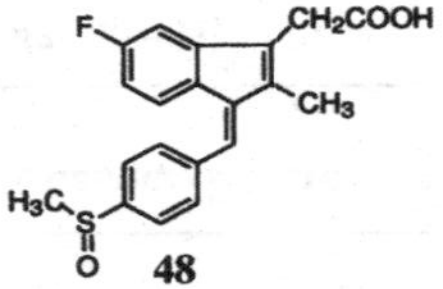 **48**	Reversible reduction to sulphide and oxidation to sulphone + unidentified metabolite	Reduction to sulphide and oxidation to sulphone (*Nocardia corallina*) Reduction to sulphide (*Arthrobacter sp.*) Oxidation to sulphone (*Aspergillus alliaceus*)	[114]

Table 2 (continued)

Drug	Pharmacological activity	Chemical structure	Mammalian metabolism	Microbial metabolism (microorganism)	Ref.
Tamoxifen	Antiestrogen Antitumor	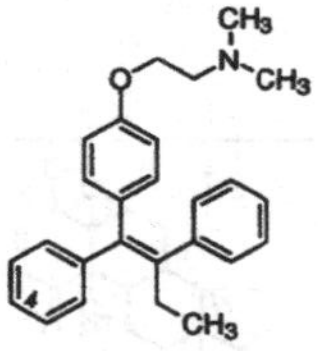 **49**	N-Demethylation and/or C-4 hydroxylation ; N-oxidation	N-Oxidation and N-demethylation (*Gliocladium roseum*, *Cunninghamella elegans*, *C. blakesleeana*) C-4 Hydroxylation (*Streptomyces rimosus*)	[115, 116]
D-Tetrandrine	Antitumor	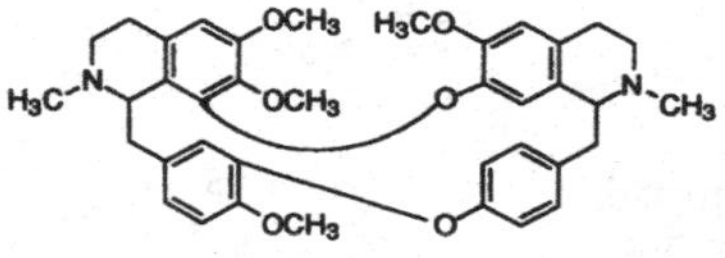 **50**		N- or N′-Demethylation (*Streptomyces griseus*, *Cunninghamella blakesleeana*)	[14, 117, 118]
Thalicarpine	Antitumor	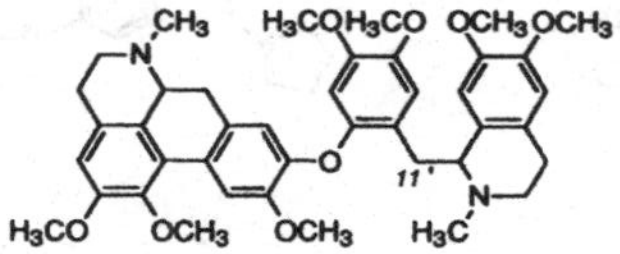 **51**		C-C Bond cleavage at C-11′ to aldehyde and reduction to alcohol (*Streptomyces punipalus*)	[14]
Thymoxamine	α-Adrenergic blocking agent	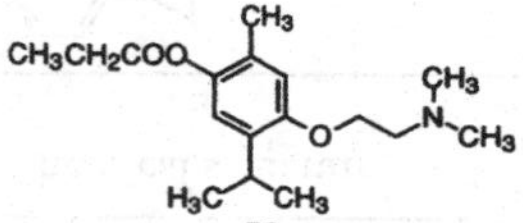 **52**	Deacetylation and N-demethylation Sulfo and glucurono-conjugation	Deacetylation and N-demethylation (*Mucor rouxii*, *Mortierella isabellina*) N-oxidation (*M. isabellina*) Sulfo and glucoconjugation (*M. isabellina*, *Actinomucor elegans*)	[119, 120]

Table 2 (continued)

Drug	Pharmacological activity	Chemical structure	Mammalian metabolism	Microbial metabolism (microorganism)	Ref.
Tranylcypromine	Antidepressant	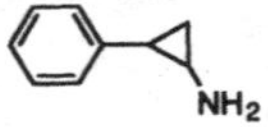 NH_2 **53**	N-Acetylation; ring-hydroxylation	N-Acetylation; C-4 ring-hydroxylation and O-acetylation (*Cunninghamella echinulata*)	[121]
Triprolidine	Antihistamine	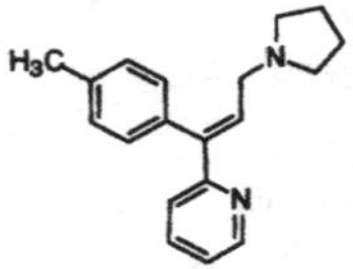 H_3C N N **54**	CH_3 oxidation to alcohol and carboxylic acid; oxidation and subsequent opening of the pyrrolidine ring; 3-hydroxylation of the pyridine ring	CH_3 oxidation to hydroxymethyl group (*Cunninghamella elegans*)	[122]
Withaferin A	Antitumor	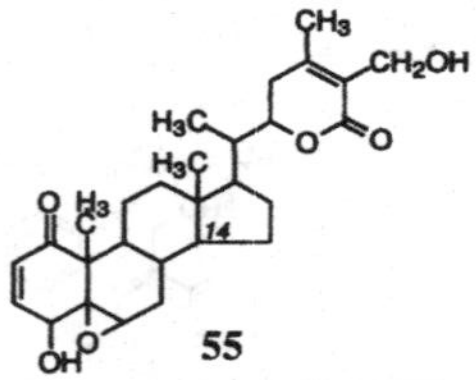 CH_3 CH_2OH H_3C H_3C O O O H_3C 14 O OH **55**		C-14α-Hydroxylation; unidentified metabolite (*Cunninghamella elegans*) 7 metabolites (*Arthrobacter simplex*),	[14, 123–125]

Table 2 (continued)

Drug	Pharmacological activity	Chemical structure	Mammalian metabolism	Microbial metabolism (microorganism)	Ref.
Zearalenone	Oestrogenic (farm animals)	OH, O, CH_3, HO, 2, 4, 10′, 8′, 6′, 1′, 2′, O; **56**	Reduction (α- or β-) of 6′-ketone; conjugation to glucuronide	Hydrolysis of the lactone ring; oxidation of 10′-OH; reduction of 6′-ketone (*Gliocladium roseum*).	[126, 127]
				Reduction (α- or β-) of 6′-ketone (*Streptomyces griseus, Mucor bainieri*)	[127]
				Reduction (α- or β-)of 6′-ketone and 1′,2′-double bond (*Aspergillus ochraceus, A. niger*)	[127]
				Reduction of 1′,2′-double bond (*Saccharomyces cerevisiae*)	[127]
				C-8′ (*S*) Hydroxylation (*Streptomyces rimosus*)	
				2,4-O-methylation and 2-O-methylation, 4-O-glucosylation (*Cunninghamella bainieri*)	[126, 127]
				4- and 2,4-O-glycosylation (*Rhizopus sp., Thamnidium elegans*)	[127–130]
				4-O-sulfate conjugation (*Rhizopus arrhizus*)	[131]

Phase II) metabolism of drugs and the privileged use of a small number of bacterial or fungal species.

Hydrolysis reactions are illustrated by the deacylation of colchicine (**10**) (amide hydrolysis), olivomycin A (**29**), Rifamycin B (**46**) or thymoxamine (**52**) (ester hydrolysis). The reduction of pentoxyfylline (**32**), zearalenone (**56**) or warfarin (**65**) are examples of the common reduction of keto groups, generally affording, with a high stereospecificity, one of the alcohol stereoisomers.

Oxidative biotransformation provides a commonly reported means of removing alkyl groups from substituted oxygen and nitrogen atoms. These reactions are believed to proceed via monooxygenase-mediated α-hydroxylation of the alkyl group to an unstable hemiacetal or hemiaminal, as outlined in Fig. 4 for an *O*-alkyl substituted compound.

O-Dealkylations have been used in the regioselective conversion of methoxyaryl groups to phenols in natural products, particularly alkaloids [47] such as papaverine (**30**), and in the production of mammalian metabolites by microbial biotransformation, as exemplified by the biotransformation of bisprolol (**4**).

N-Dealkylations by microorganisms [57] are very common reactions in drug metabolism. Microbial removal of alkyl groups from nitrogen is generally, but not exclusively, restricted to tertiary (trisubstituted) amino substrates (Fig. 5). Like the dealkylation of ethers, microbial *N*-dealkylation has been largely employed in the study of alkaloid metabolism (codeine **9**, lergotrile **22**, tetrandrine **50**, etc.). A considerable number of *N*-alkylated drugs, all of which are primarily metabolized in animals by *N*-dealkylation, have been examined using *C. elegans* ATCC 9245 or ATCC 36112 as microbial metabolizers: most arylalkylamines including *N*-alkylated amphetamines (**57**) [136–138] and pargyline (**58**) [139], deprenyl (**59**) [139], methoxyphenamine (**60**) [140] show significant *N*-dealkylation reactions; a series of antihistaminic compounds such as pheniramine (**39**) (and its halogen substituted derivatives bromo- and chloro-pheniramines **40**, **41**), pyrilamine (**61**) [73], tripelennamine (**62**), methapyrilene (**63**) and thenyldiamine (**64**) [141] are *N*-demethylated in yields ranging from 5 to 20%.

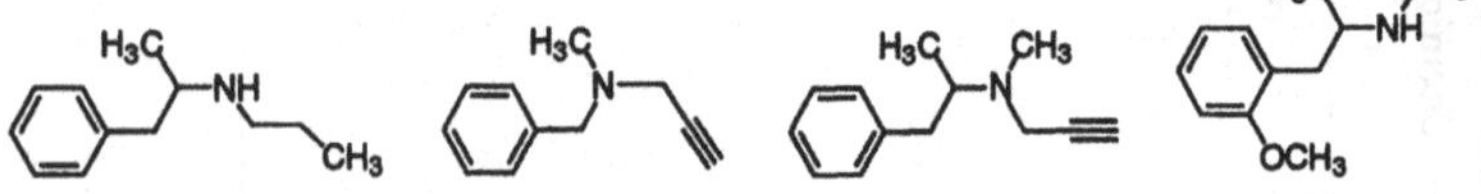

57 **58** **59** **60**

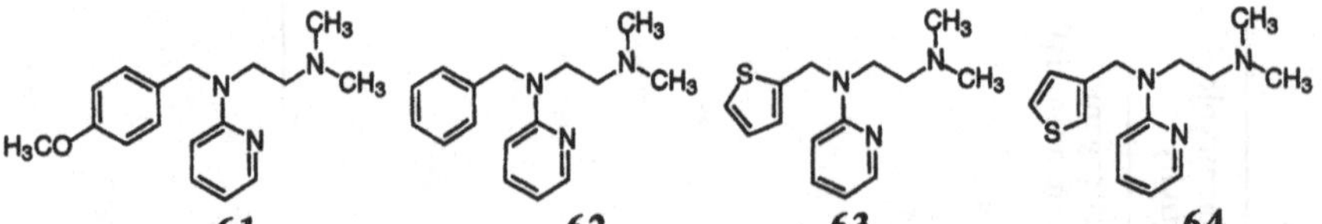

61 **62** **63** **64**

In addition, other drugs such as alosetron (**2**), cyproheptadine (**12**), diazepam (**13**) or tamoxifen (**49**) are *N*-demethylated by various microorganisms as a major metabolic route. The carboxylic acid resulting from the oxidative cleavage of the piperidine ring of phencyclidine (**36**), probably proceeding through an

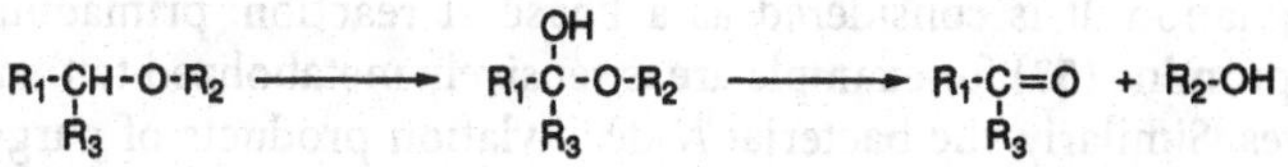

$$R_1\text{-}CH(R_3)\text{-}O\text{-}R_2 \longrightarrow R_1\text{-}C(OH)(R_3)\text{-}O\text{-}R_2 \longrightarrow R_1\text{-}C(R_3){=}O + R_2\text{-}OH$$

Fig. 4. Mechanism of an O-dealkylation reaction catalyzed by monooxygenases

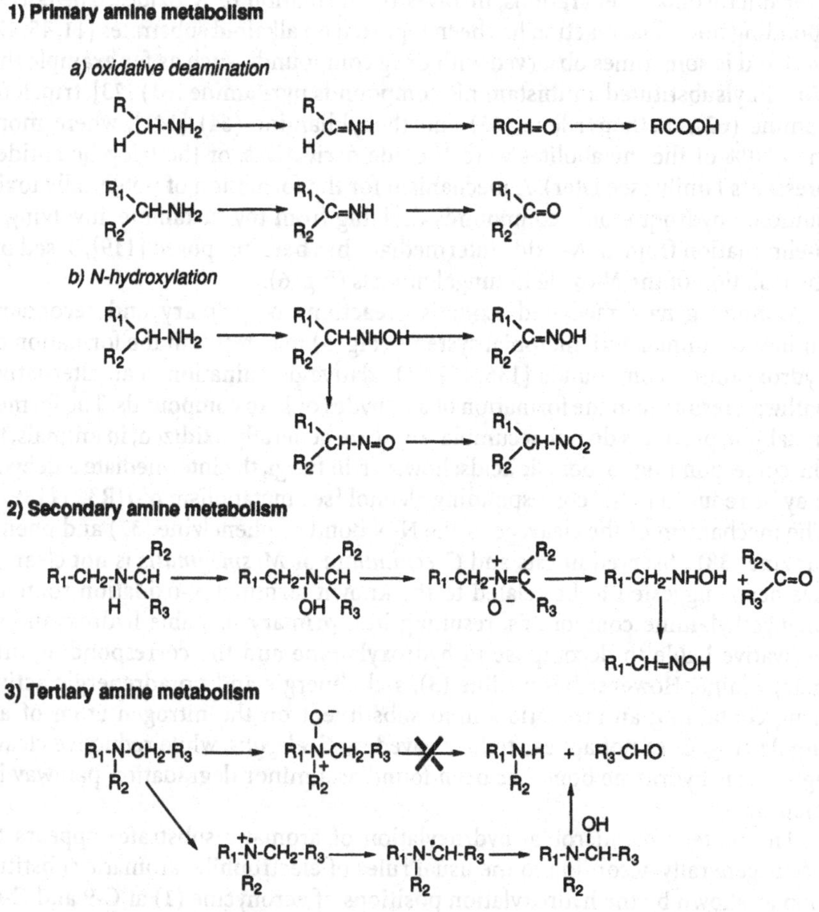

1) Primary amine metabolism

a) oxidative deamination

$$R(H)CH\text{-}NH_2 \longrightarrow R(H)C{=}NH \longrightarrow RCH{=}O \longrightarrow RCOOH$$

$$R_1R_2CH\text{-}NH_2 \longrightarrow R_1R_2C{=}NH \longrightarrow R_1R_2C{=}O$$

b) N-hydroxylation

$$R_1R_2CH\text{-}NH_2 \longrightarrow R_1R_2CH\text{-}NHOH \longrightarrow R_1R_2C{=}NOH$$

$$R_1R_2CH\text{-}NHOH \longrightarrow R_1R_2CH\text{-}N{=}O \longrightarrow R_1R_2CH\text{-}NO_2$$

2) Secondary amine metabolism

$$R_1\text{-}CH_2\text{-}N(H)\text{-}CHR_2R_3 \longrightarrow R_1\text{-}CH_2\text{-}N(OH)\text{-}CHR_2R_3 \longrightarrow R_1\text{-}CH_2\text{-}\overset{+}{N}(O^-){=}CR_2R_3 \longrightarrow R_1\text{-}CH_2\text{-}NHOH + R_2R_3C{=}O$$

$$R_1\text{-}CH_2\text{-}NHOH \longrightarrow R_1\text{-}CH{=}NOH$$

3) Tertiary amine metabolism

$$R_1\text{-}N(R_2)\text{-}CH_2\text{-}R_3 \longrightarrow R_1\text{-}\overset{+}{N}(O^-)(R_2)\text{-}CH_2\text{-}R_3 \not\longrightarrow R_1\text{-}N(R_2)\text{-}H + R_3\text{-}CHO$$

$$R_1\text{-}N(R_2)\text{-}CH_2\text{-}R_3 \longrightarrow R_1\text{-}\overset{+\bullet}{N}(R_2)\text{-}CH_2\text{-}R_3 \longrightarrow R_1\text{-}N(R_2)\text{-}\dot{C}H\text{-}R_3 \longrightarrow R_1\text{-}N(R_2)\text{-}CH(OH)\text{-}R_3 \longrightarrow R_1\text{-}N(R_2)\text{-}H + R_3\text{-}CHO$$

Fig. 5. Biological oxidation reactions of primary, secondary and tertiary amino compounds

α-hydroxylation, has been shown to be a major metabolic product in dog, human and *C. echinulata*. Some secondary *N*-dialkylated compounds have been reported to be the substrates of *N*-dealkylation reactions: propanolol (**45**) suffers the removal of the *N*-isopropyl group in low yield on incubation with *C. echinulata* ATCC 9244, and furosemide (**17**) is *N*-dealkylated by *C. elegans* ATCC 36112.

A further reaction of *N*-dealkylated amino compounds, which is also currently observed in microbial systems with drugs exhibiting a free amino group, is *N*-acetylation. It is considered as a Phase II reaction: primaquine (**42**) or tranylcypromine (**53**) for example are extensively metabolized to their *N*-acetyl derivatives. Similarly, the bacterial *N*-dealkylation products of pargyline (**58**), deprenyl (**59**) [139], phenelzine (**37**) or pheniprazine (**38**) are mostly converted to their *N*-acetylated derivatives. Another widespread reaction of tertiary and heterocyclic aromatic amino compounds in animals [142], and particularly in liver microsomal preparations, involves the formation of *N*-oxides. The corresponding microbial reaction has been reported on alkaloid substrates [11, 45, 47, 143] and is sometimes observed with drug compounds, such as for example the *N*-trialkylsubstituted antihistaminic compounds pyrilamine (**61**) [73], tripelennamine (**62**), methapyrilene (**63**) and thenyldiamine (**64**) [141], where more than 80% of the metabolites were *N*-oxide derivatives, or the tricyclic antidepressants family (see later). A mechanism for the formation of potentially toxic quinonic/hydroquinonic compounds deriving from thymoxamine, involving a β-elimination from an *N*-oxide intermediate, has been proposed [119], based on the isolation of the *N*-oxide in fungal models (Fig. 6).

Monooxygenase-mediated oxidative reactions of primary and secondary amines by animal and microbial systems (Fig. 5) may result in the formation of hydroxyamino compounds [136, 137]. Oxidative deamination is an alternative pathway, resulting in the formation of aldehydes or keto compounds. The former metabolic products do not accumulate and are generally oxidized, in animals, to the corresponding carboxylic acids; however, in fungi, the intermediate aldehyde may be reduced to the corresponding alcohol [see metabolism of HR325 (**149**)]. The mechanism of the cleavage of the N-N bond of phenelzine (**37**) and pheniprazine (**38**) observed in rats and *C. echinulata* or *M. smegmatis* is not clear: it has been suggested to be related to the known terminal *N*-oxidation route of phenyethylamine compounds, resulting in a primary unstable hydroxyamino derivative liable to decompose to hydroxylamine and the corresponding primary amine. However besipirdine (**3**), a cholinergic and noradrenergic active drug containing an aromatic amino substituent on the nitrogen atom of an indole ring, does not appear to be cleaved by *C. elegans*, while reductive cleavage of this hydrazine bond has been found as a minor degradation pathway in mammals.

The pattern of microbial hydroxylation of aromatic substrates appears to occur generally according to the usual rules of electrophilic aromatic substitution as shown by the hydroxylation positions of acronycine (**1**) at C-9 and C-11 (*para*- and *ortho*-positions to the amino group and *meta*-position to the car-

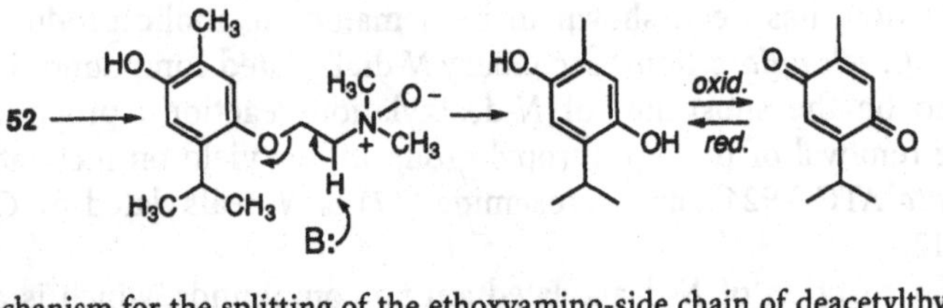

52 → [structure] → [structure] ⇌ (oxid. / red.) [structure]

Fig. 6. Mechanism for the splitting of the ethoxyamino-side chain of deacetylthymoxamine, involving an oxidative β-elinination from the N-oxide

bonyl substituent). The *para*-position is favored and constitutes frequently the unique aromatic hydroxylation position, as described for diazepam (**13**), GR117,289 (**18**), mexiletine (**24**), phenazopyridine (**35**), propanolol (**45**), tamoxifen (**49**) or tranylcypromine (**53**). A few other drugs give rise to adjacent monohydroxylated products which do not correspond to the expected hydroxylation pattern: ellipticine (**15**) for example is hydroxylated by *A. alliaceus* to give, in comparable amounts, a 8- and a 9-phenol, while a 7- and a 9-phenol are produced in mammalian metabolism. Similar results have been already observed in the C-10 and C-11 hydroxylations of yohimbine alkaloids [144] by *C. elegans* and are found again with besipirdine (**3**), cyproheptadine (**12**) or L-696,474 (**20**). The reasons for formation of such products are unclear: the operation of the NIH shift and the formation of phenols by a regioselective opening of an arene-oxide in all microbial hydroxylation reactions [11, 35] are questionable.

Moreover, a variety of other drugs of synthetic origin, containing a phenyl substituent, do not suffer any aromatic hydroxylation, neither in animal systems nor in microbial systems: they often correspond to trisubstituted amino group containing derivatives, where *N*-dealkylation is a major detoxification pathway and probably results in products exhibiting a sufficient decrease in hydrophobicity.

Epoxides, the products of monooxygenase activity on isolated or conjugated double bonds, are considered as very reactive metabolites, potentially responsible for toxic effects through reaction with nucleophilic groups present in DNA and proteins, and possibly responsible for immune reactions resulting from the modification of the macromolecules. As a rule, in animals, they have to be detoxified by enzymatic systems catalyzing their elimination by electrophilic attack on water (epoxide hydrolase) to give diols, or on thiols such as glutathione (glutathione-*S*-transferase) to give thioethers. Epoxides from isolated or conjugated double bonds are sometimes formed in microbial systems and eventually survive in the incubation and isolation conditions (see carbamazepine **7**, cyproheptadine **12** or RU27987 **79**), despite the known occurrence of epoxide hydrolase [145, 146] and glutathione transferase [147, 148] activities in fungi. Hydroxylation at saturated carbons represents a common mode of biotransformation in drug metabolism, either in animals or microorganisms. Benzylic or allylic positions (even as a methyl group) are generally favored for such a reaction, as exemplified for 10,11-dihydrocarbamazepine (**8**), codeine (**9**), lapachol (**21**), lucanthone (**23**), mycophenolic acid (**26**), novobiocin (**27**), praziquantel (**44**), rifamycin B (**46**), spironolactone (**47**) or triprolidine (**54**). Completely unactivated mono-, di- or trisubtituted carbons are also frequently hydroxylated, affording metabolites which could be difficult to access by chemical oxidation or synthesis: one of the *gem*-dimethyl group of acronycine (**1**) is hydroxylated to a hydroxymethyl group by a *Streptomyces*, whereas one of the methyl group of the *t*-butyl substituent of ebastine (**14**) is oxidized, via a preliminary hydroxylation reaction, to a carboxylic group, in high yield, by *C. blakesleeana*. Bornaprine (**6**), diazepam (**13**), GR117289 (**18**), MK954 (**25**), phencyclidine (**36**), zearalenone (**56**) are examples of hydroxylation at isolated and unactivated -CH_2- positions. Withaferin A (**55**), like many other steroid compounds [149], is hydroxylated at C-14, an unactivated tertiary carbon atom, as is ibuprofen (**19**) or dihydroquinidine (**157**, see below).

Other miscelleanous oxidation reactions involve *S*-oxidation of thioethers to sulfoxides and sulfones as observed with pergolide (**33**), the thiomethyl derivative of spironolactone (**47**), and sulindac (**48**).

Only Phase I metabolites were initially thought to be formed in the metabolism of drugs by microorganisms. As a matter of fact, a number of Phase II metabolites have been occasionally isolated and characterized in microbial models. Beside *N*-acetyl derivatives of amino compounds (see above), *O*-glucosylated derivatives of erythromycin A (**16**), furosemide (**17**), lapachol (**21**), NSC159628 (**28**), thymoxamine (**52**), zearalenone (**56**) have been isolated from incubations with fungi or *Actinomycetes*, sometimes in high yield. In addition, sulfate or phosphate conjugation of hydroxylated metabolites has been frequently observed (see erythromycin A **16**, phenazopyridine **35**, thymoxamine **52** or zearalenone **56**). It is almost certain that a reexamination of some of the literature examples where very polar metabolites were observed (and extracted with difficulty) but remained unidentified, would show the occurrence of one or several additional Phase II metabolites. However, a Phase II reaction such as glucuronidation, which is a characteristic conjugation reaction in animals, is rarely found in microorganisms: only two examples of glucuronide conjugation have been reported, one with phenol derivatives formed in the metabolism of naphthalene and biphenyl by *C. elegans* [150], a second one corresponding to the glucuronidation of the 2-nitrogen atom of the tetrazole ring of angiotensin II receptor antagonists such as MK 954 (**25**) by a *Streptomyces* strain [83, 84]. Such derivatives seem to be frequently replaced in microorganisms by D-glucose conjugation products of phenol, alcohol or carboxylate metabolites; more exotic glycosylated species can be sometimes found, as for example the 4-*O*-methyl-D-glucosides repeatedly isolated from metabolic studies with *B. bassiana* [151–155]. Another conjugation reaction, the formation of thioethers of glutathione, frequently found as Phase II metabolites in animals, has never been described, despite the fact that glutathione-S-transferases are present in low amounts in bacterial or fungal microorganisms [147, 148, 156]. The metabolism of the analgesic compound phenacetin (**34**) and its *O*-alkyl homologues by *C. elegans* yields the expected *O*-dealkylation products [97], but none of the GSH-conjugated derivatives of its putative reactive metabolites in human, such as for example the *N*-acetyl-*p*-quinoneimine, which is considered as responsible for the cytotoxic effects of the drug.

In summary, there are many advantages in the use of microbial models. They often parallel very closely transformations in humans, not differing more than other in vitro models. Their use would result in a lesser demand for animals and a greatly reduced cost for metabolic studies. Other advantages include the general occurrence of selective (or major) transformations using different strains, selected by screening, which results in easier isolation of a smaller number of metabolites in high yields and sizable amounts. Specific metabolites can thus be produced from the parent drug and made available for detailed structural studies, as standard samples for the chromatographic identification of animal metabolites, and after scaling-up, for further pharmacological and toxicological studies, without the need for tedious chemical synthesis. Manipulation of experimental parameters of culture or incubation (induction, aeration, enzyme inhibition, etc.) may be used in order to control the generation of desired metabolites. Identification of active metabolites from prodrugs is facilitated, as is the early prediction of possibly activated (and toxic) metabolites. Even when new unexpected metabolites are produced, it is possible to deduce from them likely metabolic pathways operating further in animals. New metabolites with higher activity or less toxicity may be discovered, or metabolites with new biological activities, constituting new leads for pharmacological research ("biocombinatorial chemistry").

Even at the level of Phase I metabolites, it is not realistic to expect the formation of all of the animal metabolites of a drug using a single microorganism. For some time, members of the genus *Cunninghamella* (for example *C. elegans* or *C. echinulata*) [157] or *Beauveria bassiana* [63] have been proposed to be able to mimic virtually all of the transformations performed by intact mammalian models, organ and cell systems. In fact, although these strains are known to have extended metabolic capabilities (see Table 2), it is most frequently necessary to address some other strains, selected by screening from a limited number of classical strains already used in the literature, to cover the whole pattern of animal metabolites, and especially when one wants to obtain a high-yield single metabolite preparation. Finally, it is to be noted that in a few cases, with a number of microorganisms, no metabolism (or a very slow one) occurs: this is generally related to a too high hydrophobicity of the drug, which apparently is taken up by the microorganism, but remains indefinitely trapped in a lipidic (cell membrane) compartment without access to the detoxifying activity of the cytoplasm and subcellular systems.

Accounting with so many advantages and few drawbacks, a valuable strategy in the metabolic studies associated with the development of a new drug has been recommended, consisting of investigating first its microbial metabolism, slightly before or in parallel with animal metabolic studies. This will afford in sufficient amounts for structural identification the necessary metabolite standards, at least those of Phase I, to identify by comparative chromatographic methods, eventually using minute amounts of the radiolabelled drug, most of the metabolites produced by a limited number of animals. Would some of these metabolites be needed for a detailed pharmacological or toxicological study, scaling-up of the microbial transformation with selected strains will be able, with minimal efforts, to afford even gram quantities of the desired metabolites. If there is still some doubt about the effective general predictive value of the microbial models, the observed formation of a reactive metabolite (such as an epoxide for example) by a microorganism is nevertheless indicative of a potential metabolism-induced toxicity of the parent drug in animal systems.

Although the microorganism does not always form the metabolic products of the drug in the same relative abundance as occurs in humans, it may be useful for drug interaction and drug disposition studies. The addition of quinidine and sparteine affects the rates of methoxyphenamine or propanolol metabolism in *C. bainieri* [110, 140], as it does in humans or animals. As no correlation exists in the mammalian and microbial isozymes, the mechanism involved in the microorganism is still unknown, but may be similar to that involved in animals. In addition, the fungal metabolism of xenobiotics can be affected by dose, nutrients, inducers and environmental conditions [121], indicating that it may be possible to evaluate drug toxicity in microbial models [111, 158].

3.2 Methodology Used in Microbial Model Development

As a technique, this methodology is clearly related to that which has been developed in the broader field of microbial transformation studies: drugs, as well as any other chemicals, can be metabolized and biotransformed by a variety of enzymes and the use of the microbial biocatalysts in synthetic chemistry, either as pure enzymes or cultivated whole cells, is now a well documented technique [4, 24, 26–28, 30, 40, 143, 159–172]. A general strategy for the development of microbial models of mammalian metabolism of drugs is given in Fig. 7. Three different approaches have been defined by Davis [21] for the development of microbial models of the metabolism of a given agent. i) In a "prospective" study, routes of metabolism are first explored in microbial systems, then to be extrapolated to mammals. The prediction and confirmation of metabolite production in mammals would then be markedly facilitated by the prior generation of analytical standards using the microbial model. ii) "Retrospective" microbial modeling is conducted following the exploration of mammalian routes of metabolism. Such studies can be used to confirm tentative assignments of (minor) mammalian metabolites, and serve to further document and validate biotrans-

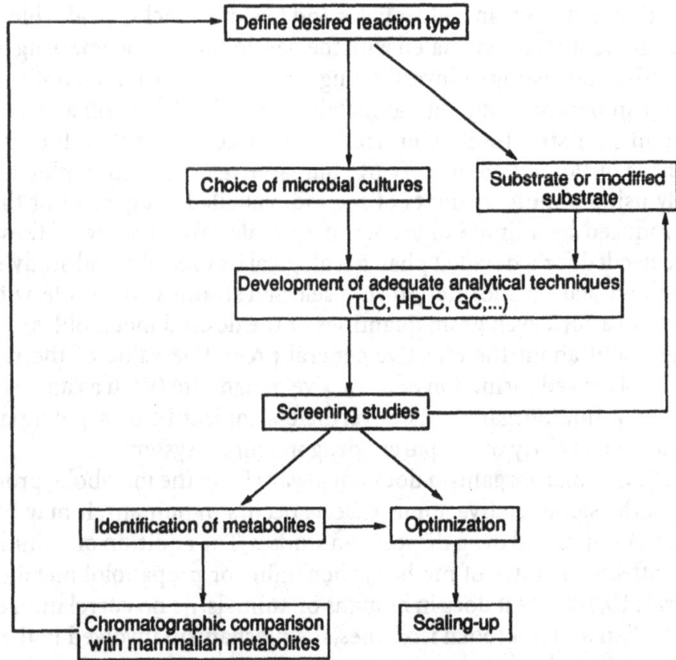

Fig. 7. General strategy for the development of a microbial model of metabolism of a drug

formation parallels between the two systems. iii) One may also conduct "parallel" (i.e. simultaneous) microbial and mammalian metabolism studies. In any approach, as discussed above, the preparative scale production of metabolites would facilitate investigations concerning their biological and toxicological evaluation, stereochemistry, or mechanism of formation.

The development of an appropriate analytical methodology is essential for success and has to precede screening studies: most generally used techniques focus on thin-layer chromatography (TLC), gas chromatography (GC), high performance liquid chromatography (HPLC), possibly coupled to mass spectrometry (GC-MS or LC MS). The recently introduced LC-NMR coupled techniques appear to be particularly adapted to the product scale and potential of microbial models [173]. The availability of identified metabolites or those suggested to occur in mammals ("retrospective" studies) is an important element for the development of adequate analytical methods. If no metabolite standards are available, a useful initial strategy is to adapt methods developed previously for the parent drug or for other agents in the same structural class, for example to develop a reverse-phase HPLC system with a retention time for the parent drug of about 30 min. Since most metabolites are more polar than the parent agent, they will generally exhibit a shorter retention time. One can thus obtain a chromatographic profile of metabolites in a reasonable analysis time.

Generally, a large number of microorganisms will be selected for a preliminary screening of their ability to metabolize a drug substrate. The selection of microorganisms is based on a number of factors including literature precedent, experience and intuition. Cultures may be isolated from soil, air, or from sewage treatment facilities, or they may be obtained in pure form from culture collections. Because of their established identity and general availability, most investigators prefer to use documented microbial cultures (i.e. strains deposited in and available from standard culture collections). It is highly advantageous to use the literature (see also Table 2) to obtain suggestions for cultures needed to perform specific types of biotransformations such as *O*- and *N*-dealkylations, dehydrogenation, hydrolyses, reductions, or hydroxylations. In our experience, the maintenance of a limited number of selected standard strains (about 30–40) may be sufficient for performing a large number of the desired biotransformations. However, workers in this field generally maintain larger collections (perhaps up to 100–400 cultures) in their own laboratories. That allows them to combine the "literature" and the "random screening" approaches.

Cultures in our collection are grown on various types of agar slants, depending on the growth requirements of the specific class of microorganism (e.g. nutrient agar for most bacteria, or ATCC Medium 5 sporulation agar for Streptomyces, YM agar for fungi) then maintained at 5 °C. Cultures are transferred to fresh slants every 1–3 months to maintain viability, and only freshly transferred strains are used for biotransformation experiments. In order to keep the initial screening process easier, a unique enriched liquid medium is generally chosen which can support the growth of a wide range of microorganisms. A selection of general use media has been given by Smith and Rosazza [11]. We prefer to use a medium consisting of: corn steep liquor, 10 g; glucose, 30 g; dihydrogen potassium phosphate, 1 g; dipotassium hydrogen phosphate, 2 g; sodium nitrate, 2 g;

potassium chloride, 0.5 g; magnesium sulfate · $7H_20$, 0.5 g; ferrous sulfate · $7H_20$, 0.02 g in 1 l of deionized water (final pH between 6.5 and 7.0). Soybean peptone (5 g/l) may be added in order to increase the obtained biomass and eventually induce some monooxygenase activities [52, 174]. *Actinomycetes* sometimes require more specific media which can be found in the literature.

A two-stage fermentation procedure has been previously recommended for obtaining high activities. However, with fungi and for initial screening studies, we prefer to use a one-stage culture in Erlenmeyer flasks (100 ml of corn steep medium in 250 ml flask), starting from freshly sporulated agar slants, and using a spore suspension for the inoculum, which generates, on incubation at 27 °C for 60–70 h in a gyrotatory shaker (200 rpm), a culture containing a large number of small mycelial pellets. Drugs are usually added to the grown cultures as solutions (1–3 vol.%) in a water-miscible nontoxic solvent such as ethanol, acetone, dimethylformamide or dimethylsulfoxide, sometimes added with a small amount of a neutral emulsifying agent, such as 0.1% Tween 80. Polyvinylpyrrolidone polymers have been successfully used for the solubilization of extremely water-unsoluble substrates [66]. An amorphous precipitate of the solute generally occurs upon addition of the organic solution to the aqueous incubation medium [13], but the finely divided suspension of substrate is generally rapidly adsorbed into the lipids of the cell membrane and becomes available for metabolization, unless crystallization takes place. For initial screening, drug substrates are added at a concentration of 200–500 mg/l of culture broth, which is generally nontoxic to the organisms. This concentration usually allows easy detection of even minor metabolites.

There are conflicting data in the literature about the importance and the effect of the pellet morphology of mould submerged cultures, related to their metabolizing activity. Some authors, for example, recommend the addition of agar (0.1–1%) or water-soluble polymers (carboxymethylcellulose, polyvinyl alcohol, etc.) to the culture medium [64] in order to reduce the pellet formation, to promote filamentous growth and to obtain higher conversion yields. Conversely, others have thoroughly investigated the nutritional [175] and physical factors (pH, shear forces, oxygenation, etc.) of the cultures to favour the formation of pellets, and the addition of microcrystalline cellulose [21] has been recommended to prevent clumping of the mycelium and promote the formation of a large number of small pellets, which would lead to optimal conversion activity. In our hands, at least at the screening step, we did not observe any significant difference in the metabolizing activities of a fungal strain, in whatever morphology it was growing.

However the pellet formation is very useful when an alternative drug incubation protocol is followed, where microbial growth is strictly separated from biotransformation, involving the use of "resting cells" or "replacement cultures": this method involves the recovery of the grown biomass from the initial culture broth, and its resuspension, after intermediate washing, in a well-defined medium, typically at a two- to three-fold higher cell concentration than in the growing culture. Clearly, pellets represent a material much easier to handle, being filterable on a simple cheese-cloth, and relatively insensitive to repeated manipulation. This incubation method offers a number of distinct advantages including the following: i) sterile conditions are not required, since growth medium is absent; ii) isolation of metabolites from a simple buffer is much easier than from a complex medium; iii) fermentation parameters (pH, aeration, biomass concentration, etc.) can be individually optimized, independently of microbial growth. However, the specific activity of the biomass is frequently slightly lower, compared to the unmanipulated mycelium.

Incubation times may extend to 7–10 days, in the absence of any added energy source, because most microorganisms used have accumulated during growth sufficient internal nutrients to maintain primary metabolic functions. During the incubation period the cultures require essentially no maintenance. Samples (1–2 ml) of the fermentation broth are withdrawn at 24 h-intervals after addition of the substrate for subsequent analysis. With fungi cultures it is difficult to obtain homogeneous samples of the whole incubation mixture and generally only the supernatant fraction can be used for quantitative measurements, thus accounting for only a fraction of the remaining substrate and, possibly, of the metabolite(s). Aqueous samples can be directly analyzed by HPLC, or repeatedly extracted with organic solvent for TLC or GC analysis. Occasionally, for difficult-to-extract solutes (Phase II metabolites), incubation samples have to be frozen and lyophilized to facilitate the isolation process. Freeze-dried samples may then be extracted with polar solvents such as alcohols or acetone.

Preparative-scale incubations for production of quantities of the metabolite(s) for structure elucidation and biological evaluation can be achieved, after optimization, either in numerous larger sized Erlenmeyer flasks holding one to several liters of medium, or in stirred bench-top fermentors. Usual fermentors can be used for mycelium forming organisms, provided the shearing forces are minimized, for example by using a helix impeller in place of the classical Rushton turbines; obtaining fungal pellets in fermentors sometimes necessitates a careful counting and proportioning of the spore inoculum (about 10^6–10^7 spores/l). Airlift (bubbles-driven) fermentors can be profitably used to favour hydroxylation reactions. Optimization through manipulation of the culture or incubation parameters [81], or the addition of inducers [16] or inhibitors [62] can be used at this stage to increase the yield of the biotransformation or orient it to a more specific production of the desired metabolite.

3.3
Selected Examples and Applications of the Microbial Metabolism of Drugs

3.3.1
Warfarin and Related Anticoagulants

Racemic warfarin (**65**), a vitamin K antagonist, has been used for decades both as an oral anticoagulant in man and as a rodenticide. The metabolism of this drug has been found to be substrate-enantioselective: 9*S*-warfarin is considered as more active than the 9*R*-antipode. In mammalian systems, warfarin undergoes a stereoselective reduction of the ketonic side chain [176, 177], affording mainly the 9*R*,11*S*-alcohol (**71**), but the major biotransformation route involves substrate-enantioselective aromatic hydroxylations at 4′-, 6-, 7- or 8-positions

(**66–69**) and aliphatic hydroxylation at 10-position (**72**), the regio- and stereoselectivity of which depend on the specific P450 isozymes present [178]. In addition, an α-diketone (**73**) resulting from the oxidative opening and decarboxylation of the lactone ring, attributed to a dioxygenase activity [179, 180], was isolated from rat microsomal systems as a minor metabolite. The detection and identification of some metabolites, and particularly the latter, was complicated by the occurence of NMR-resolvable tautomeric forms, and the formation of intramolecular cyclic hemiketals from the enol group and the ketonic side chain, as already described for warfarin itself.

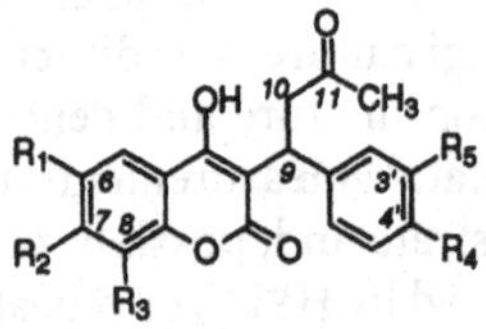

65: $R_1 = R_2 = R_3 = R_4 = R_5 = H$
66: $R_1 = OH$; $R_2 = R_3 = R_4 = R_5 = H$
67: $R_2 = OH$; $R_1 = R_3 = R_4 = R_5 = H$
68: $R_3 = OH$; $R_1 = R_2 = R_4 = R_5 = H$
69: $R_4 = OH$; $R_1 = R_2 = R_3 = R_5 = H$
70: $R_5 = OH$; $R_1 = R_2 = R_3 = R_4 = H$

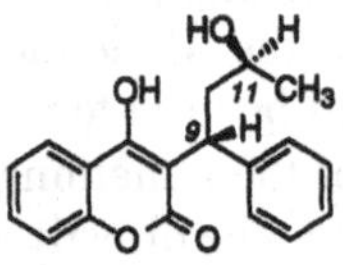

71

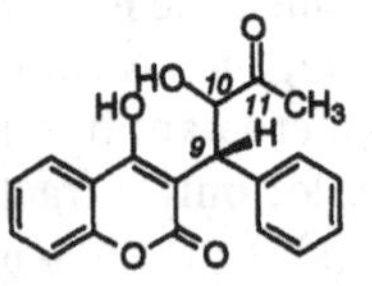

72

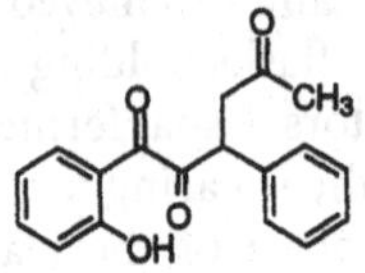

73

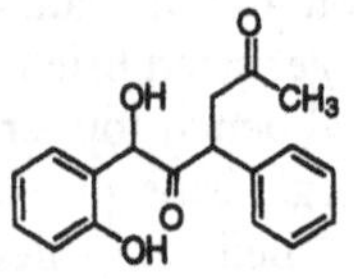

74

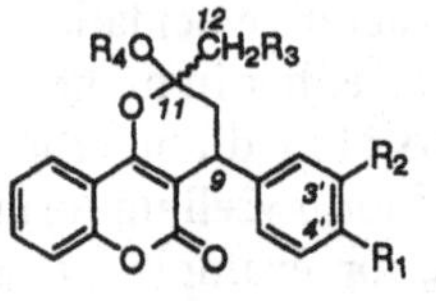

75: $R_1 = R_2 = OH$; $R_3 = R_4 = H$
76: $R_1 = R_2 = H$; $R_3 = OH$; $R_4 = H$
77: $R_1 = OH$; $R_2 = H$; $R_3 = H$; $R_4 = CH_3$

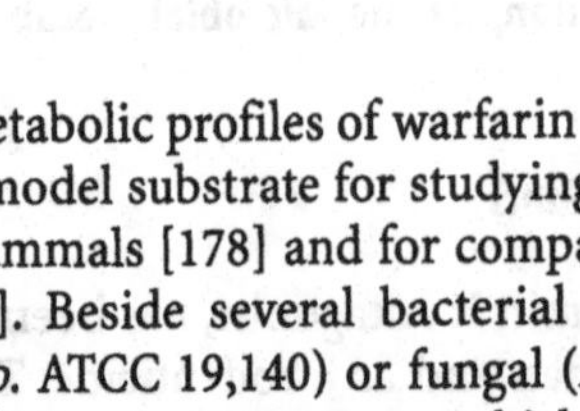

78

The diversity of the metabolic profiles of warfarin explains why this molecule has been chosen as a model substrate for studying the P450-mediated oxidative metabolism in mammals [178] and for comparing microbial and animal systems [23, 157, 181]. Beside several bacterial (*Nocardia corallina* ATCC 19,070, *Arthrobacter sp.* ATCC 19,140) or fungal (*Aspergillus niger* UI-X-172, *Cunninghamella bainieri*) microorganisms, which afforded only a fraction of the animal metabolites with various stereochemical preferences [181–183], the filamentous fungus *Cunninghamella elegans* ATCC 36,112 was found to produce all known mammalian metabolites, exhibiting the same stereoselectivity towards the 9*R*-enantiomer [21, 157]. In addition, a new metabolite,

3′-hydroxywarfarin (**70**), was isolated [184]. The microbial metabolism of warfarin was recently reinvestigated: from an extensive screening (> 50 strains), *Beauveria bassiana* IMI 12,939 was selected as an alternative source of the major mammalian metabolites, plus a diastereomeric mixture of α-ketols (**74**) [23, 63, 185]. Using direct coupling of HPLC and NMR, combined with mass spectrometry, as a detection and identification technique without prior isolation, most known metabolites have been recently identified directly in the broth incubation media from *Streptomyces rimosus* NRRL 2234 and *Beauveria bassiana* ATCC 7159 [173]. In addition, several new metabolites have been characterized: from the *Beauveria* strain, 3′,4′-dihydroxywarfarin (**75**) and the 3′-(4-*O*-methylglucoside) (**78**); from *S. rimosus*, 12-hydroxywarfarin (**76**) and 4′-hydroxy-11-methoxywarfarin (**77**). Last, it can be noticed that regioselective biotransformations of warfarin (C=O reduction, C-6 and C-10 hydroxylations) by the widely used cell suspension cultures of the plant *Catharantus roseus* have also been described [163].

A number of related coumarinic anticoagulants such as coumarin, phenprocoumon, clocoumarol, etc. have been the subject of similar investigations, with comparable results [181, 186, 187].

3.3.2
RU27987 (Trimegestone)

RU 27987 (**79**) is a synthetic 3-keto-$\Delta^{4,9}$-19-norsteroid recently developed by Hoechst-Marion-Roussel as a progestomimetic and a regulating factor of calcium assimilation in post-menopausal women, thus active as a drug for osteoporosis. This therapeutic agent is used in very low doses and, in the course of its toxicological study, identification of the metabolites produced in animals and in women was difficult owing to the very low amounts (10–100 µg) recovered. Moreover, only very few microbial conversions of such Δ-4,9(10) dienic steroids have been described up to now [188]. It was thus a good opportunity to apply the concept of microbial models of mammalian metabolism, in order to make some prediction about the nature of favored metabolites, to prepare significant amounts of the main ones for structure identification and as authentic chromatographic samples for the analysis of animal metabolites, and, if needed, to produce larger amounts of some of them for pharmacological and toxicological studies.

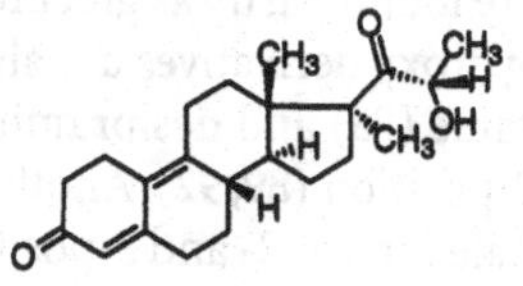

79

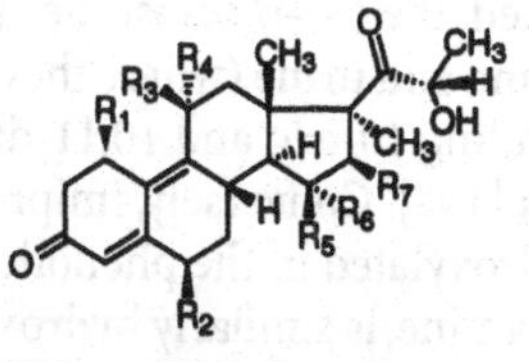

80: R_1 = OH; R_2-R_7 = H
81: R_2 = OH; R_1, R_3-R_7 = H
82: R_3 = OH; R_1, R_2, R_4-R_7 = H
83: R_4 = OH; R_1-R_3, R_5-R_7 = H
84: R_5 = OH; R_1-R_4, R_6, R_7 = H
85: R_6 = OH; R_1-R_5, R_7 = H
86: R_7 = OH; R_1-R_6 = H

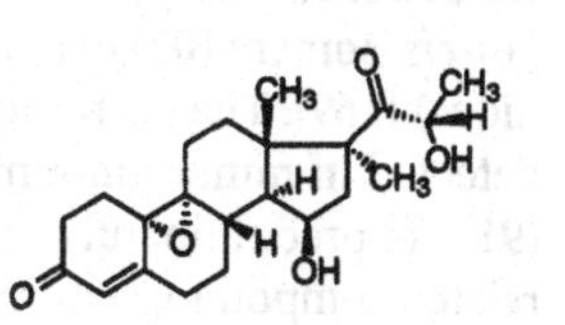

87

A screening of about 40 strains currently used for steroid hydroxylation was undertaken, using reverse-phase HPLC and UV absorption to detect and quantitate metabolite formation. Most of the strains tested were able to metabolize extensively this substrate within a 1- to 5-day period, producing in the incubation medium variable amounts of at least eight hydroxylated metabolites (**80**–**87**), essentially depending on the strain used [189].

By selecting the best strain for each metabolite, it was possible to prepare all of them in sufficient amount for structural identification (by NMR and MS) and preliminary pharmacological assays. More than 50% yields of the 11β-hydroxy derivative **82** and the 15 α-hydroxy derivative **85** were obtained using a strain of *Mucor hiemalis* or *Fusarium roseum* ATCC 14,717 respectively. Other hydroxylated metabolites were obtained in 10–30% yields using various *Absidia*, *Cunninghamella*, or *Mortierella* strains. Only metabolite **86** was poorly represented in all strains, which necessitated its isolation as a minor product from an incubation with *A. blakesleeana*. Epoxide **87** was formed only after prolonged incubation times with *C. bainieri*, indicating a secondary reaction occurring on one of the initial metabolization products. The samples obtained were compared to the minute amount of animal and human metabolites; they allowed the characterization of two of them, the 1β-hydroxy and the 6β-hydroxy derivatives (**80**, **81**), which were previously unidentified. In addition metabolite **80** (1β-hydroxytrimegestone) proved to have interesting biological activities [190].

3.3.3
Tricyclic Antidepressants

A series of tricyclic antidepressants constitutes a family of clinically widely used drugs: the first to be considered were imipramine (**88**) and amitriptyline (**95**), the pharmacological effect of which is caused, in part, by their active *N*-demethylated metabolites, desipramine (**91**) and nortriptyline (**96**) respectively. All of these drugs are metabolized in mammalian systems to aromatic hydroxylation products (position 2 and/or 8) and benzylic hydroxylation products (position 10 and/or 11), depending on the drug structure and the metabolizing organism, and giving rise to various isomeric possibilities [191]. For example, in humans, amitryptiline (**95**) and nortryptiline (**96**), lacking the central nitrogen atom, undergo benzylic hydroxylation to a major extent, while only minor amounts of phenolic metabolites are found in urine. Upon hydroxylation of nortryptiline (**96**), four 10-hydroxy stereoisomers may be formed, the (–)-*trans* isomer (**99**) predominating over the (+)-*cis* isomer (**102**) in plasma and urine (where they are found partly as glucuronides) [169]. The corresponding 10-oxo and 10,11-dihydroxy derivatives are also detected in minor amounts [192]. Conversely, imipramine (**88**) and desipramine (**91**) are predominantly hydroxylated in the phenolic 2-position (**89**, **92**). Another related compound, clomipramine, is similarly hydroxylated in the 2- and 8-positions [193]. The stereochemistry of the minor 10-hydroxylated products (**90**, **93**) has not yet been determined. Recently, the pharmacological properties of the hydroxy metabolites of the tricyclic antidepressants have been reinvestigated and, on the basis of their high efficacy and tolerability, the 10-hydroxy metabolites have been proposed as better antidepressant agents than the parent drugs [194].

The metabolism of imipramine (**88**) by microorganisms was first examined, as an early opportunity to establish the reliability of microbial systems to mimic and to allow prediction of the mammalian metabolism of a commonly used drug [19,193]. A high percentage of anaerobic or aerobic gut bacteria are capable of *N*-demethylating imipramine (**88**) to desipramine (**91**) [195]. Among a number of other microorganisms which formed one or more of the mammalian metabolites, *Cunninghamella blakesleeana* produced 2-hydroxy (**89**) and 10-hydroxyimipramine (**90**), while *Mucor griseocyanus* produced the 10-hydroxy (**90**) and the *N*-oxide (**94**) metabolites, in addition to the *N*-desmethyl derivative, desipramine (**89**) [196].

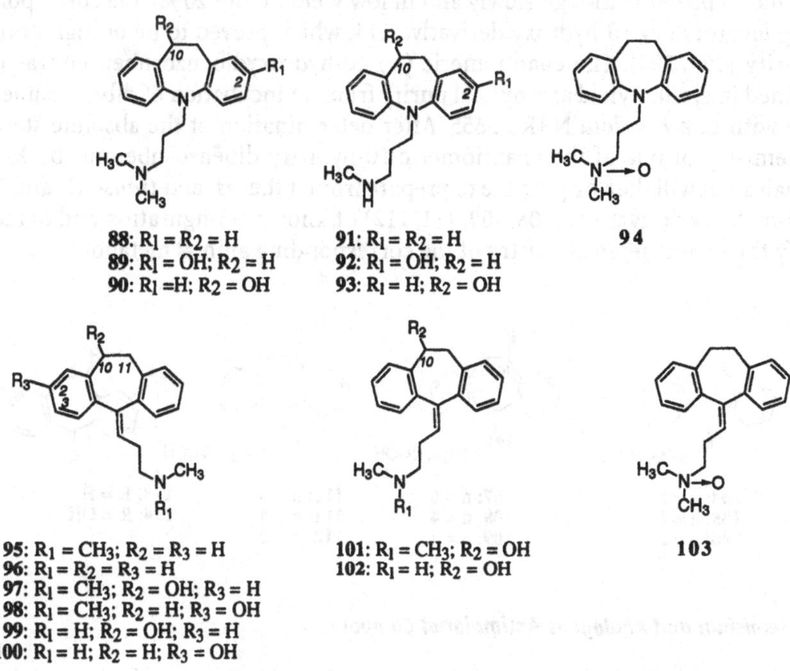

The microbial metabolism of amitriptyline was more recently investigated, using *Cunninghamella elegans* ATCC 9245 [197]: eight major metabolites were detected by HPLC and identified as *trans*- (**97**) and *cis*-10-hydroxy amitriptylines (**101**) and the corresponding *N*-demethylated nortriptyline derivatives (**99, 102**), 2- and 3-hydroxyamitriptylines, amitriptyline *N*-oxide (**103**), and nortriptyline (**96**). The *trans*-10-hydroxyamitriptyline (**97**), the stereochemistry of which was not reported, was formed as the major product (up to 35% after 72 h).

Another tricyclic antidepressant, amineptine (**104**), acting essentially by inhibition of dopamine uptake, differs from the other drugs by its aminoheptanoic side-chain. After oral administration to rat, dog or human, the main metabolites were acids with shortened C5 and C3 side-chains (**105, 106**), together with the corresponding 10-hydroxylated derivatives, the relative and absolute stereochemistries of which were not determined [198]. In an attempt to obtain significant amounts of the hydroxylated derivatives for pharmacological studies, and to elucidate their isomeric structure, the microbial metabolism of amineptine was investigated, using common fungal strains [199]. Most of them rapidly metabolize this substrate, affording mainly β-oxidation products in high yield (*Mucor plumbeus, Cunninghamella echinulata*); two of them (*Mortierella isabellina* and *Beauveria bassiana*) produced in addition two new metabolites which were identified as *cis*- (**107**) and *trans*-10-hydroxyamineptine (**110**) (the *trans* isomer predominating). Only very minor amounts of the desired C3- and C5-hydroxy derivatives were produced. However, using dibenzosuberone (**113**) as a substrate in incubation experiments, *Absidia cylindrospora* LCP 1569 was found to produce, though slowly and in low yield (about 20%), the corresponding levorotatory 10-hydroxy derivative **114**, which proved to be of high optical purity [199, 200]. The enantiomeric (+)-10-hydroxydibenzosuberone was obtained in similar yield and optical purity from an incubation of dibenzosuberone with *C. echinulata* NRRL 3655. After determination of the absolute stereochemistry of one of the enantiomeric 10-hydroxy dibenzosuberone, by X-ray analysis, it will then be possible to prepare from it the *cis*- and *trans*-C3- and C5-10-hydroxy derivatives (**108, 109, 111, 112**) of known configuration and to identify the major stereochemistry of the corresponding animal metabolites.

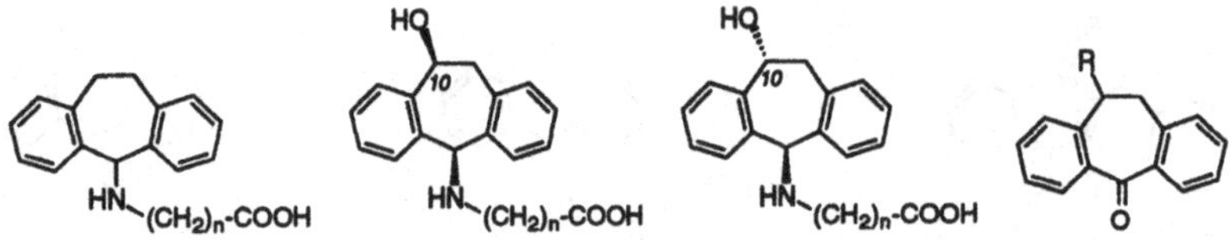

HN-$(CH_2)_n$-COOH

104: n = 6
105: n = 4
106: n = 2

107: n = 6
108: n = 4
109: n = 2

110: n = 6
111: n = 4
112: n = 2

113: R = H
114: R = OH

3.3.4
Artemisinin and Analogous Antimalarial Compounds

Artemisinin (**115**), an unusual endoperoxide sesquiterpenic lactone isolated from *Artemisia annua*, a Chinese medicinal plant, is an active antimalarial agent used as an alternative drug (Qinghaosu) against strains of *Plasmodium* resistant to classical antimalarial drugs. An ethyl ether derivative of dihydroartemisinin (**117**), arteether (**118**), has been also used for high-risk (cerebral) malarial patients. Extensive investigations on the animal and human metabolism of both drugs have been undertaken [22, 201]: they mainly showed the reduction of the endoperoxidic linkage to an ether linkage, as in deoxyartemisinin (**119**) (and consequently the loss of antimalarial activity), followed by reduction of the carbonyl group (**122**), regioselective hydroxylation reaction (**121**), or rearrangement to a tricyclic compound of tentative structure **126** [202]. The endoperoxide linkage was more resistant in arteether (**118**), as dihydroartemisinin (**117**) was observed as the major rat liver microsomes metabolite, together with deoxy- and 3α-hydroxy deoxyderivatives (**122, 123** and **124**) [202–204]. Microbial production represented the only convenient source of most of the animal metabolites (11 of them were known for arteether). Incubations of artemisinin with *Nocardia corallina* and *Penicillium chrysogenum* resulted in the production of the major animal metabolite **119**, and of a new hydroxylated metabolite **120** [202]. Arteether (**118**) was converted by *Aspergillus niger* or *Nocardia corallina* into four main deoxy metabolites: **122**, the hydroxylation products **123** and **125**, and a rearrangement product **127** [203, 204]. However dihydroartemisinin (**117**) was not identified as a microbial metabolite.

115: $R_1, R_2 = O$; $R_3 = H$
116: $R_1, R_2 = O$; $R_3 = OH$
117: $R_1 = OH$; $R_2 = R_3 = H$
118: $R_1 = OCH_2CH_3$; $R_2 = R_3 = H$

119: $R_1, R_2 = O$; $R_3 = R_4 = R_5 = H$
120: $R_1, R_2 = O$; $R_3 = R_4 = H$; $R_5 = OH$
121: $R_1 = R_2 = O$; $R_3 = R_4 = OH$; $R_5 = H$
122: $R_1 = OH$; $R_2 = R_3 = R_4 = R_5 = H$
123: $R_1 = R_5 = OH$; $R2=R3=R4=H$
124: $R_1 = OCH_2CH_3$; $R_2 = R_3 = R_4 = H$; $R_5 = OH$
125: $R_1 = OCH_2CH_3$; $R_2 = R_4 = R_5 = H$; $R_3 = OH$

126 **127**

In an attempt to generate one or more hydroxylated new products retaining the endoperoxide grouping, and starting from the known hydroxylation abilities of *Beauveria bassiana* when acting on amido group-containing chemicals [4, 205–207], Ziffer et al. [208, 209] used as a substrate for this microorganism the *N*-phenylurethane derivative of dihydroartemisinin (**128**), which was converted in low yield (10%) to the C-14 hydroxylated derivative **129**. Other artemisinin derivatives were also employed as substrates with *B. bassiana*: for example arteether (**118**) was converted in fair yields into two active new metabolites **130** and **131** containing the intact endoperoxide group [209, 210]. Other metabolites of arteether (including a few ones retaining the endoperoxide moiety) hydroxylated in positions -1α, -2α, -9α, -9β or -14, and corresponding to some minor metabolites found in rat liver microsomes incubations, have been recently prepared [211] using large scale fermentations with *Cunninghamella elegans* ATCC 9245 and *Streptomyces lavendulae* L-105.

Anhydroartemisinin (**132**), a semisynthetic derivative with very high antimalarial activity, was converted to the rearranged compound **136** and to the 9β-hydroxylated derivative **133** by *S. lavendulae* L-105, whereas a *Rhizopogon* species (ATCC 36060) formed hydroxylated metabolites **134** and **135** [212]. Preparative-scale incubations have led to the isolation of sufficient amounts of these metabolites to obtain clear structural identifications (including X-ray analysis), HPLC/MS standards for comparison with the 28 corresponding animal metabolites (the 9β-hydroxy derivative was found as the major metabolite in the rat) and antimalarial testing.

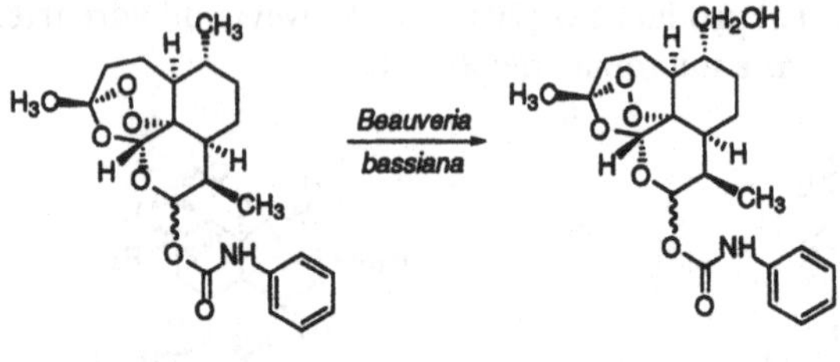

128 → *Beauveria bassiana* → 129

118 → *Beauveria bassiana* → 130 (28%) + 131 (10%) + 124 (7%)

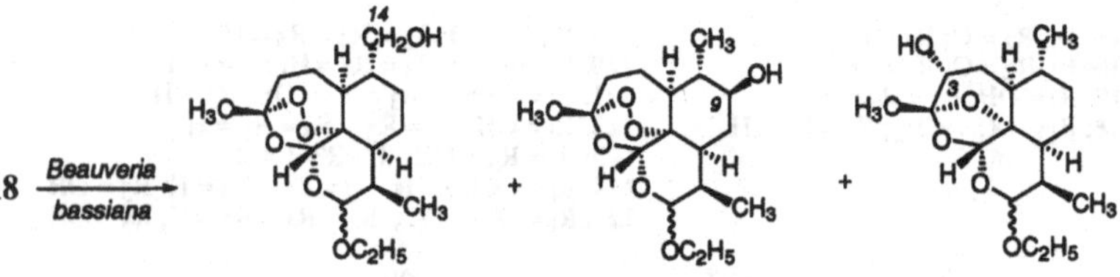

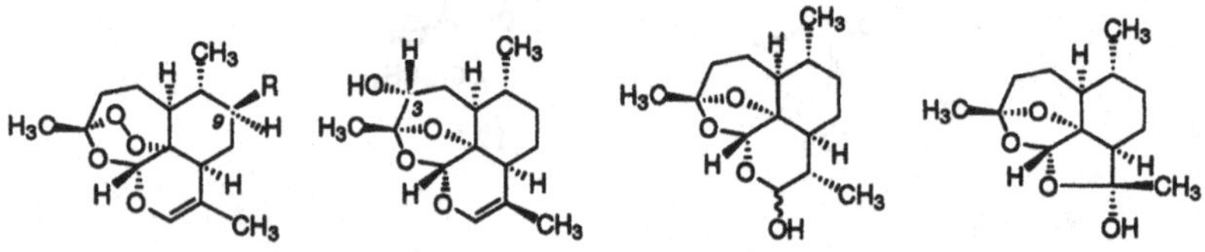

132: R = H
133: R=OH

134

135

136

3.3.5
Taxol and Related Antitumour Compounds

Taxol (Paclitaxel) **137**, a natural product derived from the bark of the Pacific yew, *Taxus brevifolia* [213–215], and the hemisynthetic analogue Docetaxel (Taxotere) **138**, two recent and promising antitumour agents, have been the matter of extensive in vivo and in vitro animal metabolic studies. The major metabolites of taxol excreted in rat bile [216] were identified as a C-4′ hydroxylated derivative on the phenyl group of the acyl side chain at C-13 (**139**), another aromatic hydroxylation product at the *meta*-position on the benzoate group at C-2 (**140**) and a C-13 deacylated metabolite (baccatin III, **142**); the structure of six minor metabolites could not be determined. The major human liver microsomal metabolite, apparently different from those formed in rat [217], has been identified as the 6α-hydroxytaxol (**141**) [218, 219]. A very similar metabolic pattern was demonstrated for Docetaxel and the structure and hemisynthesis of most of its metabolites have been reported [220,221].

The metabolism of taxol by *Eucalyptus perriniana* cell suspension cultures has been recently reported to induce hydrolyses of ester bonds at C-13, C-10 and C-2 [222]. At this moment only very few data have been published about the microbial metabolism of taxoid compounds: only site specific hydrolyses of acyl side-chains at C-13 or C-10 by extracellular and intracellular esterases of *Nocardioides albus* SC13,911 and *N. luteus* SC13,912, respectively, have been reported [223]. On the other hand, Hu et al. [224–226] have recently described some fungal biotransformations of related natural taxane diterpenes extracted from Chinese yews or their cell cultures, in order to obtain new active substances or precursors for hemisynthesis. The taxadiene **145**, a 14β-acetylated derivative

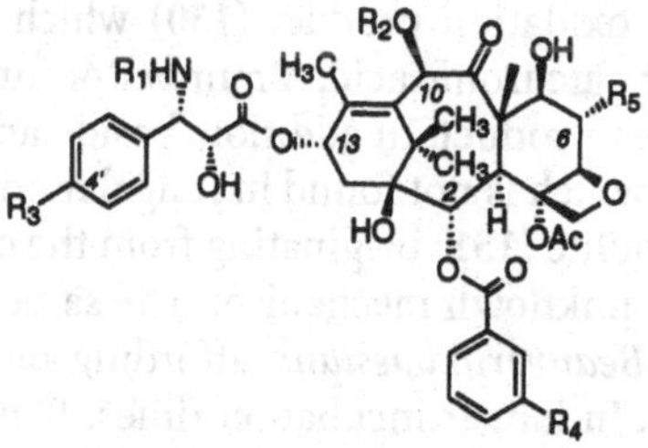

137: $R_1 = C_6H_5CO$; $R_2 = CH_3CO$; $R_3 = R_4 = R_5 = H$
138: $R_1 = (CH_3)_3COCO$; $R_2 = R_3 = R_4 = R_5 = H$
139: $R_1 = C_6H_5CO$; $R_2 = CH_3CO$; $R_3 = OH$; $R_4 = R_5 = H$
140: $R_1 = C_6H_5CO$; $R_2 = CH_3CO$; $R_3 = R_5 = H$; $R_4 = OH$
141: $R_1 = C_6H_5CO$; $R_2 = CH_3CO$; $R_3 = R_4 = H$; $R_5 = OH$

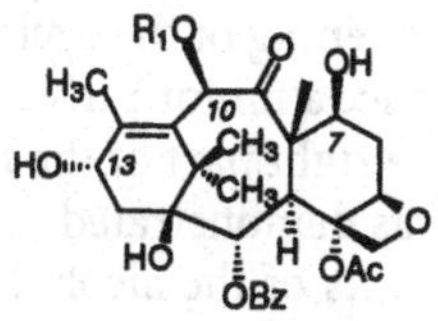

142: $R_1 = COCH_3$
143: $R_1 = H$

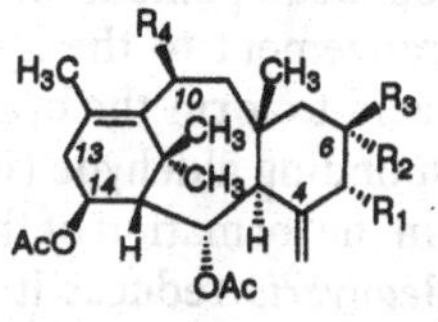

145: $R_1 = OAc$; $R_2 = R_3 = H$; $R_4 = OH$
146: $R_1 = OAc$; $R_2 = R_4 = OH$; $R_3 = H$
147: $R_1 = OAc$; $R_2 = H$; $R_3 = R_4 = OH$

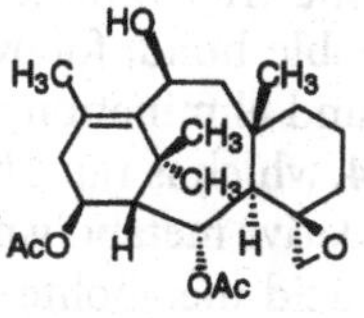

148

lacking inter alia the 13α-hydroxyacyl group and the 4(20)-oxirane ring of taxol, was extensively (55%) metabolized by *Cunninghamella echinulata* AS 3.1990 to a 6β- (major) and a 6β- (minor) hydroxylated derivatives, with hydrolytic deacetylation at C-10 (**146**,**147**). In addition, epoxide **148** was formed as a minor metabolite [224]. The C-14 hydroxyl group acylation was essential for the hydroxylation at C-6 by this strain, and hydroxylation was inhibited by the presence of the epoxide ring of **148** [226]. A comparative biotransformation study with the same strain incubated with taxol (**137**) or taxol hemisynthesis precursors such as baccatin III (**142**) and 10-deacetylbaccatin III (**143**) showed respectively no transformation or simple deacetylation and/or epimerization at the C-7 position [226]. More work would be necessary in the future to complete the study of microbial models of metabolism of complex taxoid compounds.

Surprisingly, two fungi, *Taxomyces andreanae* [227–229] and *Pestalotiopsis microspora* [230], respectively isolated from the inner bark of Pacific or Himalayan yews, were recently shown to be able to produce in culture, out of the host, small amounts of taxol. This discovery raises the possibility of a gene transfer between *Taxus* species and their corresponding endophytic fungi [230].

3.3.6
Miscelleanous Examples

Another example of the use of microbial models, applied to the elucidation of the metabolic pathways of drugs in animals, can be found in the investigation of the metabolism of HR325 (**149**), a recently developed synthetic immunomodulating drug [231, 232]. This compound is mainly metabolized [155, 233] in rats and other animals to give a benzylic oxidation product (**150**) which is then excreted as a conjugate with glucose or glucuronic acid. A number of fungi produce the same hydroxymethyl oxidation product. In addition, beside an oxidative opening of the cyclopropane ring, which is not found in fungal models, rats produce a minor carboxylic acid metabolite (**151**) originating from the cleavage of the right part of the molecule by an unknown mechanism. The same cleavage was demonstrated on HR325 using *Beauveria bassiana*, affording significant amounts of the alcohol derivative **152**. In longer incubation times, *B. bassiana* afforded the 4-*O*-methyl-D-glucosyl conjugated derivative **153** in 35% isolated yield. It was possible to demonstrate [155] in this microorganism that the cyanohydrin **154** is an intermediate in the formation of the split products and to show that the primary metabolic reaction leading to the cleavage of the right part of the HR325 molecule is a monooxygenase-mediated epoxidation of the enol double bond, followed by a spontaneous rearrangement to the hydroxy ketone and elimination of cyclopropane carboxylic acid to form the cyanohydrin **154**, which is rapidly hydrolyzed into the corresponding aldehyde (Fig. 8). The oxidative metabolism of the aldehyde may result in the formation of the carboxylic acid metabolite (**151**) in animals, whereas *Beauveria* reduces it to the alcohol **152**.

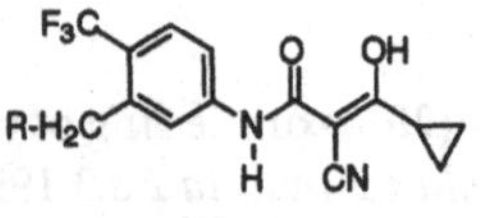

149: R = H
150: R = OH

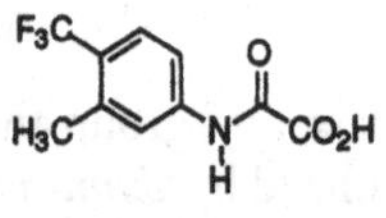

151

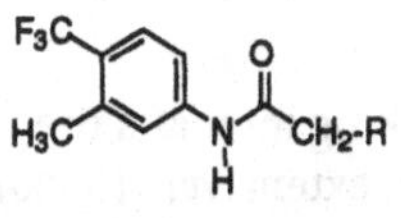

152: R = OH
153: R = β-4-OMe-glucosyl

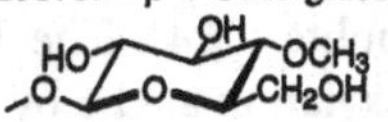

Quinidine (**155**) and dihydroquinidine (**157**) have been used for a long time for the treatment of cardiac antiarrythmia. These cinchona alkaloids (and their analogues in the quinine and cinchonidine family) are metabolized in animals and humans [234, 235] to give, among several products, the corresponding (3*S*)-3-hydroxy derivatives (**156, 159**) [236–240], which were shown to be pharmaco-

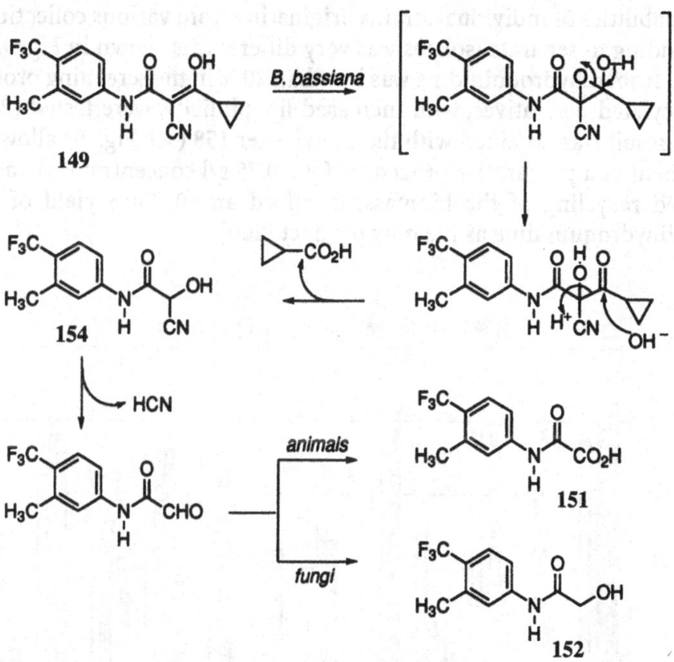

Fig. 8. Oxidative metabolism of HR325 in *Beauveria bassiana*

logically active [241] and possibly devoid of the immunotoxic effects of the natural alkaloids.

A microbiological transformation of quinidine was reported by Eckenrode [242] demonstrating its conversion into the 3-hydroxylated mammalian metabolite by various *Streptomyces* strains in low yields (<3%). A further extensive screening of *Actinomycetes* and fungi [243] showed that only 10% of the tested strains were capable of oxidizing members of this alkaloid family, in low yield (<10%), affording the 3-hydroxy derivatives together with the *N*-oxides. In order to provide a large scale preparative method, 150 fungi were tested for their ability to produce the(3*S*)-3-hydroxy derivative of dihydroquinidine (**159**) and a strain of *Mucor plumbeus* was selected for further studies [244]. Remarkably, significant activity was nearly limited to this unique fungal species but hydroxy-

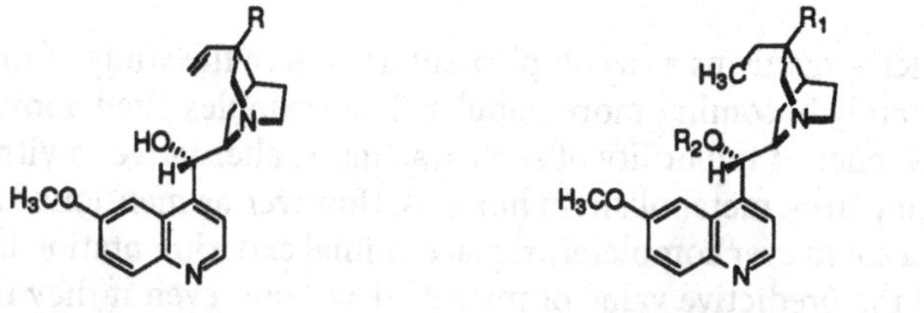

155: R = H
156: R = OH

157: $R_1 = R_2 = H$
158: $R_1 = H$; $R_2 = OCOCH_3$
159: $R_1 = OH$; $R_2 = H$

lation capabilities of individual strains originating from various collections and corresponding to separate isolates was very different, as shown in Fig. 9. As the best yield from dihydroquinidine was less than 20% in the screening procedure, several acylated derivatives, with increased lipophilicity, were tested [245]: an excellent result was obtained with the acetyl ester **158** (see Fig. 9) allowing the development of a preparative procedure (at a 0.75 g/l concentration), including a repeated recycling of the biomass, to afford an 80–90% yield of (3*S*)-3-hydroxydihydroquinidine as the only product [200].

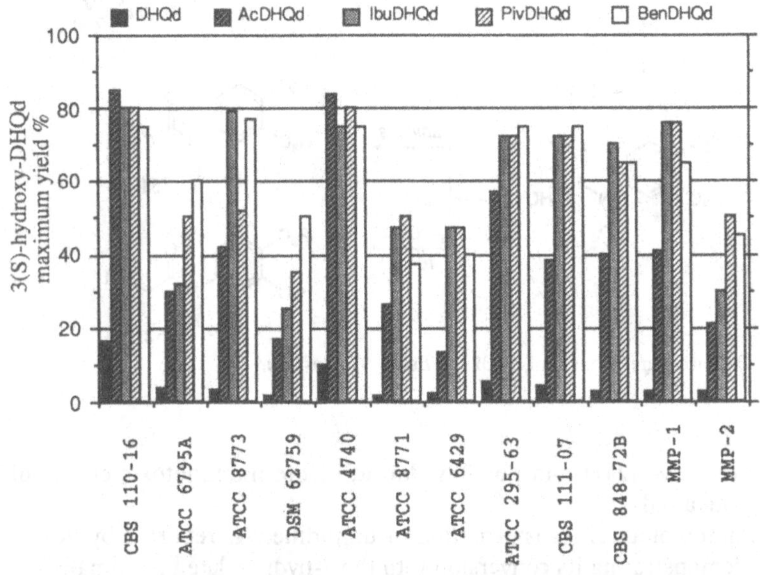

Fig. 9. Biotransformation of dihydroquinidine (and acylated derivatives) to the corresponding (3S)-3-hydroxy compounds (as % observed at the reaction plateau), using various collection strains of *Mucor plumbeus* (ATCC: American Type Culture Collection; CBS: Centraalbureau voor Schimmelcultures; DSM: Deutsche Sammlung von Mikroorganismen; MMP: Mycotheque of the Paris Museum of National History)

4
Conclusion and Perspectives

The use of microorganisms as a complementary tool in the study of drug metabolism is certainly becoming more popular. The examples cited above provide additional evidence of the utility of such systems as alternative in vitro models for investigating drug metabolism in humans. However, as mentioned above, no in vitro model could ever completely replace animal experimentation in medical research, and the predictive value of microbial systems, even if they frequently correlate with their animal counterpart, is questionable. Nevertheless, microorganisms can be used as a reliable and efficient alternative to small animals or to synthetic chemistry for obtaining sizable amounts of a number of drug derivates which facilitate the identification and the toxicological studies of most animal metabolites. Moreover, the high potential of microbial reactions can be used for the creation of molecular diversity far beyond the metabolic transformations observed in animals. The identification and isolation of drug metabolizing enzymes, and particularly cytochromes P450, in fungi and *Actinomycetes* remains to be done, just as the comparison of their activities and specificities with animal liver enzymes, in order to gain more predictivity in metabolic studies. However, the individual expression of the main representatives of most human liver families of cytochromes P450 in heterologous systems (bacteria or yeast) is now currently realized and successfully used as a predictive tool for investigating the metabolism of a potential drug. This new methodology will probably be refined in the future to cover most of the problems arising from drug metabolism, genetic polymorphism and drug interactions.

5 References

1. Velluz L, Bellet P (1959) C R Acad Sci 248:3453
2. Rosi D, Peruzotti G, Dennis EW, Berberian DA, Freele H, Archer S (1965) Nature 208:1005
3. Rosi D, Peruzotti G, Dennis EW, Berberian DA, Freele H, Tullar BF, Archer S (1967) J Med Chem 10:867
4. Fonken GS, Johnson RA (1972) Chemical oxidations with microorganisms. Marcel Dekker, New York
5. Karim A, Brown EA (1972) Steroids 20:41
6. Ozaki M, Kariya T, Kato H, Kimura T (1972) Agr Biol Chem 36:451
7. Beukers R, Marx AF, Zuidweg MHJ (1972) Microbial conversion as a tool in the preparation of drugs. In: Ariëns EJ (ed) Drug design, vol 3. Academic Press, New york, p 1
8. Marschek WJ, Karim A (1973) Appl Microbiol 25:647
9. Sebek OK (1975) Acta Microbiol 22:381
10. Smith RV, Rosazza JP (1974) Arch Biochem Biophys 161:551
11. Smith RV, Rosazza JP (1975) J Pharm Sci 64:1737
12. Smith RV, Rosazza JP (1975) Biotech Bioeng 17:785
13. Smith RV, Acosta DJ, Rosazza JP (1977) Adv Biochem Eng 5:69
14. Rosazza JP (1978) Lloydia 41:279
15. Rosazza JP, Smith RV (1979) Adv Appl Microbiol 25:169
16. Smith RV, Davis PJ (1980) Adv Biochem Eng. 14:61
17. Smith RV, Rosazza JP (1982) Microbial transformations as a mean of preparing mammalian drug metabolites. In: Rosazza JP (ed) Microbial transformations of bioactive compounds. CRC, Boca Raton, FL, p 1
18. Smith RV, Rosazza JP (1983) J Nat Prod 46:79
19. Clark AM, McChesney JD, Hufford CD (1985) Med Res Rev 5:231
20. Reighard JB, Knapp JE (1986) Pharm Int 7:92
21. Davis PJ (1988) Dev Indust Microbiol (J Indust Microbiol suppl No 3) 29:197
22. Clark AM, Hufford CD (1991) Med Res Rev 11:473
23. Griffiths DA, Best DJ, Jezequel SG (1991) Appl Microbiol Biotechnol 35:373
24. Kieslich K (1976) Microbial transformation of non-steroid cyclic compounds. Wiley Interscience, New York
25. Csuk R, Glanzer Br (199) Chem Rev 91:49
26. Santaniello E, Ferraboschi P, Grisenti P, Manzocchi A (1992) Chem Rev 92:1071
27. Sariaslani FS, Rosazza JP (1984) Enzyme Microb Technol 6:242
28. Sebek OK (1983) Mycologia 75:383

29. Servi S (1990) Synthesis 1
30. Servi S (ed) (1992) Microbial reagents in organic synthesis. Kluwer Acad Publ, Dordrecht, The Netherlands
31. Sih CJ, Chen C-S (1984) Angew Chem Int Ed Engl 23:570
32. Yamada H, Shimizu S (1988) Angew Chem Int Ed Engl 27:622
33. Peterson DH, Murray HC, Eppstein SH, Reineke LM, Weintraub A, Meister PD, Leigh HM (1952) J Am Chem Soc 74:5933
34. Peterson DH, Murray HC (1952) J Am Chem Soc 74:1871
35. Auret BJ, Boyd DR, Robinson PM, Watson CG, Daly DW, Jerina DM (1971) J Chem Soc Chem Commun p 1585
36. Ferris JP, Fasco MJ, Stylianopoulou FL, Jerina DM, Daly JW, Jeffrey AM (1973) Arch Biochem Biophys 156:97
37. Borchert HH (1991) Arch Pharm 324:401
38. Sebek OK, Perlman D (1971) Adv Appl Microbiol 14:123
39. Sebek OK (1974) Lloydia 37:115
40. Iizuka H, Naito A (1981) Microbial conversion of steroids and alkaloids. University of Tokyo Press-Springer, Berlin Heidelberg New York
41. Mahato SB, Mukherjee A (1984) Phytochemistry 23:2131
42. Mahato SB, Banerjee S (1985) Phytochemistry 24:1403
43. Mahato SB, Banerjee S, Podder S (1989) Phytochemistry 28:7
44. Mahato SB, Majumdar I (1993) Phytochemistry 34:883
45. Holland HL (1981) Microbial and in vitro enzymic transformation of alkaloids (chap. 5). In: Manske RHF, Rodrigo RGA (eds) The alkaloids, vol 18. Academic Press, New York; p 323
46. Davis PJ (1984) Natural and semi-synthetic alkaloids. In: Kieslich K (ed) Biotechnology, vol 6a. Verlag Chemie, Weinheim, p 207
47. Rosazza JPN, Duffel W (1986) The Alkaloids 27:323
48. Betts RE, Walters DE, Rosazza JP (1974) J Med Chem 17:599
49. Brannon DR, Horton HR, Svoboda GH (1974) J Med Chem 17:653
50. Knaggs AR, Cable KM, Cannell RJP, Sidebottom PJ, Wells GN, Sutherland DR (1995) Tetrahedron Lett 36:477
51. Rao GP, Davis PJ (1997) Drug Metab Dispos 25:709
52. Schwartz H, Licht R-E, Radunz H-E (1993) Appl Microbiol Biotechnol 40:382
53. Umezawa H, Takahashi Y, Fujii A, Saino T, Shirai T, Takita T (1973) J Antibiot 26:1117
54. El-Marakby SA, Clark AM, Baker JK, Hufford CD (1986) J Pharm Sci 75:614
55. Kittelmann M, Lattmann R, Ghisalba O (1993) Biosci Biotechnol Biochem 57:1589
56. Gibson M, Soper CJ, Parfitt RT, Sewell GJ (1984) Enzyme Microb Technol 6:471
57. Sewell GJ, Soper CJ, Parfitt RT (1984) Appl Microbiol Biotechnol 19:247
58. Kunz DA, Reddy GS, Vatvars A (1985) Appl Environ Microbiol 50:831
59. Zeitler H-J, Niemer H (1969) Hoppe-Seyler's Z Physiol Chem 350:366
60. Hufford CD, Collins CC, Clark AM (1979) J Pharm Sci 68:1239
61. Hufford CD, El-Sharkawy SH, Jurgens TM, Mikell JR (1992) Pharmaceut Res 9:623
62. Zhang D, Hansen EB, Deck J, Heinze TM, Henderson A, Korfmacher WA, Cerniglia CE (1997) Xenobiotica 27:301
63. Griffiths DA, Brown DE, Jezequel SG (1993) Xenobiotica 23:1085
64. Schwartz H, Liebieg-Weber A, Hochstatter H, Bottcher H (1996) Appl Microbiol Biotechnol 44:731
65. Chien MM, Rosazza JP (1979) Drug Metabol Dispos 7:211
66. Chien MM, Rosazza JP (1980) Appl Environ Microbiol 40:741
67. Marshall VP, Cialdella JI, Baczynski LA, Liggett WF, Johnson RA (1989) J Antibiot 42:132
68. Kuo M-S, Chirby DG, Argoudelis AD, Cialdella JI, Coats JH, Marshall VP (1989) Antimicrob Ag Chemother 33:2089
69. Hezari M, Davis PJ (1992) Drug Metab Dispos 20:882
70. Hezari M, Davis PJ (1992) Drug Metab Dispos 21:259
71. Cannell RJP, Knaggs AR, Dawson MJ, Manchee GR, Eddershaw PJ, Waterhouse I, Sutherland DR, Bowers GD, Sidebottom PJ (1995) Drug Metab Dispos 23:724

72. Hanlon GW, Kooloobandi A, Hutt AJ (1994) J Appl Microbiol 76:442
73. Hansen Jr EB, Cerniglia CE, Korfmacher WA, Miller DW, Heflich RH (1987) Drug Metab Dispos 15:97
74. Chen TS, Doss GA, Hsu A, Lingham RB, White RF, Monaghan RL (1993) J Nat Prod-Lloydia 56:755
75. Otten S, Rosazza JP (1978) Appl Environ Microbiol 35:554
76. Otten S, Rosazza JP (1979) Appl Environ Microbiol 38:311
77. Otten SL, Rosazza JP (1981) J Nat Prod 44:562
78. Otten SL, Rosazza JP (1983) J Biol Chem 258:1610
79. David L, Gayet J-C, Veschambre H (1985) Agr Biol Chem 49:2693
80. Davis PJ, Glade JC, Clark AM, Smith RV (1979) Appl Environ Microbiol 38:891
81. Zedan HH, El-Tayeb OM, Walash MN (1983) J Gen Microbiol 129:1035
82. Freitag DG, Foster RT, Coutts RT, Pickard MA, Pasutto FM (1997) Drug Metab Dispos 25:685
83. Chen TS, Doss G (1992) Merk and Co: Europ Patent 92/02,519
84. Chen TS, White LSR, Monaghan RL (1993) J Antibiot 46:131
85. Jones DE, Moore RH, Crowley GC (1970) J Chem Soc C 1725
86. Sebek OK, Dolak LA (1984) J Antibiot 37:136
87. Kuo MS, Yurek DA, Chirby DG, Cialdella JI, Marshall VP (1991) J Antibiot 44:1096
88. Raju MS, Wu GS, Gard A, Rosazza JP (1982) J Nat Prod 45:321
89. Schmitz H, Claridge CA (1977) J Antibiot 30:635
90. Rosazza JP, Kammer M, Youel L, Smith RV, Erhardt PW, Truong DH, Leslie SW (1977) Xenobiotica 7:133
91. Verdeil JL, Bister-Miel F, Agier C, Plat M, Guignard JL, Viel C (1985) Compt Rend Acad Sci Paris 300, Série III:305
92. Dunn GL, Gallagher G, Davis D, Hoover RE, Stedman J (1973) J Med Chem 16:996
93. Valenta JR, Dicuollo CJ, Fare LR, Miller JA, Pagano JF (1974) Appl Microbiol 28:995
94. Davis PJ, Yang S-K, Smith RV (1984) Appl Environ Microbiol 48:327
95. Davis PJ, Yang S-K, Smith RV (1985) Xenobiotica 15:1001
96. Smith RV, Davis PJ, Kerr KM (1983) J Pharm Sci p 733
97. Reddy CSG, Acosta D, Davis PJ (1990) Xenobiotica 20:1281
98. Foster BC, Thomas BH, Zamecnik J, Dawson BA, Wilson DL, Duhaime R, Solomonraj G, McGilveray IJ, Lodge BA (1991) Can J Microbiol 37:504
99. Hufford CD, Baker JK, Clark AM (1981) J Pharm Sci 70:155
100. Baker JK, Wohlford JG, Bradbury BJ, Wirth PW (1981) J Med Chem 24:666
101. Caroll FI, Brine GA, Boldt KG, Cone EJ, Yousefnejad D, Vaupel DB, Buchwald WF (1981) J Med Chem 24:1047
102. Foster GR, Coutts RT, Pasutto FM, Mozayani A (1988) Life Sci 42:285
103. Hansen EB, Cho BP, Korfmacher WA, Cerniglia CE (1995) Xenobiotica 25:1081
104. Clark AM, Hufford CD, McChesney JD (1981) Antimicrob Ag Chemother 19:337
105. Clark AM, Evans SL, Hufford CD, McChesney JD (1982) J Nat Prod 45:574
106. Hufford CD, Clark AM, Quinones JK, Baker JK, MacChesney JD (1983) J Pharm Sci 72:92
107. Clark AM, Hufford CD, Puri RK, McChesney JD (1984) Appl Environ Microbiol 47:540
108. Clark AM, Hufford CD, Gupta RC, Puri RK, McChesney JD (1984) Appl Environ Microbiol 47:537
109. Clark AM, McChesney JD, Hufford CD (1986) Pharm Res 3:170
110. Foster BC, Buttar HS, Qureshi SA, McGilveray IJ (1989) Xenobiotica 19:539
111. Foster BC, Litster DL, Wilson DL, Ormsby E, Dawson BA (1992) Xenobiotica 22:1221
112. Lancini GC, Thiemann JE, Sartori G, Sensi P (1967) Experientia 23:899
113. Banerjee UC (1993) Enzyme Microb Technol 15:1037
114. Davis PJ, Guenthner L (1985) Xenobiotica 15:845
115. El-Sharkawy SH, Abul-Hajj Y (1987) Pharmacol Res 4:353
116. El-Sharkawy SH (1991) Appl Microbiol Biotechnol 35:436
117. Davis PJ, Rosazza JP (1976) J Org Chem 41:2548
118. Davis PJ, Wiese D, Rosazza JP (1977) Lloydia 40:239

119. Moussa C, Houziaux P, Danrée B, Azerad R (1997) Drug Metab Dispos 25:301
120. Moussa C, Houziaux P, Danrée B, Azerad R (1997) Drug Metab Dispos 25:311
121. Foster GR, Litster DL, Zamecnik J, Coutts RT (1991) Can J Microbiol 37:791
122. Hansen Jr EB, Heflich RH, Korfmacher WA, Miller DW, Cerniglia CE (1988) J Pharm Sc 77:259
123. Rosazza JP, Nicholas AW, Gustafson ME (1978) Steroids 31:671
124. Fuska J, Prousek J, Rosazza JP, Budesinsky M (1982) Steroids 40:157
125. Fuska J, Khandlova A, Sturdikova M, Rosazza JP (1985) Folia Microbiol 30:427
126. El-Sharkawy S, Abul-Haii YJ (1988) Xenobiotica 18:365
127. El-Sharkawy S, Abul-Haiu YJ (1988) J Org Chem 53:515
128. Kamimura H (1986) Appl Environ Microbiol 52:515
129. El-Sharkawy S, Abul-Hajj YJ (1987) J Nat Prod 50:520
130. El-Sharkawy SH (1989) Acta Pharm Jugosl 39:303
131. El-Sharkawy SH, Selim MI, Afifi MS, Halaweish FT (1991) Appl Environ Microbiol 57:549
132. Abul-Hajj YJ (1986) J Nat Prod 49:244
133. Fiorentino A, Pinto G, Pollio A, Previtera L (1991) Bioorg Med Chem Lett 1:673
134. Pollio A, Pinto G, Della Greca M, De Maio A, Fiorentino A, Previtera L (1994) Phytochemistry 37:1269
135. Kutney JP (1993) Acc Chem Res 26:559
136. Coutts RT, Foster GR, Jones GR, Meyers GE (1979) Appl Environ Microbiol 37:429
137. Foster BC, Coutts RT, Pasutto FM (1989) Xenobiotica 19:531
138. Foster BC, Nantais LM, Wilson DL, By AW, Zamecnik J, Lodge BA (1990) Xenobiotica 20:583
139. Coutts RT, Foster GR, Pasutto FM (1981) Life Sci 29:1951
140. Foster BC, Wilson DL, McGilveray IJ (1989) Xenobiotica 19:445
141. Cerniglia CE, Hansen EBJr, Lambert KJ, Korfmacher WA, Miller DW (1988) Xenobiotica 18:301
142. Gorrod JW, Damani LA (1985) Biological oxidation of nitrogen in organic molecules. Ellis Horwood & VCH, Chichester & Weinheim
143. Holland HL (1992) Organic synthesis with oxidative enzymes. VCH, New York
144. Hartman RE, Kraus EF, Andres WW, Patterson EL (1964) Appl Microbiol 12:138
145. Chen X-J, Archelas A, Furstoss R (1993) J Org Chem 58:5528
146. Pedragosa-Moreau S, Archelas A, Furstoss R (1993) J Org Chem 58:5533
147. Sheehan D, Casey JP (1993) Comp Biochem Physiol 104B:1
148. Sheehan D, Casey JP (1993) Comp Biochem Physiol 104B:7
149. Hu S-H, Genain G, Azerad R (1995) Steroids 60:337
150. Cerniglia CE, Freeman JP, Mitchum RK (1982) Appl Environ Microbiol 43:1070
151. Petzoldt K, Kieslich K, Steinbeck H (1974) German Pat. Deutsch Offen 2,326,084 (19 May 1973)
152. Kieslich K, Vidic HJ, Petzoldt K, Hoyer JA (1976) Chem Ber 109:2259
153. Neef G, Eder U, Petzoldt K, Seeger A, Wieglep P (1982) J Chem Soc Chem Commun 366
154. Vigne B, Archelas A, Fourneron J-D, Furstoss R (1986) Tetrahedron 42:2451
155. Lacroix I, Biton J, Azerad R (1997) Bioorg Med Chem 5:1369
156. Dowd CA, Sheehan D (1993) Biochem Soc Trans 22:58 S
157. Wong YWJ, Davis PJ (1989) Pharm Res 6:982
158. Davila JC, Hseih GC, Acosta D, Davis PJ (1990) Toxicol Lett 54:23
159. Krasnobajew V (1984) Terpenoids. In: Kieslich K (ed) Biotechnology, vol 6a. Verlag Chemie, Weinheim, p 31
160. Tramper J, VanderPlas HC, Linko P (eds) (1985) Biocatalysis in organic synthesis. Elsevier, Amsterdam
161. Sedlacæk L (1988) CRC Crit Rev Biotechnol 7:187
162. Lamare V, Furstoss R (1990) Tetrahedron 46:4109
163. Hamada H, Fuchikami Y, Jansing RL, Kaminsky LS (1993) Phytochemistry 33:599
164. Faber K (1992) Biotransformations in organic chemistry. Springer, Berlin-Heidelberg New York

165. Margolin AL (1993) Enzyme Microb Technol 15:266
166. Crout DHG, Roberts SM, Jones JB (eds) (1993) Enzymes in organic synthesis. In: Tetrahedron: Asymmetry, Special Issues 4:757
167. Crout DHG, Griengl H, Faber K, Roberts SM (eds) (1994) Proceedings of the European Symposium on Biocatalysis, September 12–17, 1993, Graz, Austria. In: Biocatalysis, Special Issues 9:722
168. Wong C-H, Whitesides GM (1994) Enzymes in synthetic organic chemistry. Pergamon, Oxford
169. Dahl ML, Nordin C, Bertilsson L (1991) Ther Drug Monit 13:189
170. Drauz K, Waldmann H (eds) (1995) Enzyme catalysis in organic synthesis. VCH, Weinheim
171. Hanson JR (1995) An introduction to biotransformation in organic chemistry. WH Freeman, Oxford
172. Azerad R (1995) Bull Soc Chim 132:17
173. Cannell RJP, Rashid T, Ismail IM, Sidebottom PJ, Kaggs AR, Marshall PS (1997) Xenobiotica 27:147
174. Sariaslani FS, Kunz DA (1986) Biochem Biophys Res Commun 141:405
175. Byrne GS, Ward OP (1989) Biotechnol Bioeng 33:912
176. Hermans JJR, Thijssen HHW (1989) Biochem Pharmacol 38:3365
177. Hermans JJR, Thijssen HHW (1992) Drug Metab Dispos 20:268
178. Kaminsky LS, Zhang ZY (1997) Pharmacol Ther 73:67
179. Thonnart N, Vanhaelen M, Vanhaelen-Fastre R (1977) J Med Chem 20:604
180. Thonnart N, Vanhaelen M, Vanhaelen-Fastre R (1979) Drug Metab Dispos 7:449
181. Rizzo JD, Davis PJ (1988) Xenobiotica 18:1425
182. Davis PJ, Rizzo JD (1982) Appl Environ Microbiol 43:884
183. Rizzo JD, Davis PJ (1989) J Pharm Sci 78:183
184. Wong YWJ, Davis PJ (1991) J Pharm Sci 80:305
185. Griffiths DA, Brown DE, Jezequel SG (1992) Appl Microbiol Biotechnol 37:169
186. Mandal NC, Chakrabartty PK, Jash SS, Basu K, Bhattacharyya P (1990) Ind J Exp Biol 28:189
187. Takeshita M, Miura M, Unuma Y, Iwai S, Sato I, Arai T, Kosaka K (1995) Res Commun Mol Pathol Pharmacol 88:123
188. Holland HL, Riemland E (1985) Can J Chem 63:1121
189. Lacroix I, Biton J, Azerad R, unpublished results
190. Biton J, Azerad R, Lacroix I, Marchandeau JP (1997) Eur Pat 0,808,845 A2
191. Potter WZ, Calil HM (1981) Metabolites of tricyclic antidepressants: biological activity and clinical implications. In: Usdin E (ed) Clinical pharmacology in psychiatry. Elsevier, New York, p 311
192. Breyerpfaff U, Nill K (1995) Xenobiotica 25:1311
193. Zeugin TB, Brosen K, Meyer UA (1990) Anal Biochem 189:99
194. Nordin C, Bertilsson L (1995) Clin Pharmacokinet 28:26
195. Clark AM, Clinton RT, Baker JK, Hufford CD (1983) J Pharm Sci 72:1288
196. Hufford CD, Capiton GA, Clark AM, Baker JK (1981) J Pharm Sci 70:151
197. Zhang DL, Evans FE, Freeman JP, Duhart B, Cerniglia CE (1995) Drug Metab Dispos 23:1417
198. Grislain L, Gelé P, Bromet N, Luijten W, Volland JP, Mocaer E, Kamoun A (1990) Eur J Drug Metab Pharmacokinetics 15:339
199. Beaumal JY, Lefoulon F, Azerad R, unpublished results
200. Azerad R (1997) Hydroxylations at saturated carbons. In: Katsuki T (ed) Asymmetric oxydation reactions. A practical approach. Oxford Univ Press, in press
201. Lee IS, Hufford CD (1990) Pharmacol Therap 48:345
202. Lee I-S, ElSohly HL, Croom EM, Hufford CD (1989) J Nat Prod 52:337
203. Lee I-S, ElSohly HL, Hufford CD (1990) Pharm Res 7:199
204. Hufford CD, Lee I-S, ElSohly HL, Chi HT, Baker JK (1990) Pharm Res 7:923
205. Johnson RA, Herr ME, Murray HC, Fonken GS (1968) J Org Chem 33:3217

206. Furstoss R, Archelas A, Waegell B (1980) Tetrahedron Lett 21:451
207. Fourneron J-D, Archelas A, Vigne B, Furstoss R (1987) Tetrahedron 43:2273
208. Hu YL, Highet RJ, Marion D, Ziffer H (1991) J Chem Soc Chem Commun 1176
209. Ziffer H, Hu Y, Pu Y (1992) *Beauveria sulfurescens* mediated oxidation of dihydroartemisinin derivatives. In: Servi S (ed) Microbial reagents in organic synthesis. Kluwer Academic Publ, Dordrecht, p 361
210. Hu YL, Ziffer H, Li GY, Yeh HJC (1992) Bioorg Chem 20:148
211. Hufford CD, Khalifa SI, Orabi KI, Wiggers FT, Kumar R, Rogers RD, Campana CF (1995) J Nat Prod 58:751
212. Khalifa SI, Baker JK, Jung M, Mcchesney JD, Hufford CD (1995) Pharmaceut Res 12:1493
213. Gueritte-Voegelein F, Guénard D, Potier P (1987) J Nat Prod 50:9
214. Kingston DGI, Samaranayake G, Ivey CA (1990) J Nat Prod 53:1
215. Kingston DGI (1991) Pharmac Ther 52:1
216. Montsarrat B, Mariel E, Cros S, Garès M, Guénard D, Guéritte-Voegelein F, Wright M (1990) Drug Metab Dispos 18:895
217. Walle T, Kumar GN, Mcmillan JM, Thornburg KR, Walle UK (1993) Biochem Pharmacol 46:1661
218. Harris JW, Katki A, Anderson LW, Chmurny GN, Paukstelis JV, Collins JW (1994) J Med Chem 37:706
219. Kumar GN, Oatis JE, Thornburg KR, Heldrich FJ, Hazard ES, Walle T (1994) Drug Metab Dispos 22:177
220. Monegier B, Gaillard C, Sable S, Vuilhorgne M (1994) Tetrahedron Lett 35:3715
221. Commercon A, Bourzat JD, Bezard D, Vuilhorgne M (1994) Tetrahedron 50:10,289
222. Hamada H, Sanada K, Furuya T, Kawabe S, Jaziri M (1996) Nat Prod Lett 9:47
223. Hanson RL, Wasylyk JM, Nanduri VB, Cazzulino DL, Patel RN, Szarka LJ (1994) J Biol Chem 269:22,145
224. Hu SH, Tian XF, Zhu WH, Fang QC (1996) Tetrahedron 52:8739
225. Hu SH, Tian XF, Zhu WH, Fang QC (1998) J Nat Prod, in press
226. Hu S, Tian X, Zhu W, Fang Q (1997) Biocat Biotransform 14:241
227. Stone R (1993) Science 260:154
228. Stierle A, Strobel G, Stierle D (1993) Science 260:214
229. Stierle A, Strobel G, Stierle D, Grothaus P, Bignami G (1995) J Nat Prod-Lloydia 58:1315
230. Strobel G, Yang XS, Sears J, Kramer R, Sidhu RS, Hess WM (1996) Microbiology-UK 142:435
231. Greene S, Watanabe K, Braatz-Trulson J, Lou L (1995) Biochem Pharmacol 50:861
232. Williamson R, Yea CM, Robson PA, Curnock AP, Gadher S, Hambleton AB, Woodward K, Bruneau JM, Hambleton P, Moss D, Thomson AT, Spinella-Jaegle S, Morand P, Courtin O, Sautès C, Westwood R, Hercend T, Kuo EA, Ruuth E (1995) J Biol Chem 270:22,467
233. Dupront A (1995) personal communication
234. Brodie BB, Baer JE, Craig LC (1951) J Biol Chem 188:567
235. Palmer KH, Martin B, Baggett B, Wall ME (1969) Biochem Pharmacol 18:1845
236. Caroll FI, Smith D, Wall ME (1974) J Med Chem 17:985
237. Caroll FI, Philip A, Coleman MC (1976 Tetrahedron Lett 21:1757
238. Beerman B, Leander K, Lindstrom B (1976) Acta Chem Scand B 30:465
239. Flouvat B, Resplandy G, Roux A, Friocourt P, Viel C, Plat M (1988) Therapie 43:255
240. Caroll FI, Abraham P, Gaetano K, Mascarella SW, Wohl RA, Lind J, Petzoldt K (1991) J Chem Soc Perkin Trans 1, 3017
241. Fenard S, Koenig J-J, Jaillon P, Jarreau FX (1982) J Pharmacol 13:129
242. Eckenrode FM (1984) J Nat Prod 47:882
243. Siebers-Wolff S, Arfmann H-A, Abraham W-R, Kieslich K (1993) Biocatalysis 8:47
244. Jarreau FX, Herré L, Azerad R, Ogerau T (1985) Fr Pat No 85/11,221
245. Azerad R (1993) Chimia 47:93

Received January 1998

Subject Index

Absidia blakesleeana 204
Absidia cylindrospora 204
Acetobacter pasteurianus 116
N-Acetyl cysteine 173
Acetylation 173, 185–187, 190, 194
N-Acetyllactosamine 70
Acetyltransferases 173
Acronycine 177, 194, 195
Actinomucor elegans 189
Actinomycetes 176, 196, 200, 211, 213
Actinoplanes 176, 181, 183
Activation entropy 28
Acyl CoA:amino acid N-transferases 173
Addition, asymmetric 32
Agar slants 199, 200
Alanine 183
Alcaligenes 112–114, 120
Alcaligenes bronchisepticus 4, 61
Alcohols, chiral secondary 57
Aldehyde reductase (AR) 117, 119
Aldo-keto reductase superfamily 117
Algae 176
Alkaloids 99, 176, 192, 194, 195, 210, 211
Alkyl aryl sulfides 99
Alkyl hydroperoxides 75
Allyl alcohols, addition 64
Allyl hydroperoxides 84
Allylic hydroxylation 172, 177–183, 195
Almonds 32
Alosetron 177, 192
Amidases 172
Amineptine 205
Amines, aromatic 88
Amino acids, N-carboxyanhydride 128
Aminolysis 160
Aminopyrine 175
Amitriptyline 204, 205
Amphetanimes 192
Anhydroartemisinin 207
Antidepressants, tricyclic 204
Arene-oxide 176, 181, 185, 195
Arenes 87
Aromatic hydroxylation 172–208
Arrhenius plot 28
Arteether 206, 207
Artemisia annua 206
Arthrobacter spp. 179, 188, 190, 202
Aryl amines 98
Aryl methyl sulfides 81
Arylethanols 64
Aspergillus alliaceus 177, 180, 184–188, 195
Aspergillus flavus 187
Aspergillus niger 158, 191, 202, 207
Aspergillus ochraceus 187, 191
Aspergillus sclerotiorum 18
Aspergillus terreus 180
Asymmetrization 157
Azidolysis 160

Baeyer-Villiger reaction 138, 172
Baker's yeast 120, 157
Bassochlamys fulva 61
Beauveria bassiana 158, 180–193, 203–211
Benzoylformic acid 5
Besipiridine 177, 194, 195
Biocatalysts 97
Biotransformations 74, 75, 88, 169
Biphasic systems 45
α-Bisabolol 161
Bisprolol 177, 192
Bleomycin B-2 178
Bornaprine 178, 195
Bower's compound 163
Bromothymol blue 11
Brompheniramine 186, 192
Building blocks 110, 116
1,2-Butanediol 111, 120, 121

Candida boidinii 67
Candida magnoliae 117
Candida parapsilosis 62, 63, 65
Candida parapsilosis 120, 121
Candida tropicalis 187

Carbamazepine 178, 195
Carbamoyllation 173, 183
N-Carboxyanhydride (NCA) 128
Catalase 78
Catharanthus roseus 65, 203
Chaetomium cochloides 188
Chiral building block 111, 120
Chiral units 51
Chirality 99
Chloro-1,2-propanediol (3-CPD) 110, 112–116, 120
4-Chloro-3-hydroxybutyronitrile (CHBN) 111, 120
4-Chloro-3-hydroxybutanoate esters 110, 116
4-Chloro-3-hydroxybutanoate ethyl ester (CHBE) 117, 119, 120
4-Chloroacetoacetate esters 116
4-Chloroacetoacetate ethyl ester (CAAE) 117–119
Chloroperoxidase 78, 148
Chlorpheniramine 186, 192
Cinchonidine 210
Codeine 179, 192, 195
Colchicine 179, 192
Competitive inhibitor 13
Corynebacteria 176
Corynebacterium 115, 116, 120
Corynesporia cassiicola 67
Corynesporium cassiicola 157
Crisnatol 179
Cultures, aged 63
–, combined 60
Cunninghamella spp. 176–209
Curvularia falcata 185
Curvularia lunata 182, 188
Cyanogenesis 33
Cyanohydrin, chiral 32
Cyanohydrins 39
(R)-Cyanohydrins 40
(S)-Cyanohydrins 41, 44
trans-Cyclohexan-1,2-diol 67
Cyproheptadine 179, 192, 195
Cytochrome P450 171, 172, 202, 213

Deacetylation 184, 189, 209
Deacetylthymoxamine 194
Deamination, oxidative 172, 183, 187, 188, 193, 194
Decarboxylation reaction 2
Decomposition 33
Defatting 36
Dehydrocyanation 47
Dehydrogenases 74, 171, 172
Dehydrogenation 199
Deoxyarrtemisinin 206
Deprenyl 192, 194
Deracemization 58, 59, 157, 159
Desipramine 204, 205
Detoxification 170, 175, 176, 195
DH5 -MCR 9
Dialkyl sulfides 99
Diastereoselectivity 96
Diazepam 175, 180, 192, 195
Dibenzosuberone 206
2,3-Dichloro-1-propanol (2,3-DCP) 110–112
1,3-Dichloro-2-propanol (1,3-DCP) 111, 114–116, 120
Dihydroquinidine 195, 210–212
Diltiazem 139
trans-Diols 176
1,2-Diols, terminal 65
Diols, vicinal 146
Dioxetane 149
Dioxygenase 176, 202
Diplodia spp. 183
(+)-Disparlure 161
Docetaxel 208
Double stereodifferentiation 81
Drug disposition 197
Drug interaction 197, 213
Drug metabolism 169, 177, 212
Drug toxicity 198
Drugs, radiolabelled 197

Ebastine 180, 195
Ellipticine 180, 195
Enamines 98
Enaminones 98
Enantioconvergence 159
Enantiomeric excesses 81
Enolate 97
Enones, asymmetric epoxidation 130
Epichlorohydrin (EP) 110–112, 115, 119
Epoxidation 80, 171–182, 195, 204, 209, 210
–, CPO-catalyzed 94
–, enones 130
–, olefins 91
Epoxide hydrolase 115, 172, 195
– –, microsomal 159
Epoxide hydrolase mechanism 150
Epoxide isomerase 150
Epoxides 146
Epoxychalcones, alkylation 137
–, stereochemical assignment 135
Epoxyketones, reactions 139
Epozide hydrolase 150
Erythromycin A 180, 196
Escherichia coli 115, 119, 120

Esterases 171, 172, 209
Eucalyptus perriniana 209

Fermentors 201
Ferric protoporphyrin IX 75
Filamentous growth 200
Flavanoids 127
Flavine monooxygenase 171, 172
Flavoproteins 33
Formate dehydrogenase 61
(S)-(–)-Frontalin 162
Fungi 174, 176, 196–213
–, endophytic 210
Furosemide 181, 193, 196
Fusarium anguoides 178
Fusarium roseum 204

Geotrichum candium 63
Gliocladium deliquescens 177
Gliocladium roseum 189, 191
Glucose dehydrogenase (GDH) 118
Glucoside 173, 181, 210
Glucosylation 173, 180–184, 191, 196
Glucuronic acid 171, 210
Glucuronidation 173, 178–196
Glucuronide 173, 181, 182, 196
Glucuronyltransferases 172
Glutathione 173, 185, 195, 196
Glutathione-S-transferases 173, 195, 196
Glycerol dehydrogenase 147
Glycidic acid 116
Glycidol (GLD) 110–116
Glycine 171, 173, 183
Glycine conjugation 173
Glycosides 33
Guaiacol 81

α-Haloacid dehalogenases 147
Haloalkane dehalogenase 150
Halogenation 80, 95, 97
–, vinylic 98
Halohydrin dehydro-dehalogenase (HDDase) 113, 120
Halohydrin epoxidases 91, 148
Halohydrin hydrogen-halide lyase (H-lyase) 115, 116, 120
Halohydrins 91, 96, 97
Halonium ion 97
Haloperoxidases 91, 95, 148
Heme peroxidases 81
Heminthosporium spp. 185
Horseradish peroxidase 75
HR325 210, 211
HRP, mutants 99
HRP-catalyzed oxidation 80
Hydrogen cyanide 32
Hydrogen peroxide 75
Hydroperoxides 77, 80
α-Hydroperoxy acid 87
β-Hydroperoxy esters 84
3-Hydroxy-β-butyrolactone 119
Hydroxy-halogenation 97
Hydroxylation 87, 88
(R)-Hydroxynitrile lyase 32
Hydroxynitrile lyases 31

Ibuprofen 181, 195, 199
Imipramine 204, 205
Immobilization 48
trans-Indal-1,2-diol 68
Indane-1,1-dicarboxylic acid 26
Inversion 158

Juvenile hormone 163

Kagan sulfoxidation 99
β-Ketoesters, bioreduction 63
Ketopantoyl lactone 61
Kinetic resolution 74

Lactate dehydrogenase 147
Lapachol 182, 195, 196
Lergotrile 182, 192
Leukotriene antagonist 127
Lignin peroxidase 88
Linalool oxides 162
Lineweaver-Burk plot 17
Lipases 81
Lipooxygenase 81
Lucanthone 182, 195

Malonic acid, disubstituted 3
Mammalian metabolism 170, 174–177, 195, 198, 203
Mandelic acid 60
Mandelonitrile 32
Mandlic acid 5
*m*CPBA 88
Metapyrilene 192, 194
Methoxyphenamine 192, 197
4-Methyl-primaquine 187
Methyltransferases 173
(R)-(–)-Mevalonolactone 164
Mexiletine 183, 195
Microbial hydroxylation 174
Microbial metabolism 169, 177, 197, 201, 209
Micrococcus freudenreichii 61
Microsomal epoxide hydrolase 159
MK 954 183, 195, 196

Mono-oxygenases 148
Monoamine oxidase 171, 172
Monooxygenase 171, 174, 176, 192–195, 210
Monooxygenase induction 200
Mortierella isabellina 189, 206
MP-11 99
Mucor spp. 183, 189, 191, 204–211
Mutagenesis, site-directed 16
Mycobacterium smegmatis 186, 194
Mycophenolic acid 183, 195
Myeloperoxidase 75, 78, 88

NAD^+ 113, 120, 121
NADH 120
$NADP^+$ 118
NADPH 117, 118, 121
2-Naphthols 90
NIH shift 195
Nocardia asteroides 63
Nocardia corallina 188, 202, 207
Nocardia sp. 154
Nocardioides albus 209
Nocardioides luteus 209
Non-flavoproteins 36
Nortriptyline 204, 205
Novobiocin 183, 195
NSC159628 184, 195, 196
Nucleophiles 160

n-Octanol 175
Olefins, epoxidation 91
Olivomycin-A 184, 192
One-electron reduction 77
Optical resolution 110, 121
Optically active 110–115, 120, 121
Organic solvents 159
Organs, perfused 171
Oxidases 74
N-Oxidation 98
Oxidation, asymmetric 86
Oxidoreductases 74
Oxyfunctionalizations 80

Paclitaxel 208
Pancreatic lipase. porcine 161
D-(+)-Pantothenic acid 61
D-(–)-Pantoyl lactone 61
Papaverine 184, 192
Parbendazole 185
Pargyline 192, 194
Payne rearrangements 139
Penicillium chrysogenum 207
Penicillium notatum 182
2,4-Pentandiol 67
Pentoxifyline 185, 192
Pergolide 185, 196
Peroxidase selenosubtilisin, semisynthetic 86
Peroxidases 74, 75
Peroxygenase reaction 78
Pestalotiopsis microspora 210
Phase I reactions 170 ff
Phase II reactions 170 ff
Phenacetin 184, 196
Phenazopyridine 184, 195, 196
Phencyclidine 186, 192, 195
Phenelzine 186, 194
Pheniprazine 186, 194
Pheniramine 186, 192
Phenol-formaldehyde resins 90
Phenols 88
Photochemistry, bioorganic 80
Point mutation 12
Poly-(L)-alanine 126
Poly-(L)-leucine 126
Polyamino acid catalysed oxidations, mechanism 130
Polyamino acid catalyst regeneration 129
Polyamino acids, "diphasic" method 132
– –, "triphasic" method 132
– –, Michael-type reactions 136
– –, preparation 128
Polyleucine, oxidation of sulfide to sulfoxide 136
Porcine pancreatic lipase 161
Praziquantel 187, 195
Prelog's rule 59
Primaquine 187, 194
Prodrugs 171, 197
1,2-Propanediol 111, 120
Propanolol 188, 193, 195, 197
Protein surface 161
Proteolytic enzymes 103
Pseudomonas 112–114, 119, 176
Pseudomonas polycolor 61
Pseudomonas putida 151
pUC 19 11
Pyrilamine 192, 194
Pyrrolnitrine 99

Quinidine 197, 210, 211
Quinine 210
Quinohaemoprotein ethanol dehydrogenase 116
Quinone 188, 194

Racemization 50
Redox enzymes 57
Reductase inhibitors, NADH-linked 66
Regioselectivity 96

Renal excretion 171
Replacement cultures 200
Resolution, dynamic 58
Resting cells 200
Retention 158
Rhizopogon spp. 207
Rhizopus arrhizus 191
Rhodococcus erythropolis 62
Rhodococcus sp. 151, 157
Rhodotorula rubra 185
Rifamycin B 188, 192, 195
RU27987 195, 203

S-oxidation 172, 182, 188, 196
Saccharomyces cerevisiae 191
Scaling-up 197, 198, 201
Screening 152, 196–204
Sebekia benihana 183
Semiconductor chips 90
Sequence homology 37
Sequence similarity 152
Sharpless asymmetric dihydroxylation 142
Short-chain alcohol dehydrogenase superfamily 115
Silene alba 184
Solvent 200
Sparteine 197
Spironolactone 188, 195, 196
Sporobolomyces salmonicolor 117, 119
Stereoinversion 59, 181
–, double 57
Stereoselective degradation 121
Stereospecificity 110, 121
–, enantiocomplementary 57
Steroid biotransformation 176, 203
Steroid hydroxylation 204
Stock cultures 174
Streptococcus faecalis 61
Streptomyces griseus 177–179, 185, 189, 191
Streptomyces lavandulae 207
Streptomyces mediterranei 188
Streptomyces niveus 183
Streptomyces platensis 182
Streptomyces punipalus 189
Streptomyces rimosus 181, 187, 189, 191, 203
Streptomyces spectabilis 177, 179
Streptomyces spp. 177–211
Streptomyces vendargensis 180
Streptomyces violascens 178
Subtilisin 86
Sugar biosynthesis, stereoinversion 69
Sulfatation 173
Sulfhydryl reagent 11
Sulfotransferases 173
Sulfoxidation 80, 99, 103
Sulfoxides 99, 103
Sulfuric acid 162
Sulindac 188, 196
Sulphoconjugation 178, 189, 191
Synthesis, asymmetric 74
Synthon 110

Tamoxifen 189, 192, 195
Tandem biotransformation 61
Tartrate 151
Taxol 208–210
–, phenylisoserine side-chain 140
Taxomyces andreanae 210
Taxotere 208
Taxus brevifolia 208
TDP-4-keto-L-rhamnose 3,5-epimerase 70
TDP-6-deoxy-L-talose 70
TDP-glucose 70
TDP-L-rhamnose 70
Testosterone 175
D-Tetrandrine 189, 192
Thalicarpine 189
Thamnidium elegans 191
Thenyldiamine 192, 194
Theophylline 175
Thio ester 29
Thioelimination 188
Thiol conjugation 173
Three-dimensional structure 37
Thymoxamine 189, 192, 194–196
Transamination 179
–, enzymatic 74
Transhydrocyanation 45
Transition-metal catalysts 74
Tranylcypromine 190, 194, 195
Trimegestone 203
Tripelennamine 192, 194
Triprolidine 190, 195

UDP-glucose epimerase 69
Uni Bi mechanism 39

Verticillium lecanii 181
Vinylic halogenation 98
Viridicatum toxin 184

Warfarin 175, 192, 201
Whetzelinia sclerotiorum 184
Withaferin A 190, 195

Yohimbine 195

Zearalenone 191–196

Printing: Saladruck, Berlin
Binding: H. Stürtz AG, Würzburg